SP

Pvt. Ltd.

DEDICATIONS............

This Book is dedicated to my late parents, who have always loved me unconditionally and whose good examples have taught me to work hard for the things that I aspire to achieve. This work is also dedicated to my spouse (Richa) and to my daughter (Preet), who have been a constant source of support and encouragement during the various challenges that I had come across.

Manish Mathur

Dedicated to my husband, son and my parents without whom this book would have been constructed structured and completed in quick succession of time.

Ritu Mawar

Ritu Mawar

Dr Ritu Mawar is working as Senior Scientist, plant pathology at Central Arid Zone Research Institute Jodhpur. In a period of 15 years of research, her main area of research specialization includes management of soil borne plant pathogens particularly *Macrophomina phaseolina*, *Fusarium* and *Ganoderma*. She studied aspects like Soil solarization, use of cruciferous residues, Biological control and use of composts etc. Beside of these, she has good experience of Forage crop Protection, worked as senior scientist at Indian Grassland and Fodder Research Institute, Jhansi. In her research carrier she has published more than 60 research papers chapters in books popular articles in national and international journals like Annals of Applied Biology, Crop Protection, Phytopathologia Mediterranea, Phytoparasitica, Australasian Plant Pathology and chapters for The American Phytopathological society, Springer etc. She also published an Inventory of 10 years research of Forage Protection Technologies. Dr Mawar has been visited Thailand, Italy and Nepal. She has been recipient of 8 best paper presentation award in National Symposia. She was editor of two books and act as reviewer of National and International Journals. She acts as a Zonal Councilor of Indian Phytopathological Society of Central Zone in 2007.

About the Authors

Manish Mathur

Dr. Manish Mathur has done M.Sc. in Botany and Ph.D. in Plant Ecology from Jai Narain Vyas University, Jodhpur, India. He has ARS-NET in Plant Pathology and Crop Protection (ASRB). He is Member of Commission on Ecosystem Management (IUCN). He is also a certified person in EIA from College of Energy, Environment and Sustainability, Washington, U.S.A. Presently he is working in Central Arid Zone Research Institute. He served in Plant Ecology Section of Division of NRM, where he contributed in vegetation survey (5 districts of Rajasthan and Gujarat), ethnobotany and remote sensing works. He has 53 papers (20 in vegetation ecology, 16 in medicinal plants, 5 in mineral quantification, 6 in quantitative ethno-botany, 2 in economic assessment, 2 in agro-techniques and 2 in plant pathology) to his credit. His most cited scientific works pertaining to species level nutrient use efficiency, Population Viability Analysis of an endangered semi arid species, spatial distribution patterns of semi-arid species, invasion ecology, meteorological variables and their impacts on productivity, economic assessment and conservation priorities of arid plants, quantitative ethno-botany and herbal drugs and nano-technology and these were published in some of the reputed international journals like Journal of Arid Environment, AEER, Australian Journal of Botany, Brazilian Archive of Biology and Technology, JAST and Communication in Soil Science and Plant Analysis. He also wrote four books ***Attributes of Plant Spatial Analysis* (2014), *Ethnic-Knowledge: Documentation and their Interpretation* (2015), *Male Infertility: Factors, Consequences and Natural Solutions* (2016) and *Qualitative and Quantitative Attributes of Environmental Impact Assessments* (2016).** He is the recipient of Young Scientist Associate Award (Crop Protection) and Outstanding Research Award (2016).

Plant Disease Epidemiology

ISBN: 978-93-80012-92-6

Published by:

Studium Press (India) Pvt. Ltd.
4735/22, 2nd Floor, Prakash Deep Building
(Near Delhi Medical Association),
Ansari Road, Darya Ganj, New Delhi-110 002
Tel.: + 91-11-43240200-15 (15 lines); Fax: 91-11-43240215
E-mail: studiumpress@gmail.com

Printed at:

Printed In India

Plant Disease Epidemiology

Manish Mathur
&
Ritu Mawar

2017

Studium Press (India) Pvt. Ltd.

Preface

Plant disease epidemiology is traditionally being utilized as a highly synthesized practical tool for controlling plant disease and its attributes. Such diseases are progressively developed with-in various geographical extents and with different magnitude of damage. Literally epidemic is a Greek word denotes progress of disease in population that also deals with change in disease in time and space. A disease is said to be an epidemic when it occurs at very large geographically area and within a relatively shorter period of time. Based on their nature and spread it may be rapid, slow and negative. The Irish potato famine, the Bengal famine, the Pierce's disease of grapevine, the Karnal bunt of wheat and sudden oak death are some of the notable historical epidemic that affects various dimensions of ecological services. Plant disease epidemiology is steadily flourishing in descriptive, quantitative and comparative etiquette. Such dimension related with understanding the comportment of epidemic development, linked factors, forecasting and calculating the exact timing and type's for application of control majors. In recent past many significant contribution in the form of books have been made like *"Plant Disease: Epidemic and Control"* and *"Principal of Plant Infection"* by J.E. Vanderplank, *"Epidemiology and Plant Disease Management"* by J.C. Zadoks, *"Further Studies of Cereal Rust in India"* by K. C. Mehta and *"Plant Pathology I-III Volume"* by J.G. Horsfall and A. E. Dimons. Such efforts have provided fundamental as well as complex attributes of plant disease epidemiology. These efforts are mostly descriptive in nature with some exceptions like *"Experimental Techniques in Plant Disease Epidemiology"* by J. Kranz and J. Rotem. Be that it may, appraisal of literature revealed that there is a gap about simplification of attributes of plant disease epidemiology for graduate and post-graduate students in the form of a text book. Thus, an attempt has been made to develop a text book that may be useful for students (graduate and post graduate) and for researchers. The presented book is structured in such a way that reader can get academic advantages with respect to understand the 445 questions (short question 241; descriptive questions 26 and multiple choice questions 178). Academic gain with this book also pertains to twelve practical exercise as well as explanation of some numerical problems.

Chapter one deals with basic concepts of plant epidemic and its historical perspective. This includes elements and component of plant epidemics, host, pathogen and environmental factors related with epidemic development. Major contribution of some of the pioneer workers like J. E. Vanderplank, L. R. Jones, Van Everdingen, E. C. Large, R. A. Krause, J. C. Zadoks, J. Krans, K.C. Mehta, S. Nagajan and Laurence V. Madden have been documented. Chapter second pertains to attributes of plant epidemic like how epiphytotic develop, monocyclic, polycyclic and polyetic pathogen, simple and compound interest disease, slow and rapid epiphytotics, disease progress curve of different diseases, take off level and zero date, area under disease progress curve (AUDPC), disease incidence and severity, yield loss, threshold thermo, basic and apparent infection. Excel example for calculation of AUDPC also presented. Chapter third describes mathematical modeling for botanical epidemiology. This chapter includes methods of disease progress curve like exponential, monomolecular, logistic and Gompertz, Linked Differential Equation (LDE), Compartmental models (like SIR, SEIR, SIDR, SIX, and Stochastic SIX), role of latent period, sporulation, inoculums potential, data categorization for sample size determination, parametric and non-parametric tests like (Kruskal-Wallis test, Mann-Whitney U test, Wilcoxon Signed-Rank test, Kendall's Test of Concordance and test of association. With regression and binomial analysis, potential of the Bayesian and Imprecise Dirichlet Model were also described. Chapter four dealt with mathematical complex attributes of plant disease epidemiology *i.e.*, spatial pattern analysis. This chapter provided useful information about basics of spatial analysis, hypothesis testing, indices and their interpretations etc. Nature of spatial pattern analysis was described with dispersal gradient, spatiotemporal dynamics of airborne inoculums and basic reproduction number. Comprehensive descriptions of Complete Spatial Randomness (CSR) along with its various measurement tools were also presented. Sixteen indices related with natural sampling and eighteen indices related with arbitrary sampling, and their mathematical expression and interpretation have also been provided. In this chapter use of spatial autocorrelation, its meaning, quantification techniques, variogram and semivariogram plot for plant disease and epidemic analysis were also solved out. Chapter five related with most important aspects of plant disease epidemiology *i.e.*, forecasting and warning systems. Detail information's about forecasting meaning, its parameters, predictive system, disease pressure index, pathogen disease cycle and their role in forecasting, methods of plant disease forecasting (weather based and inoculums based), forecasting works carried out in India, computer simulation model for predicting plant disease and information about aerobiology were provide. Chapter Six related with methods for epidemic avoidance mostly with non-chemical approaches. Sanitation, crop rotation, removal of alternate and collateral hosts, roughing, soil solarization, use of host resistance, vertical and horizontal resistance, gene for gene hypothesis, boom and bust cycle, vertifolia effect, gene deployment, gene pyramiding and multiline varieties concepts are the major highlights of this chapter.

As described above the present book is formulized to gain a correct and pinpoint insight about 445 different questions. This also supported with twelve practical exercises like hypothesis testing, sample size determination, K-S test, and Mann Whitney test, Kendall test of concordance, Bay's rules, T-square index, sanitation examples and association analysis.

Manish Mathur

Acknowledgement

We would like to express our sincere gratitude to the Director, CAZRI, Dr. O. P. Yadav, and to Dr. R.K. Kaul (Principal Scientist, CAZRI), Dr. Satish Lodha (Rtd. Principal Scientist, CAZRI), and Dr. S. Sundaramoorthy (Professor, Jai Narain Vyas University, Jodhpur) for encouraging us for continuous writing. Thanks and appreciations also go to our colleague in developing the book and people who have willingly helped us out with their abilities. Further this book is a compressive effort to funnel the different scattered information's related with various aspects of Plant Disease Epidemiology and for this author's pile up the information from 1950- 2015, thus all the researchers and their works were fully acknowledge and honour in this book.

Manish Mathur
Ritu Mawar

Table of Contents

CHAPTER

1 Concept of Plant Epidemic and its Historical Perspective

1.1. INTRODUCTION

Plant diseases, weeds and environmental factors are the major threats to crop production and approximately, such factors accelerate 31 to 42 percent of worldwide crop losses. Taking into account that 14.1% of crops are lost to plant disease alone, the total worldwide crop loss from plant disease is about $220 billion dollars. These losses are in part responsible for the suffering of 800 million people who lack adequate food. Historical evidences have suggested that the Irish potato famine caused by *Phytophthora infestans* resulted in starvation death of 1.5 millions peoples, similarly brown spot of rice caused human suffering and starvation deaths to the tune of about 2 million in the Bengal (Bengal famine).

In the mid of 19th century, down mildew of grape was accidentally introduced into Europe which almost destroyed the entire vineyards in France, Germany and in Italy. In early 20th century, a canker intitiated by *Endothia parasitica* wiped out the American chestnut trees. Some other examples includes Asiatic citrus canker (*Xanthomonas axonopodis* pv citri); bacterial wilt (*Ralstonia solanacearum* race 3, biovar 2); sudden oak death (*Phytophthora ramorum*); Pierce's disease of grapevine (*Xylella fastidiosa*); Karnal bunt of wheat (*Tilletia indica*) and Asian soybean rust (*Phakopsora pachyrhizi*).

Due to the aforementioned problems it is necessary to have sufficient, economic and environmentally acceptable strategies to manage the epidemic development of plant diseases in order to decrease the crop losses and reduce their consequences.

An epidemic is a Greek word made up by *Epi* = on and *Demos* = people. It is a science of disease in a population; progress of disease in the population and also encompasses changes in disease in time and space. Basically a disease is said to be an epidemic when it occurs at a very large geographical area within a relatively shorter period of time. Based on nature and its spread, it may be rapid, slow and negative and it's largely governed by environmental factors. In modern plant pathology the general criterion's to define plant epidemic are varied. It may involve the study of factors affecting the outbreak and spread of infectious diseases, it also includes the dynamics of disease in the host population. Vanderplank (the father of epidemiology) defined it as "the science of the disease in population". On the other hand Campbell and Madden (1990) defined plant disease epidemiology as the "study of temporal and spatial changes that occur during epidemics of plant diseases that are caused by pathogen in plant population.

Epidemiology may be partitioned under descriptive, quantitative and comparative. Descriptive epidemiology describes the epidemics and factors that seem to affect it while, quantitative epidemiology quantifies, or measures various matrices of an epidemic, however, comparative epidemiology compares different patho-system in different ecosystems. Thus in holistic manner, epidemiology involves the study of the temporal, spatial, and spatial-temporal dynamics of disease in populations, and the utilization of results of experiments and surveys to describe, understand, compare, and predict epidemics.

Vanderplank (1963) mathematically showed that epidemic starts when one diseased lesion leaves behind more than one. Perpetuation of a species demands that each should produce at least one of its kinds before its death. If this is less than one, soon the population is vulnerable for extinction. He generalized that one fungal lesion must produce in turn more than one lesion and when it is >1, it can create epidemic and mathematically shown this to be as $iR_c > 1$, and this is known as the "threshold theorem. Where, R_c is the number of lesions produced from a parent lesion per day and *i,* is the infectious period of that lesion. For example, say *Puccinia graminis tritici* has an active sporulation period of 20 days (i) and every alternate day spores released produce one successful infection (R_c), then $iR_c = 20 \times 0.5 = 10$; which is >1. When the value equates 1 it means the pathogen is just managing to survive, while greater than 1 denotes an increase in the population size. Rarely the value is less than 1 which indicates that the disease is declining. Kranz (1973) denote this stage as anti-epidemic.

1.2. ELEMENT AND COMPONENT OF PLANT EPIDEMIC

The epidemic is occurring due to a timely combination of a susceptible host, virulent pathogen and favorable or conducive environment. These three elements are shown by curve, which is known as a pyramid (Fig. 1). Interactions of the 3 main components are described by the disease triangle. Some pathologists have elaborated on the disease triangle by adding one or more parameters. The additional parameters include humans, vector and time.

Time is regarded as fourth component which is shown by a tetrahedron. While, human is its fifth component (Fig. 1). Human activities like crop type (annual or perennial), density of crop, resistant types (horizontal or vertical). In general, in tetrahedron the hymen is located at the top whose base is a triangle and height is the time length (Fig. 1).

Based on the magnitude, availability and timely combination of elements of the epidemic, the geographical extents of diseases can be pandemic, endemic, epidemic and sporadic (Fig. 2).

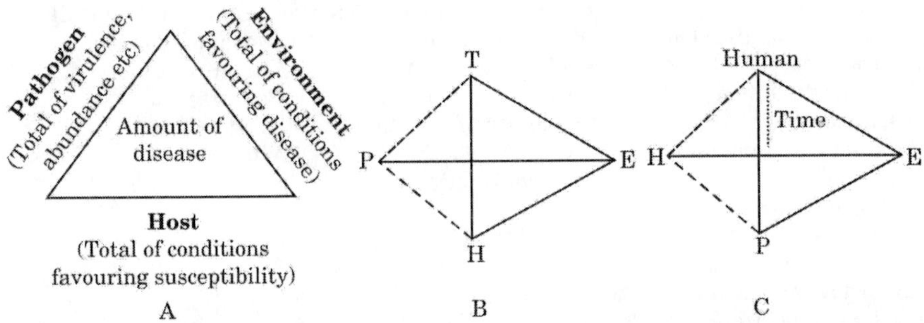

Fig. 1: A to C Elements of Epidemics, disease triangle (A); tetrahedron (B) role of human (C). (Author Original Drawing)

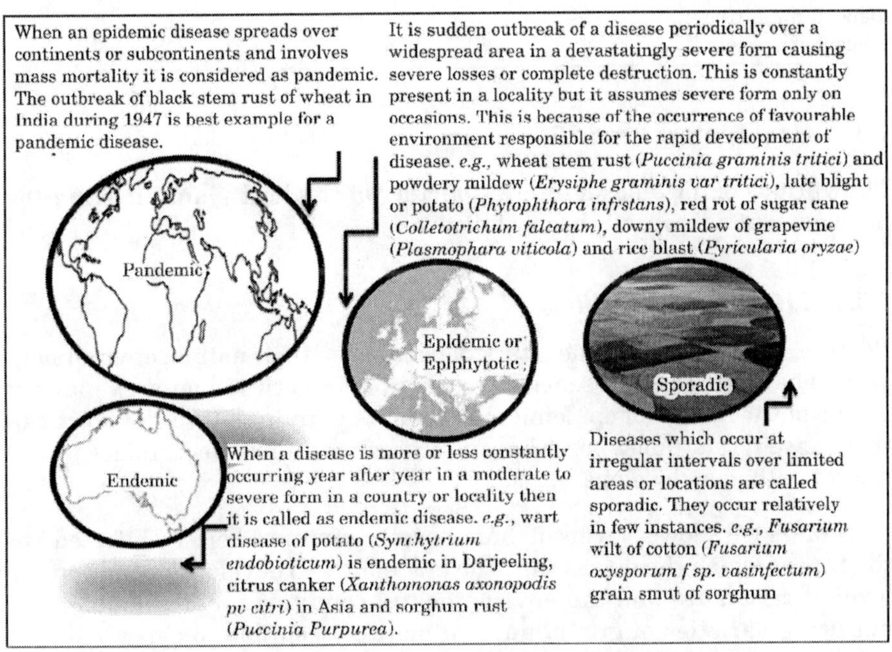

Fig. 2: Geographical extents of diseases. (Author Original Drawing)

1.3. PRIMARY ELEMENTS ESSENTIAL FOR EPIDEMIC

As described above attributes related with host, pathogen and conducive environment related with development of epiphytotic are as follows:

Host	*Pathogen*	*Environment*
Susceptibility of the host, Spatial distribution of susceptible hosts Introduction of new Hosts/variety/strain Presence of collateral or alternate host Age of host (Ontogenic resistance) Types of host (annual or perennial) Distance from source of inoculums Degree of genetic uniformity of host (vertical or horizontal resistant) Host density Plant nutrient availability like silica deposition, nitrogen, phosphorus, potassium	Introduction of a new pathogen Pathogen in their virulent form Birth rate of the pathogen (R or K selection) Mode of spread (wind, water, insect) Capability of resistant breaking Types of pathogen	Temperature and humidity Moisture and humidity Rainfall and other climatic conditions like sunshine duration, humidity etc. Soil pH Shade effect Wind

1.3.1. Host Factors Leads Epidemic

The various external and internal factors of the host plant affecting the development of epidemic are as follows

1.3.1.1. *Host susceptibility*

Plants are predisposed to the attack depending on their nature, environment and stage of growth. Presence of susceptible varieties in an area may act as one of the causes of epidemic. Host plants with high levels of (vertical) resistance do not allow a pathogen to become established, unless a new race of pathogen develops.

Plants with lower levels of (horizontal) resistance become infected but the rate at which the disease and the epidemic will develop depends on the level of resistance and the environmental conditions. For example, late maturing varieties of groundnut are more susceptible to early leaf spot (*Cercospora arachidicola*) and late leaf spot (*Phaeoisariopsis*) than the early maturing varieties. Similarly late maturing varieties of wheat are susceptible to loose smut (*Ustilago nuda tritici*) than the early maturing varieties.

Early sown sugarcane varieties of sugarcane are more susceptible to leaf rust in Deccan canals in Bombay area than the late sown varieties.

Abundance of susceptible hosts in an area is one of the major causes of the spread of epidemics.

Continuous cultivation of susceptible variety or varieties in an area, that too in a large contiguous area help in the build up of inoculums and improve the chances of epidemics. Under the above conditions the pathogen increases the rate of multiplication of its propagules and repeats the disease cycles in a short span. Destructive epidemic of early and late leaf spots of groundnut in Bombay area (Gujarat and Maharashtra States) during 1912-1913 was mainly the result of cultivation of local varieties in a larger area.

Panama wilt (*Fusarium oxysporum* f.sp. *vasinfectum*) susceptible table variety, 'Son' in banana was responsible for the destructive epidemic in parts of Bombay area (Gujarat and Maharashtra) during 1936 – 1940.

Countrywide cultivation of red rot (*Colletotrichum falcatum*) susceptible sugarcane varieties (local varieties like *Pundya, Khajuria* etc.) practically made their cultivation impossible in Bombay area.

Plant susceptibility to disease also oscillates with the age of host. Ontogenic or age-related resistance has been noted in many pathosystems. In such types of system the host is susceptible only at a particular growth stage or age and once such particular stage is passes, than a even susceptible host behave like a resistant one's. For example *Pythium* damping-off and root rots, downy mildews, peach leaf curl, systematic smuts and bacterial blight, the hosts are susceptible only during the growth period and becoming resistant when adult.

While in disease like rusts and viral infections, plant parts are resistant while still very young, become more susceptible later in their growth stages. In some other disease, like blossoms or fruit infections by *Botrytis, Penicillium, Monilinia* and *Glomerella*, plant parts are resistant during growth and early adult period but become susceptible near ripening.

1.3.1.2. *Degree of genetic uniformity of host*

Large geographical extent of genetically uniform host varieties (vertical resistance), accelerates the possibility of the appearance of pathogen new race/biotype and thus, leads to epidemic (Boom and Bust Cycle phenomena will be discussed in subsequent chapters). Examples of such conditions include *Cochliobolus* blight on the Victoria oat and in the southern corn leaf blight on corn.

For the similar reason of genetic uniformity, the height rates of epidemic development generally observe on vegetatively propagated varieties, intermediates rates in self-pollinated crops and lowest rates in cross pollinated crops.

1.3.1.3. *Availability of collateral and alternate host's*

Alternate hosts are those plants on which the heteroecious pathogens pass part of their life cycles. Similarly, collateral hosts are some wild plants in which the pathogen survives when primary host is not available. Both alternate and collateral hosts are important in building up the primary inoculum to the next crop.

They determine the course and intensity of an epidemic. Grass hosts (collateral hosts) of *Sclerospora sacchari, S. philippinensis* (downy mildews), *Pyricularia oryzae* (rice blast), *Ustilago scitaminea* (sugarcane smut) may produce abundant inoculum which aid in building up of epidemics. Outbreak of heteroecious blister rust of pine (*Cronartium ribicola*) in Europe and the U.S.A happened due to import or introduction of *Pinus strobus* from the USA. In general collateral host's belongs to the same family of the main crops and mostly they are wild grasses, while, alternate host belongs to different family. Collateral and alternate hosts of black, brown/orange and yellow rust of wheat are as follows (Table 1).

Table 1: Various collateral and alternate host's of wheat rusts

Casual organism	***Black rust***	***Brown or orange rust***	***Yellow rust***
Collateral Hosts	*Puccinia graminis tritici* *Bromus coloratus, B. carinatus, B. japonicus, B. mollis, B. patulus, Hordeum distichon, H. murinum, Lolium, perenne, Hilaria jamesii, Aegilops squarrosa, A. ventricosa, A. trinecilis*	*Puccinia recondita* *Bromus catharticus, B. japonicas, Hordeum murinum, Muhlenbergia huegelii*	*Puccinia striiformis* *Aegilops species*
Alternate Hosts	*Berberis species, Mahonia, Mahoberberis*	*Not Known*	*Isopyrum fumarioides, Thalictrum* species, *Clematis* species, *Anchusa* species

1.3.1.4. *Crop types*

In general, annual crops like corn, wheat, vegetables, cotton etc., are more prone for rapid epiphytotic compare to perennial woody species. These two phenomenon's are correlated with nature of pathogen (systemic/non systemic), mode of transmission and inoculums potential.

1.3.1.5. *Predisposing factors*

Predisposing factors are generally associated with the host and these factors make the host more vulnerable for disease. These factors may include intrinsic characters of host, environmental variables or cultural practices.

In fact these factors favors' the pathogen in term of their virulence, growth, reproduction, germination or physiological or chemical stimulation.

Predisposing factors of few fungal and bacterial diseases are as follows:

1. ***Barley Rust***

   A. *Black or Stem Rust- Puccinia graminis* f. sp. *tritici*
      - Warm humid weather conditions with intermittent rains.
      - Hot days 25-30°C and mild nights (15-20°C)
      - Leaf wetness from rain or dew

   B. *Brown or Leaf Rust- P. hordei*
      - Development of the disease is most rapid during warm (15-20°C) moist (rain or dew) weather.
      - Warm humid weather conditions with intermittent rains.
      - Hot days 25-30°C and mild nights (15-20°C)
      - Leaf wetness from rain or dew

   C. *Yellow or Stripe Rust- P. striiformis*
      - Cool humid weather conditions with intermittent rains.
      - Cooler climates (10-16°C)
      - Leaf wetness from rain or dew
      - Heavy dew or intermittent rains can accelerate the spread of the disease

2. ***Downy Mildew of Sorghum: Peronosclerospora sorghi***
   - Cool - wet - humid weather
   - Intermittent rains
   - The rapid growth of fungal pathogen is favoured by rainy and humid environment.

3. ***Ergot or Sugary Disease of Sorghum: Claviceps purpurea***
   - Relative humidity greater than 80%, and
   - Temperatures between 20 to 30°C.
4. ***Downy Mildew or Green Ear Disease of Bajara: Sclerospora graminicola***
   - Warm- wet - humid weather
   - Intermittent rains
   - The atmospheric temperatures between 25-30°C and above 85 per cent relative humidity favours high level of secondary infection.
5. ***Bacterial Disease of Apricot: Pseudomonas syringae pv. syringae***
   - Cultivar susceptibility
   - Sandy and clay soils
   - Winter pruning
6. ***Bacterial Disease of Hazelnut: Pseudomonas avellanae***
   - Acidic soils (pH<5.5)
   - Beest *et al.* (2008) have quantified the disease-weather relationships for powdery mildew (*Blumeria graminis* f. sp. *tritici* Em. Marchal) and yellow rust (*Puccinia striiformis* Westend. f. sp. *Tritici*) on winter wheat. They have provided predisposing factors for both diseases as well as their model for prediction of disease severity.

A. For Powdery Mildew

Factor	*Value*	*Model*
Wind speed	Number of days ≥ 200 km	Wind x 0.299-17.07
Sunshine	Number of days ≤ 2 hours	Sunshine x 0.674-24.07
Radiation	Accumulation below 4^0C	Radiation x 0.156+13.61
Humidity	Number of Consecutive days ≥ 95%	Humidty xx 0.794 + rain x 0.150
Rain	Accumulation above 10 mm	+ Maximum temperature x
Maximum temperature	Number of days ≥ 20^0C	0.716 – 7.16

B. For Yellow Rust

Factor	*Value*	*Model*
Vapor	Accumulation below 6 mbar	Vapor x 0.0383 + Mean temperature
Mean temperature	Accumulation above 4^0C	x –0.00214-19.2
Maximum temperature	Accumulation above 16^0C	Maximum temperature x 0.5 + rain x-4.08 + 2.54
Rain	Number of days ≥ 8 mm	

Based on their models they have reported that wind in December to February is the most influential factor for powdery mildew epidemic and disease severity can be best identified by model with temperature, humidity, and rain in April to June. Similarly for yellow rust, the temperature in February to June is most influential factor for epidemic as well as for disease severity.

1.3.2. Pathogen Factors Leads Epidemic

1.3.2.1. *Virulence level*

The virulent pathogen infect host rapidly and produce large amount of inoclum at faster rates.

All strains of a pathogen are not aggressive. Only the aggressive strains are capable of causing infectious diseases which spread as epidemic.

They are characterised by rapid cycle of infection and causing successful infection in new hosts.

Rapid cycle of infection is essential for successful infection and it happens only by aggressive strain of the pathogen. *e.g.*, *Puccinia graminis tritici* (wheat black rust) in India, stripe rust, bunt and loose smut of wheat in the U.S.A. and Europe.

The possibility of outbreak of epidemics increases with the number of physiologic forms or pathogenic strains of the pathogen present in a locality.

1.3.2.2. *Inoculum potential*

Any part of the pathogen which comes into contact with host called inoculum and contact phenomenon called inoculation. In general pathogen produces two types of inoculums, primary and secondary.

Primary inoculum are the survival tool, generally they are sexual spore and cause primary infection, while, secondary inoculum develops from primary one, they are asexual spore and spread the disease in host population. Inoculum potential refers to "amount of energy available at host surface for infection and its multiplication factor of inoculum density and inoculum capacity". Non-systemic pathogen generally produces large inoculum and thus, causes rapid epidemic.

1.3.2.3. *Reproduction pattern of pathogen*

Pathogen with high reproductive capacity and capable of rapid dissemination over wide areas mostly causes epidemics. The fungal members causing powdery mildews, downy mildews, rusts, blasts, blights etc., produce enormous amount of spores.

Such pathogens have short reproduction cycles and thus can produce many generations in a single growing season. These are polycyclic in nature. They have high birth rate, causing rapid epidemics as the diseases spread in a compound interest manner (compound interest diseases). These spores are easily dispersed by using water or insects and cause infections to new plants. The high degree of fecundity and the enormous amounts of inoculum produced by some common plant pathogens are as follows:

Pathogen	*Fecundity Number*
Stem rust of wheat (*Puccinia graminis tritici*)	Twenty five trillion uredospores in one hectare of wheat crops and 64,000 million aeciospores from aecial cupsin a single barberry bush
Cedar rust apple	Two billion teliospores
A. grapevine infected by downy mildew	32,000 sporangia per sq.cm
Bunt of wheat	6 to 12 million smut spores in a single kernel
Chestnut blight	150,000,000 spores in a single spore horn

Epiphytotics may also be caused by low death rate diseases. These diseases are caused by agents of systemic nature which are protected by plant tissues. As they are protected by plant tissues the chances of high mortality is reduced to the minimum. In these diseases the chief source for accumulation of inoculum for epiphytotics is the diseased plant organ used for vegetative propagation (corms, setts, tubers, etc,).

Soil fungi like *Fusarium* species, *Verticillium* species and most nematodes usually have one to few reproductive cycles per crop season. They often cause more localised, slower epidemics. In such cases inoculum builds up from one to many years. They are low death rate pathogen and cause simple interest disease. Cedar rust apple and white pine blister rust are the example of such epidemic.

1.3.2.4. *Introduction of new pathogen*

Some pathogens, epidemic in certain area, may become quite aggressive and outbreak as epidemic when introduced to new area. For example late blight of potato caused by *Phytophthora infestans* was epidemic in South America. This disease became epidemic when the infected tubers were introduced in Europe (in 1843-45). Fire blight (*Erwinia amylovora* in North America is endemic.

Fire blight spread to Pacific coast fruit-growing areas of the U.S.A in 1884 and subsequently it reached Canada. It reached New Zealand in 1919 and it appeared in England in 1957. The mode of introduction had been through fruit boxes.

Coffee rust (*Hemileia vastatrix*) is indigenous in Ethiopia, where *Coffea arabica* is native. The disease spread to Sri Lanka in 1869, India in 1870, Sumatra in 1876, Java in 1878 and the Philippines in 1889. It also spread from Kenya to the Congo by 1918 and reached the Cameroons. From 1950 onwards, it spread to the reminder of West Africa.

Dutch elm diseases (*Ceratocystis ulmi*) first reported in 1919 in Holland, spread throughout Europe and reached Great Britain in 1927. It was introduced to the eastern United States on elm logs imported from Europe.

1.3.2.5. *Autoinfection and alloinfection*

Knowledge of pathogen dispersal patterns is crucial to a better understanding of the local increase of disease and its subsequent spread. An important aspect in this regard is the concept of autoinfection *versus* alloinfection (Mundt, 2009).

These terms originally were proposed by Robinson (1976), who considered autoinfection to be "infection in which the donor (infector) host individual is the same as the receptor (infected) host individual." Conversely, alloinfection was considered to be "infection in which the donor (or infector) host is a different individual from the recipient (or infected) host individual."

The concept has since been extended to both larger and smaller spatial scales. For example, in studying mosaics of host genotypes, autoinfection was considered to be infection resulting from propagules produced from the same unit of contiguous plants of the same genotype, while alloinfection was considered to be infections resulting from propagules produced on other genotype units in the population (Skelsey *et al.*, 2005).

The importance of autoinfection to disease epidemiology and pathogen evolution in mixtures of host genotypes has been given some emphasis, especially in models (Tellier and Brown, 2008).

High autoinfection rates were reported for stem rust of ryegrass (*Lolium perenne*), caused by *Puccinia graminis* subsp. *graminicola*. In this case, the pathogen was found to sporulate on the inside surface of inoculated flag leaf sheaths, resulting in infection of emerging stems and inflorescences (Pfender, 2004). Autoinfection *via* this mechanism can cause greater than 50% of the total stem rust infection in a perennial ryegrass crop (Pfender, 2006); additional autoinfections will certainly occur on exposed plant parts *via* aerial transmission.

The mathematical expression of auto and allo-infection and their role in plant disease epidemiology are described in details by Mundt (2009).

1.3.3. Environmental Factors Leads Epidemic

The environmental conditions such as temperature, relative humidity, rainfall, duration and intensity of light, etc. play very important role in causing epidemics. These are actually the deciding factors and influence almost all the stages of disease cycle.

Favourable environmental conditions are needed for sporulation, liberation of spores and dissemination of pathogen, germination, infection and establishment of pathogen in the host. The environment may affect each component of the epidemic *i.e.* host, pathogen as well as vectors.

The environment may affect the availability, growth stage, succulence and genetic susceptibility of the host. It may also affect the survival, vigour, multiplication rate, sporulation and dispersal. Moisture affects the host, pathogen as well as vectors.

The term 'moisture' covers rainfall (duration and intensity), relative humidity and dew and leaf wetness (duration and intensity). The influence of rain splash and running water on dispersal of pathogen propagules is also important. Moisture facilitates spore release by many fungi and the oozing of bacteria, while in viral disease the moisture affects epidemic indirectly. The spores of certain pathogens (e.g. wheat stripe rust, *Puccinia striiformis*) may not establish effective contact with a leaf that is too wet. In general, germination and infection by powdery mildews is favored more by high humidity than by leaf wetness.

The period of leaf wetness required for maximum infection is affected by temperature. Often the leaf wetness period required for a certain degree of infection is longer in cool than in warm weather. Germination and infection are usually accelerated by warmer conditions. A good example of how to quantify the effect of temperature on plant disease is seen in a widely used method of predicting the incidence of apple scab (*Venturia inaequalis*).

Temperature affects the incubation or latent period (the time between infection and the first appearance of disease symptoms), the generation time (the time between infection and sporulation) and the infectious period (the time during which the pathogen continues producing propagules). At higher temperatures the disease cycle is speeded up with the result that epidemics develop faster. Under cooler conditions, epidemic progress is usually slower so disease incidence and severity may not reach the threshold levels necessary to cause significant crop loss. With stripe rust of wheat, cool temperate rust, disease development slows at excessively high temperatures in late spring/early summer. Importance of thermal time in assessment and modeling of growth and development of plant-pathogen have been studied for *Mycosphaerella graminicola, Gaeumannomyces graminis* in wheat (Schoeny and Lucas, 1999 and Lovell *et al.*, 2004) and *Xanthomonas campestris* pv. *dieffenbachiae* (Fukui *et al.*, 1999), *Podosphaera leucotricha* (powdery mildew) in apple shoot (Xu, 1999), *Puccinia graminis* sub species, *graminicola* (stem rust) in perennial ryegrass (Pfender, 2003) and *Sclerotinia sclerotiorum* in lettuce (Clarkson *et al.*, 2004) For modeling plant disease epidemiology with temperature, Lovell *et al.* (2004) have provided basic equation incorporated latent period (l), apparent infection rate r and thermal time (T) after infection; p (T) = 1/ 1{1+b exp[-r(t-l)]}.

In addition, the adult plant resistance of many wheat cultivars is expressed more strongly at higher temperatures, which also reduces the rate of epidemic development.

Fungal spores, vectors, and to a lesser extent, bacteria and nematodes can be dispersed by the action of wind. This will enable pathogen to move from infected plant to healthy one.

Wind can also injure hosts, thus, providing an avenue for infection and or increasing susceptibility to some pathogens. In addition to these, there are many other factors including light intensity, light quality, soil pH, fertility and soil type that influence disease development. Soil type can affect plant growth and also development of some pathogens. Light sandy soil, low in organic matter favors the growth of many types of nematodes. Damping off disease is commonly seen or favored by heavy and water logged soils. Soil pH affects pathogen development. Excessive nitrogen fertilizer can increase susceptibility to pathogen attack.

In addition to host, pathogen and environmental factors, human activities like, site selection, crop types, density, variety types (horizontal or resistant variety), cultural practices and distance from primary source of inoculum also affects the epidemic.

1.4. HISTORICAL PERSPECTIVE OF PLANT EPIDEMIOLOGY

In general there are two different schools defining conceptual nature of theoretical epidemiology while, and other consider epidemiology to have an early origin in efforts made long ago to control disease in populations of plants.

The birth of modern quantitative plant epidemiology is generally endorsed to the early work of J. E. Vanderplank (Vanderplank, 1963) who popularized the concept of modelling disease change (epidemics) in time.

Plant diseases have largely been documented in our ancient texts including the Bible, and in the Sanskrit literatures.

In Roman culture during each spring there was a celebration of the Robigalia carried out to satisfy and the 'God of Rust' thus stop casing cereal rust epidemics. Some of the epidemics that had significant socio-economic impacts on mankind are presented in Table 2.

Table 2: Historical epidemics and their geographical extents

Epidemic name	*Casual organism*	*Occur at*	*Geographical extent*
Ergot of rye	*Claviceps purpurea*	10th-19th century	Europe
Fire blight of pear and pome fruit	*Erwinia amylovora*	1780	USA and UK
Peach yellow	Peach yellow virus	1791	Eastern USA
Potato blight	*Phytophthora infestans*	1845	Ireland
Coffee rust	*Hemileia vastatrix*	1869	Sri Lanka
Vine mildew	*Plasmopara viticola*	1878-85	France
Panama disease of banana	*Fusarium oxysporum f. sp. Cubense*	1880-1959	Central and South America
Chestnut blight	*Cryphonectria parasitica*	1904	Eastern USA
Dutch elm disease	*Ceratocystis ulmi*	1920-1970	Netherland, USA and UK
Brown spot of rice (Bengal Famine)	*Cochliobolus miyabeanus*	1942	India
Maize leaf blight	*Helminthosporium maydis*	1970	Southern USA
Leaf rust	*Puccinia recondita*	1970s	Northern India
Karnal bunt	*Tilletia indica*	1970s	Northern India

Initially he works of Bourke (1953) and Cox and Large (1960) concerned with epidemic and their relationship with weather data. Bourke (1970) had worked on *"use of weather information in the prediction of plant disease epiphytotics"*. Miller and O'Brein (1957) on "Prediction of plant disease epidemics, Waggoner (1960) on "Forecasting epidemics and ground level climate and disease forecasting", Krause and Massie (1975) on "Predictive systems-modern approaches to disease control", Shrum (1978) on "Forecasting of epidemics, Madden and Hughes (1995) "Plant diseases incidence: distribution, heterogeneity and temporal analysis", McCartney (1996) "The influence of environment on the development and control of disease.

From the Indian perspective Shanta (1960) studied the epidemic and spread of the tikka disease of groundnut caused by *Cercospora personata* and *Cercospora arachidicola*, he reported that severity and its subsequent spread depend upon the prevalence of high humidity and low temperature during crop season.

Further, Mallaiah and Rao (1980) have added that air-borne conidial concentration were high when temperature approaches 29-31^{0}C with 76% relative humidity.

Padmanabhan (1965) examined the meteorological factors associated with the epiphytotics of *Helminthosporium oryzae* caused the "Great Bengal Famin" of 1942, he revealed that heavy rainfall in September accompanied by a favourable temperature of 25 to 30^{0}C, followed by cloudy weather in subsequent months with low solar radiation availability lead outbreak of brown leaf spot of rice. For predicting the outbreaks of blast of rice (*Pyricularia oryzae*). Padmanabhan (1965) used climatic parameters like nycto-temperature (from 20-24^{0}C) in association with 90% relative humidity for a week during any susceptible phase of crop brought an epidemic of blast.

Payak and Renfro (1967) have observed that the brown stripe downy mildew of maize (*Sclerophthora rayssiae* var. zeae) was sever in areas where mean precipitation was more than 190 cm.

Saxena and Lal (1984) developed a linear equation based on the sunshine hours and rainfall for prediction of *Erwinia* stalk rot of maize.

Saharan and Singh (1980) observed that the latent period of *Melampsora lini* varied greatly and the variation can be predicted by relative humidity.

Mayee (1988) developed a forecasting system based on temperature for groundnut rust epidemic in central India. He along with Ekbote (1983) reported that the inoculums of rust was available throughout the year, however, maximum infection occurred only during rainy season. They further reported that factors like mean temperature, relative humidity and number of rainy days can predict not only the sub-phases of infection chains but also the disease development.

Muralidharan and Rao (1981) developed a rice blast warning with the help of trap plots.

Shukla and Anjaneyulu (1981) correlated the rice tungro development with vector population and initial infection.

Bhattacharya *et al.* (1982) had developed rules for late blight of potato under the Indian conditions and according to them, under Indian hill conditions, rainfall is the primary factor for the generation of conducive condition for late blight.

In modern research perspective, there is an unspoken understanding that most research attributed to this sub-discipline of plant pathology is quantitative in nature and frequently characterized by models of varying complexity. Authors like Drenth (2004) described the significance of spatial structure in epidemic progress, however, he also visualized that theory has been widely mistreated in disease management.

Jeger (2004) suggested that, "Surprisingly, there has been little experiential evaluation of the amount to which improved disease management results directly from improved understanding other than in specific areas such as seed health management, the alignment of epidemiology with population genetics, or where climate is a critical influence on disease epidemics".

Similarly, Scherm *et al.* (2006) argued that, "the image of theoretical epidemiology within the larger field of plant pathology could benefit from a clearer documentation of its impact on practical disease management".

Plant epidemiology and its relationships and modelling with climate data and climate changes have been documented by various workers particularly with disease like rice leaf blast caused by *Pyricularia oryzae*, oak disease caused by *Phytophthora cinnamomi*, grape downy mildew caused by *Plasmopara viticola* etc.

Garrett *et al.* (2006) reviewed the subject and developed a framework for studying the effects of climate change on disease. In 1969, the first computer simulation programme called EPIDEM was developed and resulted from modelling each stage of life cycle of a pathogen as a function of the environment. Example of some of the computer simulation is as follow:

Programme name	***Pathogen***
EPIDEM	Early blight of tomato and potato caused by *Alternaria solani*
MYCOS	*Mycosphaerella blight of chrysanthemum*
EPIVEN	*Venturia inaequalis* for apple scab
EPICORN	*Helminthosporium maydis*
BLIGHTCAST (Krause *et al.*, 1985)	*Phytophthora infestans* (Late blight of potato)
TOM-CAST (Pitbaldo, 1992)	*Early blight of tomato*
PLSMO (Rosa *et al.*, 1993)	*Plasmopara viticola* Downy mildew of grapes
LATEBLIFHT (Jorge *et al.*, 2005)	*Phytophthora infestans* on potato foliage.
STEMRUST_G (Pfender and Upper, 2015)	*Puccinia graminis* subsp. *graminicola* (Stem rust of grasses)

Principals and details about computer simulation programmes will be discuss in subsequent chapters

1.4.1. Pioneers Workers of Plant Epidemiology

1.4.1.1. *J. E. Vanderplank*

James Edward Vanderplank was born in 1908 in Eshowe, Zululand. He regarded as father of plant epidemiology. His major contributions are as follows:

- Provided Protein for Protein hypothesis
- Wrote books
  - *Plant Diseases: Epidemics and Control* (1963)
  - *Principles of Plant Infection* (1975)
  - *Disease Resistance in Plants* (1968)
- Provided the vertifolia effect
- Elaborate Disease progress Curve and Area Under the Disease Progress Curve (AUDPC)
- He also provided the essential conditional behavior for the "Gene for Gene" hypothesis. According to him two corresponding gens of host and pathogen are essential for Gene for Gene interaction.
- Provided threshold thermo ($IR_c > 1$) for plant epidemic

1.4.1.2. *L. R. Jones*

Studied the effects of soil moisture and temperature and other environmental factors upon the epidemiology of diseases.

1.4.1.3. *Van Everdingen*

He worked on late blight of potato and published his work in form of "The relation between weather conditions and potato blight (*Phytophthora infestans*). Tijdschrift over Plantenziekten, 1926: 32, 129-140".

He developed forecasting system for this epidemic and formulated the Dutch Rules, which are based on four weather parameters (Figure 3). Letter in 1947 A. Beaumont developed 'Beaumont-humidity rule for forecasting late blight of potato.

1.4.1.4. *E. C. Large*

He worked and interpreted the progress curves for late blight of potato and other plant disease.

He had published "battle against the blight" in 1959. He assessed the incidence/severity of disease and plotted disease progress curves and crop losses.

1.4.1.5. *R. A. Krause*

He and his colloquies developed the computerized forecasting system for potato late blight **BLITECAST** - It is a combination of two late blight prediction models.

When all these conditions are satisfied under field conditions continuously for a few days, then in the Netherland, control measures are recommended

Fig. 3: Dutch rules for late blight of potato proposed by Van Everdingen (Author Original Drawing). Beaumont combined the cloud cover; rainfall and dew requirements for a period of high humidity and his rules were that were there should be at least two days with: (a) minimum temperature not less than 10°C and (b) relative humidity not below 75%.

The first part of the model forecasts the occurrence of LB at 7–14 days after an accumulated total of 10 favorable rainy days according to the model of Hyre (1954, 1955), or the accumulation of 18 severity units according to the model of Wallin (1962).

The second part of the model recommends fungicide application based on the number of favorable rainy days and the severity values accumulated over the 7 previous days.

The rainy days are based on accumulated precipitation. Severity ratings relate the duration of RH above 90% and the average temperature during these periods of high humidity.

1.4.1.6. *J. C. Zadoks*

He studied different aspects of plant disease epidemics and along-with Schein wrote a book "*Epidemiology and Plant Disease Management*".

The simulation model developed (1971) by him provided a numerical integration of Van der Plank's seminal equation of botanical epidemiology.

His famous works pertains with system analysis and the dynamics of epidemics, methodology in epidemiological research, role of epidemiology in modern phyto-pathology and *On The Political Economy of Plant Disease Epidemics: Capita Selecta in Historical Epidemiology".*

Further Zadoks and his associates have shown that within canopy spread of fungal spores is more dependent on micro-meteorology. He found that four uredial generations are needed to establish a metre area of infected pocket.

1.4.1.7. *J. Kranz*

He carried out the comparison of plant epidemics and system analysis. Edited a book *"Epidemics of Plant Diseases – Mathematical Analysis and Modeling".*

He also introduced the concept of anti-epidemic.

1.4.1.8. *K. C. Mehta*

He did his famous works on physiology and epidemiology wheat and barley rust in India. He carried out the role of barberry and Mahonia species in annual recurrence of wheat rust in India.

He established three laboratories in the Agra, Almora and in the Shimla for rust studies.

He concluded that the rust inoculums are reintroduced into the plains of the India year after year from some other sources. He had published *"Further Studies in Cereals Rust in India"* (part one 1940, and part two 1952).

He carried out the aerobiology works in the India and he suggested that rust inoculums not survive in Indian plain, their annual reintroduction carried out through Sothern hilly region. He showed that collateral and alternate host were epidemiologically insignificant in India. And uredospores spread from the Nilgiri and Pani hills in south India, and from central Nepal in the Himalayas.

Initially, Mehta had shown the movement of leaf rust both from north and south Indian hills. Later, Joshi and his team supported this view and further demonstrated that leaf rust (*P. recondita*) spreads both from southern and northern hills. It is introduced from Nilgiri and Palini hills and is established in the plains of Karnataka and Tamil Nadu in South India. The rust population from southern foci moves northwards towards Maharashtra and Madhya Pradesh. The spread of leaf rust over the Indo-

Gangetic plains is predominantly from the warmer north-eastern region and is influenced by the number of rainfall during winter month.

He conclusively proved that due to the prevalence of the high temperature in the plain of India during summer months followed by rainy season, both uredinospores and teleutospores are completely destroyed. And as a result there is no local source of infection in the plains from one season to the next season. Whereas rusts can survive in the hills in the form of urdiospores throughout the year on volunteer plants (self sown), ratoon tillers and also on the summer crop grown in the hills. He concluded that every year rust spores are blown down from the hills to plains where they re-establish infection on the rabi crop.

1.4.1.9. *S. Nagrajan*

He along with Singh (Nagarajan and Singh, 1975) provided the "Indian Stem Rust Rules" to explain the south to north transportation and deposition of stem rust spores along with rain and consequently formulated set of three upper air rules.

These rules explain satisfactorily whether the primary inoculums from South India would travel tens of hundreds of kilometres before its deposition in central India. The stem rust prediction methods proposed by them consist of the following steps. *A*. Check if there were any rains during November in central India together with southerly wind. *B*. Check the urediniospore content of the rain. *C*. Check if the Indian stem rules are satisfied. If a tropical cyclone occurred during November disease would appear near the area where cyclone dissipated. If the inoculums arrive early the disease can develop into severe proportions. If rainfall satisfying the weather conditions of rule b or c occurred during January or February, only traces of the disease would appear. *D*. Check the weather over the area of urediniospore deposition. *E*. check the viability of the transported spores, that is if the trajectory for 700-850 k bar passed near the south foci within 120 hours, spores would be viable and *F*. check if the ground level conditions following the deposition by rain, are favourable for infection or not. If all conditions are satisfied, the first prominent infection could occur in 25-30 days at location closed to where the cyclone dissipated. They further suggested that once infections have occurred in central India, their further spread can be predicted following the same rules with the removal of steps A and C.

He also developed forecasting system for leaf rust and according to Nagarajan *et al.* (1980) epidemic of leaf rust (*Puccinia recondita*) of wheat is likely to develop if (a) rust infections are observed in Uttar Pradesh and Bihar between 15th and 20th January and 5-6 different locations separated from each other by a minimum of 25 ± 5 km (b) the number of rainy days during January to mid April over northwest India is at least twice the normal that is 12 days and (c) the weekly mean maximum temperature during March to mid April over northwest India, is within 1^0 C of the normal temperature (26^0C). If the first two conditions exist fully, then a severe epidemic of leaf rust is likely to occur.

He along with Singh (1973) and with Joshi (1980 and 1985) monitored the satellite data particularly, related with tropical cyclone and their studies revealed that Nilgiris and Palni hills are the main source for the spread of stem rust urediniospores.

They further summarized that though stem rust can survive in the northern hills during all the summer months, however, due to prevalence of low temperature Himalayas have little role to play in the recurrence of stem rust epiphytotics in northern India.

The stem rust uredinospores that are available in South Indian hills get carried to north by tropical cyclones and are deposited along with rains over the central India where the infections are established on one month old wheat crop. They provided that subsequent spread of stem rust from central India to other parts of Northern India is favoured by 'Western Disturbances' linking both area (Fig. 4). The associated winter rainfall permits nationwide appearance and built up of stem rust. Set of these three weather situation (cyclone from the Bengal, western disturbance and winter rainfall) which favours north-west movement of stem rust urdinospore also known as the "**Puccinia Path**"

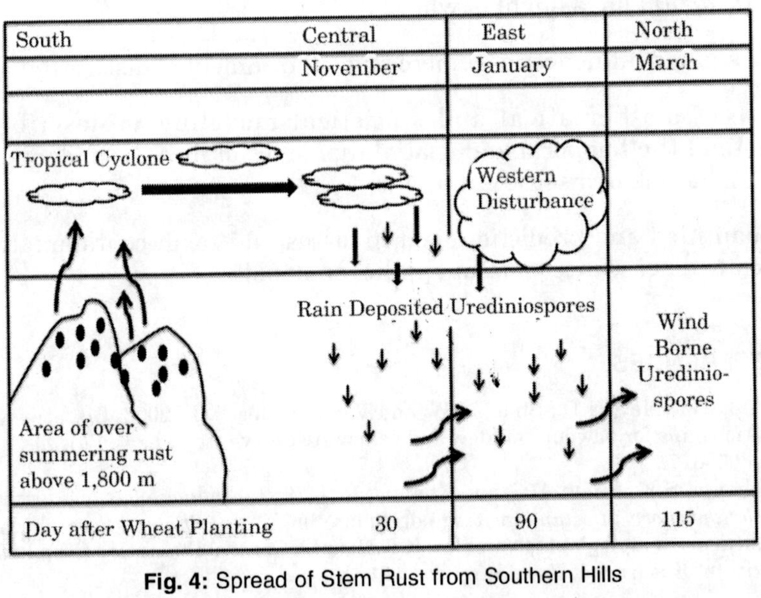

Fig. 4: Spread of Stem Rust from Southern Hills

His famous contribution *"Plant Disease Epidemology"* described the various attributes of plant epidemiology like pathogen and the disease, pathogen variability and fitness, the role of host and its population, effect of weather, climate and plant disease, pathometry, systems approach and epidemic avoidance by mean other than chemicals.

1.4.1.10. *Laurence (Larry) V. Madden*

His research is focused on understanding and describing plant diseases at the population level and relating disease intensity to crop losses.

He used mathematics and statistics to characterize disease epidemics in time and space with the goal of determining which biological and physical factors best predict: disease invasion, persistence, outbreak and the magnitude of disease spread.

He developed approaches for assessing threat of plant pathogen as bioterrorism weapons against United State crops.

He also conducted research on the understand of how fungal spores moves within fields and between fields, and has shown how to better control some economically important disease.

He developed risk and epidemiological models for *Fusarium* head blight.

He contributed in plant disease modeling, statistical analysis and crop loss assessment which were primarily aimed at oomycetes diseases of fruit crop, *Fusarium* head blight of wheat.

He studied influences of climate and environment of disease dynamics.

Use of mathematical and statistical modeling to describe and understand the temporal and spatial dynamics of disease epidemics and forecast the risk of disease outbreaks.

Quantified and prediction of crop losses and toxin contamination in relation to disease intensity and pathogen density

1.5. REFERENCES

Beest, D.E., Paveley, N.D., Shae, M.W. and Van den Bosch, F. 2008. Disease-weather relationships for powdery mildew and yellow rust on winter wheat. *Phytopathology*, 98: 609–617.

Bhattacharya, S.K., Singh, D.S. and Khanna, R.N. 1982. Epidemiology and forecasting of *Phytophthora infestans* inciting potato late blight. *In*: *Phytophthora* diseases of tropical cultivated plants, Nambiar, K.K.N. (*ed.*). Central Plantation Crop Research Institute, Kasaragod, Kerala, pp. 94–101

Bourke, P.M.A. 1953. Potato blight and the weather in Ireland. Technical Note Number 15. *In*: Irish Metrological Services, Dublin, p. 37

Bourke, P.M.A. 1970. Use of weather information in the prediction of plant disease epiphytotics. *Annual Review of Phytopathology*, 8: 345–370.

Clarkson, J.P., Phelps, K., Whipps, J.M., Young, C.S., Smith, J.A. and Watling, M. 2004. Forecasting *Sclerotinia* disease on lettuce: Towards developing a prediction model for carpogenic germination of sclerotia. *Phytopathology*, 94: 268–79.

Cox, A.E. and Large, E.C. 1960. Potato blight epidemics through the world. Agricultural Handbook, Number 174: Department of Agricultural, Washington, D.C., p. 230.

Drenth, A. 2004. Fungal epidemics – does spatial structure matter? *New Phytologist*, 163: 4–7.

Fukui, R., Fukui, H. and Alvarez, A.M. 1999. Effect of temperature on the incubation period and leaf colonisation in bacterial blight of anthurium. *Phytopathology*, 89: 1007–1014.

Garrett, K.A., Dendy, S.P., Frank, E.E., Rouse, M.N. and Travers, S.E. 2006. Climate change effects on plant disease: Genomes to ecosystems. *Annual Review of Phytopathology*, 44: 20.1–20.21.

Jeger, M.J. 2004. Analysis of disease progress as a basis for evaluating disease management practices. *Annual Review of Phytopathology*, 42: 61–82.

Jorge, L., Andrade-Piedra, Gregory, A., Forbes, Dani, S., Niklaus, J., Griinwal, M.G., Chacon, M.V., Robert, J.H. and William, E.F. 2005. Qualification of plant disease simulation model: Performance of the LATEBLIGHT model across a broad range of environments. *Phytopathology*, pp. 1412–1422.

Kranz, J. 1973. Epidemiology, Concepts and Scope. *In*: Current trends in plant pathology, Raychoudhury, S.P. and Verma, J.P. (*eds.*). Lucknow University, Lucknow. India, pp. 26–32.

Krause, R.A. and Massie, L.B. 1975. Predictive system: Modern approaches to disease control. *Annual Review of Phytopathology*, 13: 31–47.

Lovell, D.J., Hunter, T., Powers, S.J., Parker, S.R. and Van den Bosch, F. 2004a. Effect of temperature on latent period of septoria leaf blotch on winter wheat under outdoor conditions. *Plant Pathology*, 53: 170–81.

Lovell, D.J., Powers, S.J., Welham, S.J. and Parker, S.R. 2004b. A perspective on the measurement of time in plant disease epidemiology. *Plant Pathology*, 53: 705–712.

Madden, L.V. and Hughes, G. 1995. Plant disease incidence: Distribution, heterogeneity and temporal analysis. *Annual Review of Phytopathology*, 33: 529–565.

Mallaiah, K.V. and Rao, A.S. 1980. Aerobiology of two species of *Cercospora* pathogenic to groundnut. *Proceeding of Indian National Science Academy*, 2: 215–222.

Mayee, C.D. 1988. Epidemiology and management of groundnut rust. *In*: Vistas in plant pathology. Verma, A. and Verma, J.P. (*eds.*), Malhotra Publication House, New Delhi, pp. 305–320.

McCartney, A. 1996. The influence of environment on the development and control of disease. *In*: Environmentally safe approaches to crop disease control. Rechcigi, N.A. and Rechcigi, J.E. (*eds.*). CRC Lewis Publishing, Boca Raton, pp. 1–31.

Mundt, C.C. 2009. Importance of autoinfection to the epidemiology of polycylic foliar disease. *Phytopathology*, 1110–1116.

Muralidhara, K. and Rao, G.V. 1981. A simple method for forecasting outbreak of rice blast. *Indian Phytopathology*, 33(4): 560–564.

Nagarajan, S. and Singh, H. 1975. Indian stem rust rules-an epidemiological concept on the spread of stem rust of wheat. *Plant Diseases Reporter*, 59: 133–136.

Nagarajan, S., Joshi, L.M., Srivastava, K.D. and Singh, D.V. 1980. Epidemiology of brown and yellow rust of wheat over north India III. Impact of varietal change. *Plant Disease Reporter*, 62: 694–698.

Padmanabhan, S.Y. 1963. Helminthosporium disease of rice-VII: A study of the meteorological factor associated with the epipytotic of 1942 in Bengal. *Oryza,* 1(2): 101–110.

Padmanabhan, S.Y. 1965. Studies on forecasting outbreaks of blasts disease of rice influence of meteorological factor on blast incidence at Cuttack. *Proceeding of Indian Academy of Science*, 62(B): 117–129.

Payak, M.M. and Renfro, B.L. 1967. A new downy mildew disease of maize. *Phytopathology,* 57: 394–397.

Pfender, W.F. 2003. Prediction of stem rust infection favourability, by means of degree-hour wetness duration, for perennial ryegrass seed crops. *Phytopathology,* 93: 467–77.

Pfender, W.F. 2004. Role of phenology in host susceptibility and within plant spread of stem rust during reproductive development of perennial ryegrass. *Phytopathology*, 94: 308–316.

Pfender, W.F. 2006. Interaction of fungicide physical modes of action and plant phenology in control of stem rust of perennial ryegrass grown for seed. *Plant Disease*, 90: 1225–1232.

Pfender, W.F. and Upper, D. 2015. A simulation model for epidemics of stem rust in ryegrass seed crops. *Phytopathology*, 105(1): 45–56.

Pitbaldo, R.E. 1992. The development and implementation of TOM-CAST. Ontario Ministry of Agricultural and Food, Toronto, Canada.

Robinson, R.A. 1976. Plant Pathosystems. Springer-Verlag, Berlin.

Rosa, M., Genesio, R., Gozzini, B., Maracchi, G. and Orlandini, S. 1993. PLASMO: A computer programme for grapevine downy mildew development forecasting. *Computer Electronic Agricultural*, 12: 311.

Saharan, G.S. and Singh, B.M. 1980. Influence of environmental factors in the latent period of *Melampsora lini* under field conditions. *Indian Phytopathology*, 33: 270–274.

Saxena, S.C. and Lal, S. 1984. Use of meteorological factors in prediction of *Erwinia* stalk rot of Maize. *Tropical Pest Management*, 30: 82–83.

Scherm, H., Ngugi, H. and Ojiambo, P. 2006. Trends in theoretical plant epidemiology. *European Journal of Plant Pathology*, 115: 61–73.

Schoeny, A. and Lucas, P. 1999. Modeling of take-all epidemics to evaluate the efficacy of a new seed-treatment fungicide on wheat. *Phytopathology*, 89: 954–61.

Shanta, P. 1960. Studies on *Cercospora* leafspot of groundnuts (*Arachis hypogea* L.) Effect of the environmental factors on disease incidence and on survival of the pathogen: *Journal of Madras University*, 30: 167–177.

Shrum, R.D. 1978. Forecasting of epidemics, *In*: Plant disease-AN advance treatise. Horsfall, J.G. and Cowling, E.B. (*eds.*). Academic Press New York, San Francisco and London, II: pp. 223–238.

Shukla, V.D. and Anjaneyulu, A. 1981. Plant spacing to reduce rice tungro incidence. *Plant Disease*, 65: 584–586.

Skelsey, P., Rossing, W.A.H., Kessel, G.J.T, Powell, J. and Van der Werf, W. 2005. Influence of host diversity on development of epidemics: An evaluation and elaboration of mixture theory. *Phytopathology*, 95: 328–338.

Tellier, A. and Brown, J.K.M. 2008. The relationship of host-mediated induced resistance to polymorphism in gene-for-gene relationships. *Phytopathology*, 98: 128–136.

Vanderplank, J.E. 1963. Plant disease: Epidemics and Control. Academic Press, New York, p. 349

Waggoner, P.E. 1960. Forecasting epidemics. *In*: Plant Pathology, Horsfall, J.G. and Diamond, A.E. (*eds.*), Academic Press, New York and London, III: pp. 291–312.

Xu, X.M. 1999. Modeling and forecasting epidemics of apple powdery mildew (*Podosphaera leucotricha*). *Plant Pathology*, 48: 462–71.

CHAPTER

2 Attributes of Plant Epidemic

2.1. PATHOGEN TYPES AND THEIR RELATIONSHIPS WITH EPIDEMIC

Appearance of a disease in a large number of individual over large areas in relatively short time is known as epidemic.

Epidemiology deals with the study of the development of the disease in population and this entire event is a cyclical phenomenon, that is, they consist of repeated cycles of pathogen development in relation to the host and the environment. In general, the eight major events are involved in disease development (Figure 1).

The inoculum, which might consist of fungal spores, bacterial cells, nematodes, viruses within an aphid vector, or some other propagules of a pathogen, gains entry into and establishment within the host tissues through the process of infection.

The pathogen develops within the host and eventually begins to produce new inoculum, which, in turn, can be dispersed to new susceptible sites to initiate new infections. Based on stages of pathogenesis the inoculums may be primary or secondary type (as described in chapter 1).

Pathogens that produce only one cycle of development (one infection cycle) per crop cycle are called **monocyclic**, while pathogens that produce more than one infection cycle per crop cycle are called **polycyclic.**

For some diseases it is important to consider an epidemic over a period of many growing seasons. This is particularly true for perennial plants (forages, pastures, lawns, orchards, forests, etc.) or for annual crops that are grown in monoculture year after year. In these situations the inoculum produced in one growing season carries over to the next, and there could actually be a build-up of inoculum over a period of years. Such epidemics are refer to epidemics as **polyetic** epidemics, regardless of whether the pathogen is monocyclic or polycyclic within each season.

2.1.1. How the Epiphytotics Develop

In plant population the epiphytotics constructed in the similar manner as the rise in human population.

The human population rises due to (a) high birth rate, (b) low death rate, (c) low birth rate but still lower death rate or (d) high death rate but still higher birth rate. Examples of the high birth rate diseases includes late blight of potato (*Phytophthora infestans*), apple scan (*Venturia inaequalis*), rust and powdery mildews.

In such types of pathogen having high multiplication rates and rapid dispersal however, they are effectively managed by fungicidal sprays.

Low death pathogen are mostly systemic in nature, such disease are common in fruit trees or in perennial species and in few crops like sugarcane.

Such diseases can be checked only be removal of the source of inoculums through sanitary practices and other cultural operations. Such disease should be identified and remove timely because once they established, it becomes difficult to eradicate the primary inoculums sources.

Sugarcane ratoon stunting (*Clavibacter xyli* subspecies *xyli*), citrus greening (*Liberobacter asiqticum*), phony peach disease (*Xylella fastidiosa*) spread slowly but are very persistent unless the plant harbouring the pathogens in xylem or phloem is destroyed. Thus, the build up of an epidemic is actually manifestation of balance between two opposing forces.

2.1.2. Monocyclic Pathogens

Monocyclic pathogens complete part or all of one cycle of pathogenesis in one cropping season, and therefore have a maximum of one generation (one cycle of pathogenesis) per season.

Monocyclic disease is also known as Simple interest disease. The term simple interest coined by Vanderplank (1963). Monocyclic disease denotes such a disease which increases mathematically analogous to simple interest in money. In this type of disease there is a common source of inoculum. Generally called primary inoculum, from which develops only one generation of infection in the life of a host crop in one season. This is because secondary infection rarely occurs during the season.

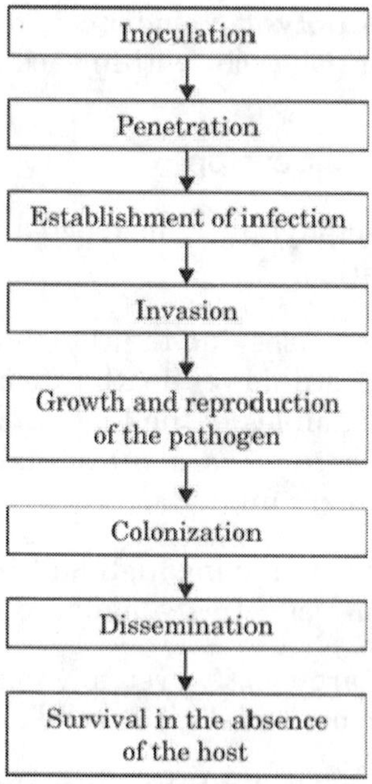

Fig. 1: Events for disease development

The pathogen causing a simple interest disease (or monocyclic disease) is called monocyclic pathogen. Thus, in monocyclic disease, the amount of inoculums does not increase significantly during the season. Therefore the rate of disease increase is affected only by the inherent ability of the casual organism as well as by environmental and cultural practices.

Basic characters of monocyclic plant disease epidemic are:

- ***Rate of increases*:** Like simple interest in terms of money.
- ***Reproduction*:** Reproduction cycle long (reproduce only one generation per growing season.)
- ***Birth rate*:** Low.
- ***Survival*:** Inoculum long lived.
- Pathogen systematic in nature
- Host survive for longer duration after infection
- ***Death rate*:** Low
- Control mostly through sanitation practices, and
- ***Example*:** Soil borne and some seed borne diseases

Monocylic disease are exemplified by loose smut of wheat (*Ustilago nuda*), cover smut smuut of barley (*U. hordei*) and grain smut of sorghum

(*Sphacelotheca sorghi*), vascular wilt diseases caused by *Fusarium oxysporum.*

Monocyclic nature of the pathogen is either determined by environmental factors or pathogen life cycle requirements that are inherent in their nature. Some pathogens are monocyclic because environmental or physical factors prevent more than one cycle of pathogenesis.

- For example, *Verticillium dahliae*, which induces a wilt of cotton and many other plants, is monocyclic primarily because of environmental constraints. The fungus survives from one season to the next as microsclerotia (compact groups of thick-walled resistant cells) in decaying stems and roots. These are eventually liberated to the soil. At the beginning of the season, the population of *V. dahliae* (initial inoculum) exists mainly as microsclerotia in soil. When roots of cotton plants grow sufficiently close to microsclerotia, exudates from the roots stimulate germination and, if the soil environment is favourable, *V. dahliae* penetrates the plant. The fungus first invades cortical tissue, then selectively invades xylem. It invades the plant systemically *via* hyphal growth and passive transport of conidia through the xylem. Infections can occur any time during the season when healthy roots come sufficiently close to microsclerotia. Late in the season, the fungus produces microsclerotia in dying and dead cotton tissues. These function as inoculum only if stimulated to germinate near susceptible roots of the next crop, usually after they are dispersed during tillage and cultivation. Thus, *V. dahliae* normally does not produce inoculum that functions during the same season to induce disease in a different plant. Consequently there is only one generation per season because of physical or environmental conditions.
- Many soil-borne pathogens (*i.e.*, *Fusarium solani* f. sp. *phaseoli*) are usually restricted to one generation per season. The incidence (number) of infected plants may increase during the season because roots of healthy plants contact inoculum, present since the beginning of the season, as they grow through soil. Additionally, some pathogens that rot stored produce (*i.e. Penicillium expansum*, which rots apples) are effectively monocyclic because storage conditions prevent their dispersal.
- In other cases, a short duration of favourable weather may permit only one cycle each season. Some pathogens are monocyclic because their life cycle requires at least one season and there are not repeating stages. *Gymnosporangium juniperi-virginianae* (which induces rust of apple) is an example. This fungus is a short-cycled rust whose alternate hosts are red cedar and apple. The fungus overwinters in galls on red cedar that take about 21 months to develop to maturity. Although the life cycle requires 2 years, the overlapping of generations assures that

inoculum from cedars is produced each year. In addition to *Gymnosporangium juniperi-virginianae* there are many other pathogens that are monocyclic because their life cycles require at least one season.

- Many smuts (*i.e.*, *Ustilago nuda*, which is loose smut of wheat) and other short-cycled rusts have no asexual repeating stage and completion of the life cycle requires an entire season, or it may be synchronized with the life cycle of its annual host plant.
- Time required for some nematodes (*i.e. Globodera rostochiensis*, the golden nematode of potato) to complete their life cycles may be about the same as a growing season.
- Inoculum in the current crop is not disseminated until the end of the season. Many soilborne pathogens fall into this catergory. For example, White rot of peas (*Sclerotinia sclerotiorum*) produces sclerota which remain in the soil until stimulated to germinate to exudates of the host. During the season fungus may multiply by producing more scllerotia but they will not be distributed in the soil until harvest.
- A limited time is available for infection for pathogen. Example includes, pearl millet is only susceptible to infection by *Claviceps fusiformis* for a few days only after pollination.

Examples of monocyclic diseases are as follows

***Soil borne*:**

- Wilt of Cotton (*Gossypium spp.*) = *Fusarium oxysporum f sp. vasinfectum.*
- Wilt of Arhar (*Cajanus cajan*) = *F. oxysporum f sp. udum.*
- Wilt of Linseed (*Linum usitatissimum*) = *F. oxysporum f sp. lini.*
- Panama disease of Banana (*Musa paradisiaca*) = *F. oxysporum f sp. cubense.*
- Head smut of Jowar = *Sphacelotheca reiliana*

***Seed borne*:**

- Loose smut of Wheat = *Ustilago tritici.*
- Covered smut of Barley = *Ustilago hordei.*
- Flag smut of Wheat = *Urocystis tritici.*
- Bunt or Stinking smut of Wheat = *Tilletia tritici.*

2.1.3. Polycyclic Pathogens

Polycyclic pathogens have more than one generation per cropping season.

Some have many generations per year and therefore produce propagules nearly continuously.

Polycyclic pathogens are commonly dispersed by air. Relatively few soil-borne pathogens are polycyclic; but some (*e.g.*, certain nematodes, fungi that cause damping-off) may have short generation times and may migrate short distances to healthy roots during the season to be polycyclic.

The generation time is influenced greatly by the host population and the environment. The generation time may approach its theoretical minimum time (pathogen development approaches its maximum rate) in a susceptible host population during weather favorable to the pathogen. Conversely, in unfavorable weather and in a resistant host population, pathogen development may be very slow.

Because environment and host population are so influential, it is clear that the number of generations of a polycyclic pathogen is likely to vary among seasons and crops. In situations especially unfavorable to their development, some polycyclic pathogens may have only one cycle per season.

- *Phytophthora infestans* causes potato late blight, can have many generations per season. *Phytophthora infestans* survives from one season to the next in infected potato tubers left behind in the harvest or discarded in cull piles. The primary inoculum consists of sporangia produced on these infected tubers. The sporangia are disseminated by air currents. If they alight on susceptible tissue in the presence of free water at a suitable temperature, they germinate (either by a germ tube or by zoospores) and penetrate the host to produce lesions. Under ideal temperature and humidity, the fungus produces sporangia from infected tissue in 4-6 days. The fungus in the lesion continues growth, causing the lesion to enlarge, and producing new sporangia. These are readily dispersed by wind and splashing water to initiate secondary cycles. The fungus in a single lesion can produce more than one hundred thousand sporangia. Lesions develop so rapidly that the fungus may sporulate from "progeny" lesions while the "parent" lesion is still supporting sporulation. Thus, there can be an overlapping of generations. Sporangial production is governed primarily by weather rather than by discrete generations. Even if only a small proportion of these sporangia establish new infections, the increase of inoculum is tremendous, and late blight often appears to "explode overnight." Population increase may be exponential.
- Diverse pathogens may be polycyclic. Members of each taxonomic group of biotic plant pathogens are polycyclic. The time required for a generation (one cycle of pathogenesis) depends on each pathogen. For example, the nematode *Trichodorus christei* progresses through

relatively few generations each season because of its long generation time (20–30 days).

- On the other hand, the generation time for *Xanthomonas phaseoli*, which induces common blight of beans, may be as short as 2-3 days. When infected leaves are wet, bacteria multiply rapidly and produce ooze on external surfaces. Droplets of ooze containing bacteria are dispersed to susceptible tissue by splashing rain or wind-driven water drops. Penetration and bacterial replication may occur so rapidly that ooze is again produced on the newly infected leaf within 2-3 days.
- In order for an epidemic to be considered polycyclic, there must be repeated complete infection cycles, that is, infection followed by pathogen development, new inoculum production, dispersal to new susceptible sites, and new infections, all within a single crop cycle. A good example is potato late blight, where a single cycle of infection, lesion development, sporulation, sporangium dispersal, and new infection can occur in as little as five days, and many overlapping cycles occur simultaneously during periods of favourable weather. Each cycle can produce more than a ten-fold increase in numbers of sporangia landing on susceptible sites, and an explosive epidemic results. Cereal rusts behave similarly; a single urediniospore can infect to produce a pustule from which hundreds of new urediniospores can be released to infect and produce hundreds of new pustules repeatedly throughout the season. Most plant diseases caused by bacteria are polycyclic, and many plant viruses, with the aid of their insect vectors, also can produce repeated cycles of infection in one season.

2.1.4. Combinations of Monocyclic and Polycyclic Epidemics

Not every plant disease epidemic is clearly either monocyclic or polycyclic. Epidemics produced by fungi with two spore stages can have elements of both, sometimes in distinct phases, and sometimes occurring simultaneously.

For example, the fungus *Venturia inaequalis*, causal agent of apple scab, produces ascospores on the dead, infected leaves that have overwintered from the previous season. These ascospores are released over a period of six to eight weeks in the spring and infect the newly expanding apple leaves. Since no new ascospores are produced until the following spring, this component of the epidemic might be considered monocyclic. However, each leaf lesion within about ten days produces a second type of spores, conidia, that also can infect newly expanding leaves. Thus, for the early part of the season, a polycyclic epidemic is superimposed on a monocyclic epidemic. Since lesions produced by conidia cannot be distinguished from those produced by ascospores, the net effect appears to be a rapidly growing polycyclic epidemic.

An epidemic of tan spot on wheat, caused by *Pyrenophora tritici-repentis*, also comprises both polycyclic and monocyclic elements.

The initial inoculum occurs on infected seed, and the epidemic starts as these seeds germinate to yield seedlings bearing leaf lesions. Conidia are disseminated by wind and splashing rain in repeated cycles of leaf infection throughout the period of crop development (the polycyclic phase).

By the time flowering begins, the inoculum for the second phase of the epidemic has built up on the upper leaves of the plant, principally the flag leaf. The flowers and developing grains are susceptible for a relatively brief period to infection by the inoculum extant at the time of flowering, and the lesions on the glumes of the developing grains do not produce inoculum capable of infecting more grains. Seed infection, therefore, is a monocyclic phenomenon. It is these infected seeds that provide a major means of carryover of this pathogen from one crop to the next and the initial inoculum for the new crop.

2.1.5. Polyetic Pathogens

If a pathogen is capable of surviving long periods in large proportion, then inoculum at the end of one season is closely related to inoculum at the beginning of the next season *i.e.* carry forward of previous inoculums in new gneration.

If inoculum increases from year to year, pathogen development over a number of seasons becomes exponential regardless of its development within a season. Such pathogens are ***polyetic***.

In polytene plant disease the inoculums build up in several years and they are compound interest disease with low r. Mostly occur in perennial host, and example of such diseases are Chestnut blight and apple powdery mildew.

- The Dutch elm disease is an example of a monocyclic pathogen giving rise to a polyetic epidemic. Note that while there is only one cycle of infection each year and disease progress within each year is roughly linear, the incidence of infected trees increases at an increasing rate from year to year. *Ceratocystis ulmi* (which induces Dutch elm disease) is monocyclic within a given season, but Dutch elm disease increases exponentially when considered over several seasons. The fungus survives the winter in infected and recently killed elms. Elm bark beetles become contaminated with conidia of *C. ulmi* when they leave brood galleries in these elms, and then inoculate healthy elms during their subsequent feeding. Most successful inoculations occur in the late spring and early summer when the pathogen readily colonizes rapidly growing tissues. The dynamics of Dutch elm disease over several years are definitely exponential.

- Apple powdery mildew is an example of a polyetic epidemic caused by a polycyclic pathogen.
- Sigatoka leaf spot on bananas decreases somewhat during the dry season but otherwise produces more or less continuously repeated cycles of infection. The host, in this case, consists of a population of plants of different ages developing continuously over a long period of time.

Not all pathogens survive from one season to the next in large proportions, and for these, pathogen development is unlikely to be exponential when considered over several seasons.

For example, in the absence of barberry plants in the Midwestern United States, *Puccinia graminis* f. sp. *tritici* (stem rust in wheat) does not overwinter there in large proportion, and the amount of stem rust in any year may be unrelated to the amount in the previous year.

2.2. COMPARISON OF DISEASE SEVERITY

Both monocyclic and polycyclic pathogens induce diseases that are important and severe. Although it may appear initially that polycyclic pathogens ought to be more bothersome in agriculture because of their many cycles, this simplification is not sustained by experience.

For example, wheat is affected by many pathogens, and although rusts (polycyclic) are probably the most notorious however, diseases induced by soil-borne pathogens frequently cause greater yield losses. Diseases of foliage frequently attract attention because of their visibility and because polycyclic pathogens frequently increase in population dramatically *via* rapid reproduction (rapid birth rates).

Conversely, diseases of roots are less noticed because roots are less visible and the pathogens tend to have longer generation times.

Populations of monocyclic pathogens increase because of longevity of reproductive units (slow death rates), and populations increase over a long time period.

Severe disease results from a large pathogen population regardless of the speed at which the population increased. Thus, severe disease is induced by both monocyclic and polycyclic pathogens.

Measurement of the quantity of inoculums provide a good assessment of the epidemic progress, but direct observation of inoculums is not always feasible or even possible.

With most plant disease some kind of symptoms produce or appear that allow us to count the number of plants infected, but there is a time lag between infection and the appearance of visible symptoms, furthermore, the symptoms are also affected by environmental conditions.

If the disease progresses in monocyclic fashion, the epidemic is linear and the slop of the disease progress curve is proportional to the amount of initial inoculums. Therefore, we can describe a monocyclic epidemic with linear disease progress using differential equation, *e.g.*, Flax wilt

$$X = QRT$$

Here, Q is the amount of initial inoculums; R is the rate of disease progress per unit of inoculums and t is the time.

Q = 0.57 CFU

R = 0.0013 and T is 60

Then, X =QRT; 0.57 × 0.0013 × 60 = 0.044 or 4%

If the disease progresses in polycyclic fashion the rate of change in disease or amount of disease is proportional to the amount of inoculums present at any point of time, *e.g.*, Holo blight

$$X = X_0e^{rt}$$

Here, X_0 is the proportion of disease at the start of the epidemic; e is the base of natural logarithm and r is "apparent infection rate" and its based on appearance of disease symptoms

Final incidence = X_0 = 25% or 0.25; T = 90 and R = 0.12

Then, $X = X_0e^{rt}$

$0.25 = X_0 \exp^{(0.124 \times 90)}$; $X_0 = 0.25/70262 = 0.000036$

Thus, epidemic can be control through three ways:

1. Reduce initial inoculums amount = Q for Mono and X_0 for Polycyclic
2. Reduce rate of infection = R for Mono and r for Polycyclic
3. Reduce duration of epidemic = t.
   - In monocyclic model Q, R and T have equal effect on X
   - While, in polycyclic disease, reducing r has a relatively greater effect on epidemic than reducing X_0. Thus, reducing X_0 make good strategies sense only if *r* is low.
4. Selecting the best model to estimate initial inoculums amount, *r* and *t*

2.2.1. Reducing the Amount of Initial Inoculum (x_1)

The initial amount of inoculum can be kept at zero by avoiding the pathogen altogether. This can be achieved by growing crops in regions where the pathogen does not occur, by choosing pathogen-free planting sites within a particular area, by choosing planting dates that ensure that the crop is not at a susceptible stage of growth when the pathogen is active or by using disease-free planting material.

The amount of initial inoculum can also be kept at zero by excluding the pathogen from a region.

Strategies used to prevent the introduction of pathogens into new areas include quarantine regulations, sanitation, use of planting material certified to be free from disease, and treatment of planting material to kill any pathogens on or in it.

Destruction of vectors, such as flying insects that may carry pathogens into an area will also keep the level of initial inoculum at zero.

If the pathogen is already present in an area, the level of inoculum can be decreased by various forms of sanitation such as crop rotation, destruction of diseased debris, removal of alternative and alternate hosts, removal and destruction of diseased plants or plant parts and soil fumigation.

Pathogens may be eliminated from planting material by heat or chemical treatments. Plant propagules can be treated with chemicals or micro-organisms to protect against infection.

In some virus diseases, inoculation with a benign or mild strain of the virus protects plants against subsequent infection by more virulent forms of the virus.

Specific genes for resistance in plant populations prevent some or all of the available inoculum from infecting a crop. If a new pathotype arises which can attack plants containing the resistance gene, the disease epidemic will develop as it did before the resistance gene was introduced.

2.2.2. Reducing the Time Available for the Epidemic to Develop (t)

The time available for an epidemic to develop can be reduced in some instances by using early-maturing cultivars which are harvested before the disease has developed to epidemic proportions.

Similarly, with some fruit crops such as bananas, the fruit is picked early (before it is ripe and before postharvest diseases associated with ripening have had time to develop) and later artificially ripened by exposure to ethylene under environmental conditions that suppress disease development.

With pastures and forage crops the plants can be grazed or harvested early (*e.g.,* to make hay) if there is a risk of an epidemic developing. In many crops however, it is not possible to alter the time available for epidemic development

2.2.3. Reducing the Rate of Disease Increase (r)

The rate of disease increase can be reduced in a number of ways. Slow rusting, slow mildewing and slow blighting types of resistance reduce the rate of disease increase, as do multilines or cultivar mixtures in which not all individuals are susceptible to a given strain of a pathogen.

Chemical treatments can also reduce the rate of disease development. Protectant fungicides reduce the level of spore germination which results in fewer lesions being formed and less inoculum being produced.

Systemic fungicides which destroy established infections can also influence the rate of disease development as well as the level of initial inoculum.

The rate of disease increase is also influenced by climate. Apart from the expensive process of growing plants in greenhouses, the environment to which a crop is exposed can be altered by changing planting dates, by changing the density of the host population, by modifying irrigation practices or by growing crops in different climatic areas.

Nutritional status of plants also influences their susceptibility to disease. Sanitation practices such as removing diseased plants or eliminating vectors can also reduce the rate of epidemic development.

2.2.4. Selecting the Best Model to Estimate Initial Inoculums Amount, *r* and *t*

Madden (1980) and Nutter (1997) has advocated that the section of a temporal model suitably describe the temporal progress of plant disease epidemics which involve subjective as well as objective criteria.

Monomolecular, exponential, logistic and Gompertz population models are the commonly use temporal population growth models and each models have certain specific steps which will be discussed in next chapter. Basic comparisons between these models are presented in Table 1.

Table 1: Basic model for population growth models.

Monomolecular	*Exponential model*	*The logistic model*	*Gompertz model*
This model suitably describes the plant disease pathosystems particularly related with single cycle during growing season. The monomolecular population growth model is applied to epidemics for which there is no spread from plant to -plant, *i.e.* there is no secondary spread within a growing season	This also known as logarithmic, geometric, or Malthusian model and according to Van der Plank (1963) it is best for compound interest disease. Therefore, +3this model implies that the carrying capacity of the host crop does not limit the absolute rate of disease increase. Disease (or pathogen) progress curves are typically J-shaped and lack an inflection point	It is the second type of compound interest model proposed by Van der Plank (1963) and may be appropriate for most plant disease epidemics, where there is plant-to-plant spread within the crop. Disease (and pathogen) progress curves typically have a characteristic sigmoid (S-shaped) form, with an inflection point at the time when disease intensity reaches a proportion of 0.5 (50%).	This model would be appropriate for polycyclic (compound interest) diseases as an alternative to the logistic model.

2.3. SIMPLE INTEREST AND COMPOUND INTEREST DISEASES

Simple interest diseases	*Compound interest diseases*
• In simple interest diseases the rate of increase is mathematically analogous to simple interest in money. • There is only one generation of the pathogen in the life of the affected crop. • The primary inoculum is seed-borne or soil-borne. The secondary infection rarely occurs during the crop season. That is, the pathogens do not spread from plant to plant in one growing season. • The dispersal of propagules is restricted by climatic and biotic conditions. • Simple interest diseases are caused by seed-or soil-borne smuts, like loose smut of wheat, covered smut of barley and soil borne fungi which attack roots, like wilt (*Fusarium oxysporum*) and root rot (*Rhizoctonia* spp.) diseases. • Most of the smuts infect the seedlings, grow along with the growth of the plant and produce spores in the inflorescence on maturity of the crop. There is no secondary spread from the smutted heads.	• In compound interest diseases the rate of increase is mathematically analogous to compound interest in money. • The pathogen produces enormous amount of spores at a very rapid rate. • These spores are disseminated rapidly by wind and infect the other plants. • Both the inoculation and sporulation period are short so that the pathogen spreads from plant to plant during the same growing season. • New crop of spores is produced, disseminated and the cycle is repeated fast. Thus more generations of the pathogen are produced in the life of a crop. *e.g.*, late blight of potato, powdery mildews and rust diseases. If we consider wheat stem rust caused by *Puccinia graminis tritici* as an example, the fungus produces uredospores in very large numbers (50,000 to 4,00,000 uredospores per uredosorus). These spores are spread by wind and infect other plants. Each of

Table: (*Contd...*)

Table: (*Contd...*)

These smut diseases are mostly systemic in nature. They do not produce propagules external to the host during the active season of the crop. Dispersal of propagules of these fungi is restricted by existing climatic and biotic conditions. • Simple interest epidemics can be described as: dx/dt = QR (1-X) x = proportion of disease tissue or plants at time t; Q = initial inoculums; and R = rate of infection.	the freshly infected wheat plant produces uredo pustules within 5 to 7 days at 24°C. Thus within a week of appearance of the first pustule in the crop several thousand new pustules are formed which could repeat the process within a week.

2.4. COURSE OF EPIDEMIC

The course of epidemic follows two distinct phases *viz.*, (a) Progressively destructive phase and (b) the decline phase

Progressively destructive phase	Decline phase
• Some epidemics develop slowly (tardive) while others develop rapidly. • Slow epidemics (or epiphytotics) usually occur among population caused by systemic pathogens. The pathogen multiplies slowly following the characters of simple interest disease. • They belong to low death rate category and have less incubation period and sporulation period. However, the rapid epiphytotics are greatly influenced by environmental factors.	• During early stage, an epidemic spreads vigorously causing diseases in new hosts. After development of a saturation stage it shows a decline by itself. • No epidemics may be due to non-availability of suspects non-availability of susceptible stages of the crop, unfavorable weather conditions and reduction in aggressiveness of the pathogens. • Generally the hosts are prone to the disease at a specific developing stage. Once this stage is crossed in a plant it's proneness to infections is reduced or completely lost. Under the conditions the epidemic declines. • The decline in the epidemic may also be due to unfavorable weather conditions for disease development. As a result future spread of the disease will be checked and the epidemic will decline. • Wheat crop in Northern India usually gets the attack of rusts in January to March. Epidemics develop during these months. Although the plant remains prone to attack afterwards also, further development of the disease is checked because of rise in temperature which is favorable for the pathogen. • Due to the above mentioned and other causes, the aggressiveness of the pathogen may be reduced. • When all susceptible individuals are destroyed by the pathogen, it may try to parasitize the remaining resistant individuals of the same

Table: (*Contd...*)

Table: (*Contd...*)

	species. In these adverse conditions, the pathogen may lose its power of successful infection, its reproduction may slow down and the pathogen becomes less aggressive.

2.5. SLOW AND RAPID EPIPHYTOTICS

The form of epidemic is decided by the nature of the pathogen, host and the weather. Epidemic may develop slowly and is called 'tardive'; while, epidemic which develops rapidly is called 'explosive'. In between these intermediate forms of epidemic may occur.

• Slow epiphytotics occur among perennial (tree) populations. • Infected host survives for several years before dying. • Most of the characters of a simple interest disease are found in slow epiphytotics. The causal agent is mostly systemic. • The pathogen multiplies slowly. • Their movement from plant to plant is much slower. They are low death rate pathogen. • In slow epiphytotics, crop sanitation is the best method. *e.g.*, Swollen shoot of cocoa. This disease spreads very slowly from tree to tree and still less from one garden to another garden. For instance, the incidence of 31 % swollen shoot incr-eased to 75 % over a period of 2.5 years. • As stated by Van der Plank (1959) the rate of multiplication of a systemic disease of trees is about tenfold a year whereas it is 10,000 fold in respect of herbaceous plants and it is of higher rates for local lesion pathogens *e.g.*, late blight of potato, wheat stem rust, *etc*.	• Rapid epiphytotics occur among annual crops. • It is caused by non-systemic pathogens with high birth rate. • Several generations of the pathogen is produced within a short time. • Rapid epiphytotics are largely governed by environmental factors compared to slow epiphytotics. • Disease increase is rapid and the disease rises to a beak in short time and then show sharp decline when the weather turns unfavorable or when the host becomes resistant due to maturity or due to restricted dispersal of propagules of pathogen. *e.g.*, apple scab. • This type of epiphytotic is controlled by protective spraying or dusting with chemicals.

2.6 DISEASE PROGRESS CURVE

Patterns of epidemic show by a curve which is known as "Disease Progress Curve". This curve denotes the progress of epidemic in population and its shape is depending on amount of initial inoculum and on pathogen growth rate.

The progress of an epidemic measured in terms of the numbers of lesions/ the amount of diseased tissue, or the numbers of diseased plants plotted over time is called the disease–progress curve.

Thus, a disease progress curve is simply a graph of the amount of disease vs. time.

Basically, this curve provides the information's about

- Recurrent weather conditions
- Appearance time
- Amount of initial inoculums, and
- Changes in host susceptibility during infection.
- Information's about disease dynamics
- Impact of management strategies on disease
- Help in disease forecasting, and

For monocyclic disease it's of saturation type (Fig. 2A), for polycyclic it's of sigmoid type (Fig. 2B), while for disease in which multiple plant parts are infected it's of bimodal type (Fig. 2C).

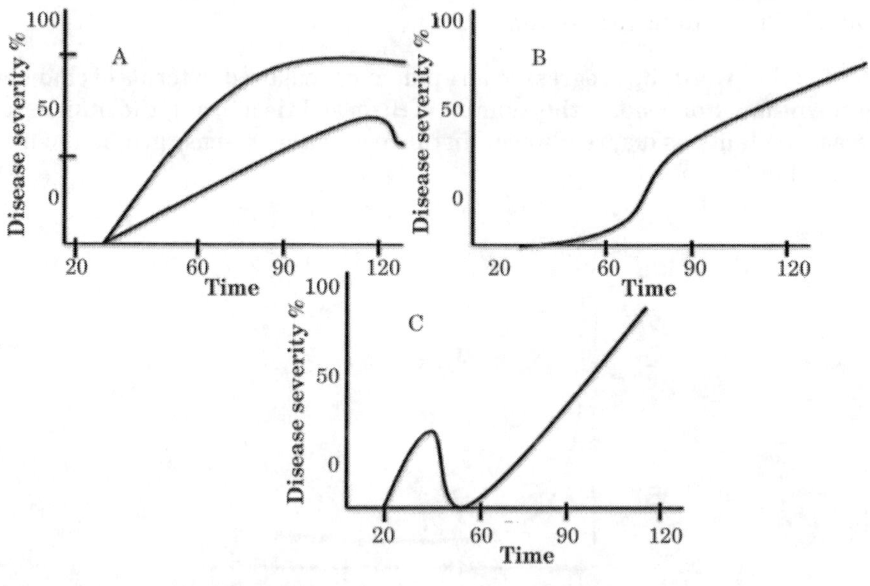

Fig. 2A-2C: DPC curves of various types of diseases

2.6.1. DPC Shape of Few Diseases

In *Phytophthora* blight of pepper seedlings (*Phytophthora capsici*) and *Fusarium* kernel root (*Fusarium moniliforme*) of maize, DPC is linear

On the other hand in bean rust (*Uromyces phaseoli*) and grey leaf spot of corn (*Cercospora zeae-maydis*), the DPC is exponential.

Practically, plant disease cannot continue to increase forever, and as the level of disease approaches 100%, the DPC gradually flattens out. For example, in epidemics such as the infection of beans by *Sclerotium rolfsii* or the infection of tobacco by *Phytophthora parasitica* var. *nicotianae,* disease progress starts out looking linear but slow down as it approach maximum.

Similarly, the DPC of *Puccinia graminis* subsp. *graminicola* on ryegrass is exponential at first, but as time goes on and the incidence and severity of disease approach 100%, the rate of disease gradually slow to zero, giving both curves a somewhat sigmoid shape. Thus, plant disease epidemics tend to be either roughly linear or exponential in the early stages, and they tend to level off as they approach some limit.

2.7. DISEASE GRADIENT CURVE

In and out movement of the pathogen form the host population called Disease Gradient Curve. This curve shows high disease intensity near the source of inoculums (Figure 3).

In other word the progress of an epidemic measured in terms of changes in the number of lesions/ the amount of diseased tissue, and the number of diseased plants as it spreads over distance, is called disease gradient curve (spatial pattern).

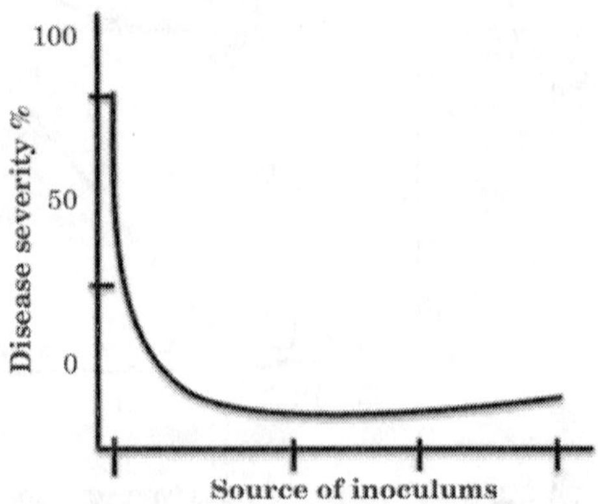

Fig. 3: Disease Gradient Curve

Mechanisms of dispersal differ widely among plant pathogens, including the following mechanisms including

- Spore ejection
- Wind

- Insect and other arthropod vectors
- Nematode vectors
- Rain
- Water runoff
- Movement of infected plant material
- Human influences such as movement of propagules in farm machinery

Dispersal and disease gradients corresponding to these mechanisms are often estimated. Dispersal gradients represent the frequency distribution of the distances travelled by all individuals in a population and the application of dispersal gradients has been useful for characterizing unidirectional dispersal.

Disease gradients take into account all events leading to the spread of disease, including release of inoculum, transport, and deposition, as well as the presence of susceptible hosts in a disease-conducive environment.

For plant pathogens, the primary sources of inoculum are often one of three general types: point, line, or area. A point source typically has a diameter smaller than 1% of the gradient length, while line or area sources do not necessarily have a fixed size.

For example, a line source may be a row of diseased plants, and pathogen/disease measures are made at increasing distances away from this source.

Furthermore, there are two types of dispersal gradients that need consideration: primary and secondary. Primary disease gradients indicate the dispersal potential of a pathogen (inoculum) in a single disease cycle (single source). Secondary disease gradients occur when inoculum is moved from lesions (plants) that had been infected during the primary dispersal event.

In addition to measurements of dispersal within fields, long distance dispersal (LDD) is important for many plant pathogens (*e.g.*, Asian soybean rust, tobacco blue mold).

Successful transmission of plant diseases over long distances often depends on: the reproductive rate of the pathogen; the carrying capacity of the source locality; the role of atmospheric turbulence, stability, and wind speed; and the survival of spores during exposure to inhospitable temperature and humidity and to UVB radiation from the sun (Mundt, 2002; Tack *et al.*, 2013 and Papaix *et al.*, 2014).

2.8. TAKE OFF LEVEL AND ZERO DATE

Switch over point on which disease development is rapidly increased known as take off level. For stem rust of wheat it's between 0.52-0.72. While, the "time" when epidemic reach at take off level known as zero date.

Basic features of zero date are as follows

- It's a date when epidemic rate (r) is very high.
- Zero date blur the impact of inoculums amount and by this the epidemic tend to be more weather sensitive.
- In general, the Zero date of more susceptible variety appear early.
- Zero date largely utilize in forecasting practices.
- Fungicidal application is generally advisable after the zero date

2.9. EPIDEMIC RATE OF DISEASE (R)

The epidemic rate of disease is denoted by r and it represents the rate of growth of epidemic. In general r is the amount of increase of disease per unit of time (per day, week, or year) in the plant population.

Shape of epidemic rate curve varies with different diseases for example for the late blight of potato, the rate curves are symmetrical (bell shaped) while for apple scab and most downy and powdery mildews, the rate curves are asymmetrical, with the r being greater during early growing season because of the higher proportion of susceptible tissues.

In contrast, for *Alternaria* leaf blight and *Verticillium* wilts the rate curve is asymmetrical but r being greater during later period of season.

These diseases start slowly but accelerated markedly as host susceptibility increases late in the season. In contrast to monocyclic disease, the epidemic rate(r) for polycyclic disease is generally calculated per day or per week rather than per year.

Basically the epidemic rate (r_p) for polycyclic diseases is much greater than the r for monocyclic diseases (r_m), *i.e.* $r = r_p > r_{m.}$.

- For example, the r_m for *Verticillium* wilt of cotton is 0.02 units per day and is 1.60 units per year for *Phymatotrichum* root rot of cotton.
- In contrast, the epidemic rate r for potato late blight is 0.3-0.5 units per day, is 0.3-0.6 units per day for wheat stem rust and 0.15 units per day for cucumber mosaic virus.
- In case of polyetic epidemic, because of the perennial nature of their host, they behave basically as polycyclic diseases with a lower r. This

is occur because there are many disease trees and almost as much inoculums at the beginning of a year as the end of the previous one, and both increase over the years, causing slower it as severe epidemic. Most significant example of polyetic epidemics are chestnut bight (r = 0.3–1.2 units) and elm yellows (*Phytoplasma*, r =0.6 units per year).

2.10. AREA UNDER DISEASE PROGRESS CURVE (AUDPC)

Disease severity in any plant-pathosystem can be assessed either once at the peak of the epidemic or several times at some intervals starting from disease initiation until the end of the epidemic.

The former method measures the cumulative effects of all the factors operating during the course of epidemic *viz.* the terminal disease severity scores (TDS), while the latter can be used to estimate different parameters like the area under the disease progress curves (AUDPC), relative area under the disease progress curve (RAUDPC), logistic (r) and Gompertz (k) apparent infection rates, the time required for the disease to reach a specific level of severity in logistic and Gompertz models, and the genotype-score on PC-1 and PC-2 estimated through principal component analysis (Mukherjee *et al.*, 2010).

In general, estimation of percentage infection is, nonetheless, subject to several sources of error (Forbes and Jeger, 1987).

- Underestimation occurs when evaluation is made only on that portion of the disease with visible symptoms that is still on the plant. Infected leaflets eventually fall off and the timing of the fall is probably cultivar-dependent. Diseased tissue can be green and symptomless and therefore overlooked during the evaluation. Infected green tissue can also be sporulating, but this is not always visible unless one is very close to the leaflet. This level of scrutiny is generally not employed in routine cultivar evaluations.
- Human error can occur when proportional differences are evaluated. Research has shown that people estimate disease more precisely at lower and higher levels of severity than at intermediate levels (Forbes *et al.*, 2014). The use of logarithmically-based scales does not necessarily correct this bias and thus simple percentage infected area is advised (Forbes and Korva, 1994).
- In general, it is better if all readings for a trial are taken by an evaluator with expertise in disease assessment in order to maintain the same degree of accuracy and precision in the disease assessments (Campbell and Madden, 1990). It is also best to record readings independently (that is, without knowing the value given at the previous reading) by

having someone else register the values in the data sheet or by using a digital recorder which can hide the previous reading. If paper is used to take disease readings, the color should be chosen to avoid reflection of sunlight. The best time of the day to evaluate disease severity is in the morning or in cloudy conditions.

The area under the disease progress curve (AUDPC) is a useful quantitative summary of disease intensity over time, for comparison across years, locations, or management tactics. AUDPC models have been used for epidemic of relatively short duration, which are late in crop growth cycle and where yield is accumulated over a short period of time.

The AUDPC is a variable which estimates the amount of disease across the season. The AUDPC is expressed in % days (that is, the accumulation of daily percentage infection values) and is interpreted directly without transformation (Forbes *et al.*, 2014).

The higher the AUDPC, the more disease in the genotype. It is often helpful to plot the percentage leaf area infected versus the evaluation date ("disease progress curve") to get a better idea of how genotypes perform in the experiment. The most commonly used method for estimating the AUDPC, the trapezoidal method, is to discretize the time variable (hours, days, weeks, months, or years) and calculate the average disease intensity between each pair of adjacent time points (Madden *et al.*, 2007).

We can consider the sample time points in a sequence $\{t_i\}$, where the time interval between two time points may be consistent or may vary, and we also have associated measures of the disease level $\{y_i\}$. We can define $y(0) = y_0$ as the initial infection or the disease level at $t = 0$ (*i.e.*, the first disease severity observation in our study). $A(t_k)$, the AUDPC at $t = t_k$, is the total accumulated disease until $t = t_k$, given by

$$A_k = \sum_{i=1}^{N_t - 1} \frac{(y_i + y_{i+1})}{2} (t_{i+1} - t_i)$$

In similar words N is the number of assessment time; y = disease measurement and t is the time (days). AUDPC models make two assumptions

- Injury to the host is proportional to the amount of tissue infected, and
- Injury is proportional to the duration of the disease.

Rearrangement of above AUDPC formula is calculated by multiplying each assessment value by its respective weight (Simko and Hans-Peter, 2012).

$$AUDPC = \left[y_i \, x \, \frac{t_2 - t_1}{2}\right] + \left[\sum_{i-2}^{n-1}\left(y_1 \, x \frac{t_{i+1} - t_{i-1}}{2}\right)\right] + \left[y_n \, x \, \frac{t_n - t_{1n-1}}{2}\right]$$

Where y_i and y_n are the assessments at the first and last observations, respectively, and t_1, t_2, t_{n-1} and t_n are the times of the first, second, penultimate, and last observation. The weight of each assessment is the duration between the midpoint of the previous time interval to the midpoint of the subsequent time interval (Simko and Hans-Peter, 2012).

$$(v_i = \frac{t_{i+1} - t_{i-1}}{2} + \frac{t_{i+1} - t_i}{2} + \frac{t_i - t_{i-1}}{2})$$

Thus, according to (Smiko and Hans-Peter, 2012), all weights are calculated by extrapolating in two directions, with the exception of the first and the last assessments for which weight is extrapolated in one direction only, This leads to a substantial undervaluation of the first and the last assessments in comparison to all other assessments. It is possible that in some special cases this lower relative weight may be desirable, but for the majority of evaluations there is no added benefit in giving only half a weight to assessments at these two observations. To give the appropriate importance to the first and the last assessments, weights for these assessments should also be extrapolated in both directions, therefore, extrapolating a weight in the missing direction using half of the average interval duration between observations. Smiko and Hans-Peter (2012) have suggested a new method that is used to calculate total area (including adjusted weight for the first and last assessment) and termed it as Area under the Disease Progress Stairs (AUDPS). The mathematical formula for AUDPS is

$$\begin{aligned} AUDPS &= \left[y_1 \, x \left\{\frac{t_2 - t_1}{2} + \frac{D}{2(n-1)}\right\}\right] + \left[\sum_{i=2}^{n-1}\left(\frac{t_{i+1} - t_{i-1}}{2}\right)\right] \\ &+ \left[y_n \, x \left\{\frac{t_n - t_{n-1}}{2} + \frac{D}{2(n-1)}\right\}\right] \\ &= \left[\sum_{i=1}^{n-1} \frac{y_1 + y_{i+1}}{2} \, x(t_{i+1} - t_1)\right] + \left[\frac{y_1 + y_n}{2} \, x \frac{D}{n-1}\right] \\ &= AUDPC + \left[\frac{y_i + y_n}{2} \, x \frac{D}{n-1}\right] \end{aligned}$$

The AUDPC values, as well as percentage infection values, can be analyzed using analysis of variance (ANOVA) after exploration of the data through simple statistics such as means, standard errors, frequency distributions and box plots. Multiple means comparisons tests (*e.g.*, Dunnett) can also be applied. The analysis of residuals is recommended to test the validity of the model and assess homogeneity of the variances (Upton and Cook, 2008).

The AUDPC and percentage infection are considered pseudo-quantitative variables with hierarchy and can be analyzed without transformation (Jeger and Vilijanen-Rollinson, 2001). If yield has been evaluated in addition to the AUDPC, the correlation between yield and the AUDPC can be calculated with the Spearman method. The Spearman method works by ranking each observation within each replication and the Spearman rank correlation is calculated on the ranked values. A coefficient close to 1.0 indicates a good correlation between yield and the AUDPC (*i.e.*, overall disease severity).

Among these parameters, AUDPC has been widely used in two areas of plant pathological research *viz.* crop loss assessment (Ferrandino and Elmer 1992) and assessment of partial or quantitative resistance.

The estimation of AUDPC is based on simple formulae developed by Wilcoxson *et al.* (1975), Shaner and Finney (1977), Bjarko and Line (1988), Das *et al.* (1992), Chen and Line (1995), Miedaner and Sperling (1995) and Broers *et al.* (1996), following the trapezoidal rule for calculation. Mohapatra (2002) and Mukherjee *et al.* (2005) recognized AUDPC, RAUDPC (Relative Area Under Disease Progress Curve), *r* (Logistic apparent infection rate), k (Gompertz apparent infection rate) and genotype-score on PC-1 as top ranking parameters for identification of slow-blasting resistance in rice. The assessment of rate reducing resistance in polycyclic diseases covers a range of sampling frequencies at regular intervals, which is time consuming, labour intensive and costly.

Vander Plank (1963) proposed estimating the logistic apparent infection rate (r) using a two-point method *i.e.* initial disease severity and final disease severity data, which was adopted by Berger (1981) through Gompertz transformations. Jeger and Viljanen-Rollinson (2001) reported that estimation of AUDPC from two data points on the disease progress curves for stripe rust resistance in wheat provided an equivalent amount of information as those from all points. This was followed by estimation of AUDPC from two data points for assessment of late blight resistance in potato by Haynes and Weingartner (2004), who suggested that the estimation from two data points provide the information similar to those from all points, thereby saving time and economic resources.

AUDPC can be quantified at two levels viz., all point method and two point method. According to Shaner and Finney (1977) AUDPC through all point method can be quantify with following equation

$$AUDPC_{AP} = \sum_{i=1}^{n-1} \left[\frac{X_{i+1} + X_i}{2}\right] [t_{i+1} - t_i]$$

Where X_i is the proportion of host tissue damaged at the ith day; t_i is the time in days after appearance of disease at the ith day and n is the total number of observations. In case of two data point *i.e.,* initial and final assessment, Jeger and Viljanen-Rollinson (2001) have provided method which was based on the integration of disease progress models.

According to them for any real continuous function y = f (t) (with v>0), the area under the function is simply the definite integral evaluated between the limit of integration t_0 and T, where $T>t_0$, thus y =f (t) = $1/(1+A_e^{-rt})$, where f(t) is the logistic function with $A=(1-y_0)/y_0$, y is the amount of disease in the numerical scale of 0-1, y_0 is the value of y at t_0 = 0, and r is the logistic rate parameter. Therefore, following the above criteria, the AUDPC can be described as.

$$AUDPC = \int \left\{\frac{dt}{(1 + Ae^{-rt})}\right\}$$

and this can be evaluated between the limits t=0 and T, and substituting in $y = T + [\ln\{(y_0/y_T)\}/[\ln[y_T/(1-y_T)\}-\ln\{y_0/(1-y_0)\}]/T$. Thus, only two assessments of diseases are necessary, one at the start of epidemic and the other at the end of the epidemic or at a critical growth stage.

The apparent infection rates in the logistic (r, Van der Plank, 1963) and the Gompertz (k model, Berger, 1981) can be quantified by following formulas, respectively.

$$r_{TP} = \left| \frac{\log_e\left\{\frac{X_2}{1 - X_2}\right\} - \log_e\left\{\frac{X_1}{1 - X_1}\right\}}{t_2 - t_1} \right|$$

$$k_{TP} = \left| \frac{gompit\left\{\frac{X_2}{1 - X_2}\right\} - gompit\left\{\frac{X_1}{1 - X_1}\right\}}{t_2 - t_1} \right|$$

Where, X_1 and X_2 are disease severity scores on the first and last day of observation, and t_1 and t_2 are the initial and final day of observation.

2.10.1. Demerits of AUDPC

International Potato Center have provided following problems and their solutions associated with AUDPC (Forbes *et al.*, 2014).

Problem	*Solution*
Timing and reading: The selection of resistant genotypes is frequently based on the AUDPC. However, the same AUDPC value can result from an early-starting, slow progressing infection or from a late-starting, fast-progressing infection; therefore, AUDPC values do not provide information on the type of resistance present in the genotypes, nor on their potential durability. For this reason it is also advisable to examine disease progress curves	It is important to start the readings when disease begins, ideally even before the first symptoms appear. Generally, it is not relevant to continue the readings after the susceptible materials reach 100% disease severity. While an AUDPC value can be calculated from two data points, greater accuracy is achieved with each additional observation, although under some conditions very accurate AUDPC values can be derived from two data points
Incomparability across experiments: The AUDPC perse should not be used for comparing species genotypes across experiments. Furthermore, the AUDPC units as indicators of resistance or susceptibility are not easily interpretable. For example, an AUDPC value of 2043 may arise from a moderately resistant genotype grown under conditions conducive to severe infection, or it could also arise from a highly susceptible genotype grown under conditions not conducive to severe infection.	In an effort to standardize the AUDPC, researchers often use the relative AUDPC (rAUDPC) The rAUDPC is calculated by dividing the AUDPC by the "maximum potential AUDPC." The maximum potential AUDPC is simply the AUDPC a genotype would have if it had 100% infection at all readings. The maximum potential AUDPC can be calculated by multiplying the total number of days between the first and last readings by 100. The rAUDPC is also not the best measure for comparing results across different experiments for the same reasons explained above for AUDPC

2.10.2. Working example for Calculation of AUDPC and rAUDPC

For example an experiment have five new varieties and their susceptibility will recorded after 15,25,45,65,85,110 and 135 days. For their AUDPC following formula can be use (Fig. 4).

```
=((D2+C2)/2)*($D$1-$C$1)+((E2+D2)/2)*($E$1-$D$1)+((F2+E2)/2)*($F$1-$E$1)+((G2+F2)/2)*($G$1-$F$1)+((H2+G2)/2)*($H$1-$G$1)+((I2+H2)/2)*($I$1-$H$1)
```

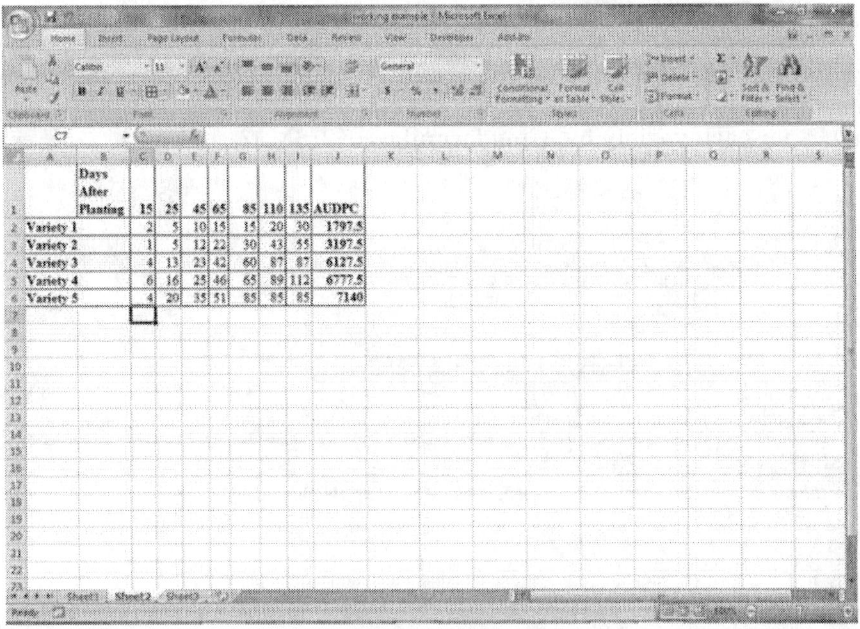

	Days After Planting	15	25	45	65	85	110	135	AUDPC
Variety 1		2	5	10	15	15	20	30	1797.5
Variety 2		1	5	12	22	30	43	55	3197.5
Variety 3		4	13	23	42	60	87	87	6127.5
Variety 4		6	16	25	46	65	89	112	6777.5
Variety 5		4	20	35	51	85	85	85	7140

Fig. 4: Excel sheet for AUDPC

For rAUDPC we can use the formula =J2/((I1-C1)*100); here C1 is the first date of recording and I1 is the last day of recording (Fig. 5).

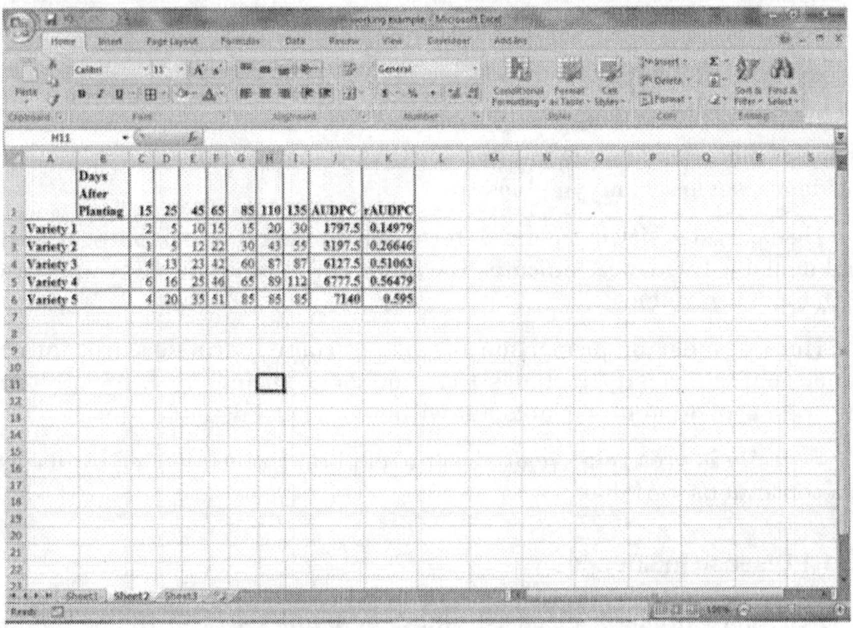

	Days After Planting	15	25	45	65	85	110	135	AUDPC	rAUDPC
Variety 1		2	5	10	15	15	20	30	1797.5	0.14979
Variety 2		1	5	12	22	30	43	55	3197.5	0.26646
Variety 3		4	13	23	42	60	87	87	6127.5	0.51063
Variety 4		6	16	25	46	65	89	112	6777.5	0.56479
Variety 5		4	20	35	51	85	85	85	7140	0.595

Fig. 5: Excel sheet for AUDPC

Now assigned the scale value for susceptible variety, in this example we have assigned 8 as scale value and by using =L7/K7 formula we get constant (11.34; Fig. 6). After this we can calculate ssusceptibility scale values for each variety by using formula==K2*M7.

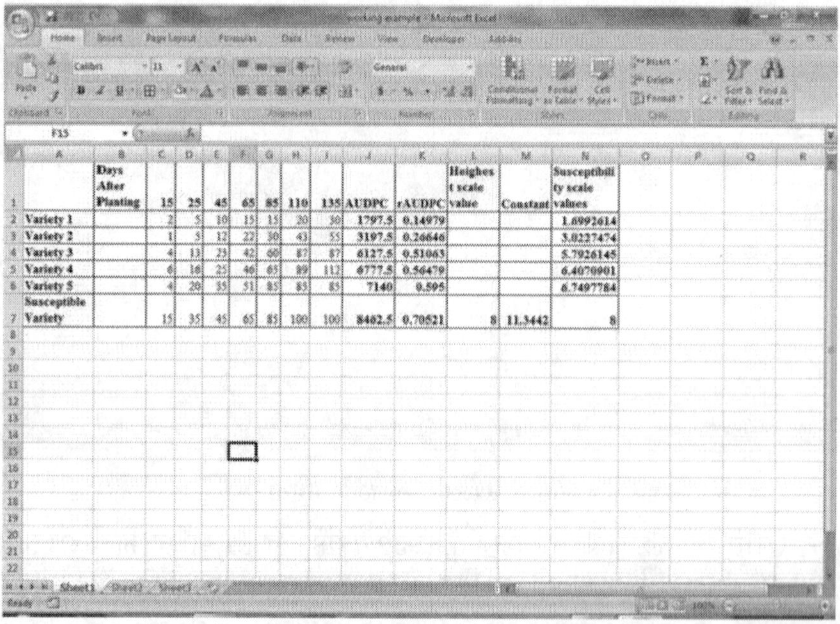

	Days After Planting	15	25	45	65	85	110	135	AUDPC	rAUDPC	Heighest scale value	Constant	Susceptibility scale values
Variety 1		2	5	10	15	15	20	30	1797.5	0.14979			1.6992614
Variety 2		1	5	12	22	30	43	55	3197.5	0.26646			3.0227474
Variety 3		4	13	23	42	60	87	87	6127.5	0.51063			5.7926145
Variety 4		6	18	25	46	65	89	112	6777.5	0.56479			6.4070901
Variety 5		4	20	35	51	85	85	85	7140	0.595			6.7497784
Susceptible Variety		15	35	45	65	85	100	100	8462.5	0.70521	8	11.3442	8

Fig. 6: Excel sheet for AUDPC

2.11. MEASURING DISEASE IN A POPULATION

Disease measurement is related with three different aspects, disease incidence, severity and yield loss.

Disease intensity is a general term, which includes incidence, severity and density. Incidence indicates the status of individual plant units and with binary variables.

Here number or proportion disease is count variable with natural denominator. Density includes the number of lesions or other units of infection is also count variable but without natural denominator.

Severity is area relative or absolute of plant tissue affected by disease is a continuous variable.

2.11.1 Disease Incidence

It is the percentage of diseased plants or plant parts in the sample or population, irrespective of their severity.

- It is suitable for the early stages of the epidemic.
- It applies mainly to disease which affects entire plants (virus, wilts, smuts or spots on fruit) if one lesion makes the latter unfit for sale.
- It can be easily accessed by simple counting diseased plants *vs* healthy sampling unit. It is less affected by errors then disease severity.
- Sampling unit must be defined diseased only if the severity is beyond a certain level (physiological damage threshold).
- Thus, it's the number or portion of plant units that are disease (fruit, leaves number) in relation to the total number of units examined, and it can be described from following formula;

$$\frac{Total\ Number of\ Infected\ Unit}{Total\ Number\ of\ Plant\ Unit\ Examined}\ X\ 100$$

Measuring disease incidence is relatively quick and most commonly used in epidemiology study to measure the spread of disease through a field, region or country. In diseases like cereal smuts, brown rot of stone fruit, vascular wilt of annual, the disease incidence is directly related with severity of the disease and yield loss because each diseased plant or fruit loss.

2.11.2. Disease Severity

It is the percentage of the relevant host tissues or organs covered by symptoms of the disease. This is the results of the combination from number and size of the lesions.

- It is appropriate for rust, downy and powdery mildew, leaf spots.
- It is essential to distinguish between symptom area directly caused by pathogen and the other indirect symptoms (chlorosis, necrosis, leaf shedding, etc.).
- Also the accelerated senescence must be taken into account.
- These differences between direct and indirect damages are important for epidemiological studies, but they are not important for crop loss assessment.
- Thus, it's the amount of plant that is diseased, which can be quantified through following methods and by graphical illustration (Fig. 7).

$$\frac{Area\ of\ Diseased\ Tissue}{Total\ Tissue\ Area}\ X\ 100$$

Disease severity is usually expressed as the percentage or plant area destroyed by a pathogen. More often disease assessment scales ranges from 1 to 10 or 1 to 4.

Both disease incidence and severity are generally assessed to compare effect of treatments/management practices.

In general during the initial stage of the epidemic, the incidence is high while, disease severity is low.

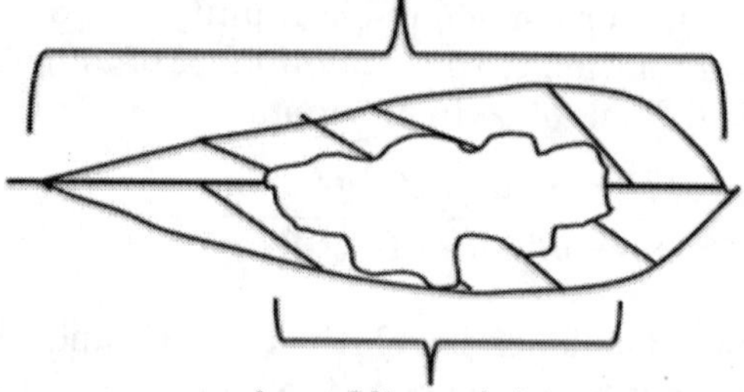

Fig. 7: Graphical illustration of disease severity

For quantification of disease development and decline of epiphytotics, the disease severity can also be express through using inoculums potential, disease potential, inoculums density, capacity and proneness and susceptibility and their relationships can be express by following equation.

$$= [\textit{Inoculum Potential}]\, X\ \textit{Disease Potential}$$

$$= [\textit{Inoculum density X Capacity}]\, X\ \{\textit{Proneness X Susceptibility}\}$$

Inoculum potential of the pathogen is a function of the number of infective propagules (inoculums density) and their pathogenic capacity (inoculums capacity).

Disease potential is concerned with the condition of the host. High proneness and high inherent susceptibility are ideal for development of an epidemic in the plant pathogen.

Both the above factors, are affected by environment which turn them favourable or unfavourable for development of epiphytotics.

International Rice Research Institute (IRRI, 2002) had developed a method to quantified the panical blast severity in rice (PBS), which is as follows (Pasha *et al.*, 2013):

$$PBS = \frac{(10\ x\ N1) + (20\ x\ N3) + (40\ x\ N5) + (70\ x\ N7) + (100\ x\ N9)}{Total\ number\ of\ panicles\ observed}$$

N1-N9 are the number of panicles with score 1-9 (0= no incidence; 1= less than 5%; 3=5-10%; 5=11-25%; 7= 26-50% and 9 more than 50%.

By using disease severity and disease incidence, we can also quantify the disease intensity index (Trapero-Casas and Jimenez Diaz, 1985 and Gatachew *et al.*, 2014).

$$Disease\ Intensity\ Index = \frac{(I\ x\ S)}{4}$$

$$Disease\ Severity\ (\%) = \frac{n\ x\ v}{4N} x\ 100$$

Where n = number of plants in each category, (v)= numerical values of symptoms category; N = total number of plants and 4 = maximum numerical value of symptoms category

$$Disease\ Incidence\ (\%) = \frac{Number\ of\ Infected\ Plant\ Units\ x\ 100}{Total\ Number\ of\ Units\ Assessed}$$

2.11.3. Disease Density

Number of lesions or other units of infection per plant or per unit area of plant tissue. It is often related with disease severity.

- It is only applicable for certain types of diseases and is relevant when there are lesion, pustules, colonies etc.
- If number of lesions is n, and the actual area of each lesion is known then one could determine mean area (A) and then severity would be Y=A x n.
- As stated earlier disease density is a count variable but without a natural denominator. It is a type of discrete variable but may be approximated as continuous variable for some purposes.

2.11.4. Decision Making for Epidemic Control

Control of different diseases requires different control procedures. Economic considerations are crucial in the process leading to decision for against action.

In view of economics, disease management can be defined as the total actions, intentional or not that serve to regulate disease levels so that they remain below the economic threshold level; the actions can be directed toward a single disease or toward all diseases threatening a crop.

Various action's (for example a= sanitation; b= changes of planting time; c = potential resistance; d= treatment with eradicate fungicide; e= treatments with protective fungicide; f= residual resistance in the adult stage or regular treatment with fungicides) and their timing are illustrated in Figure 8

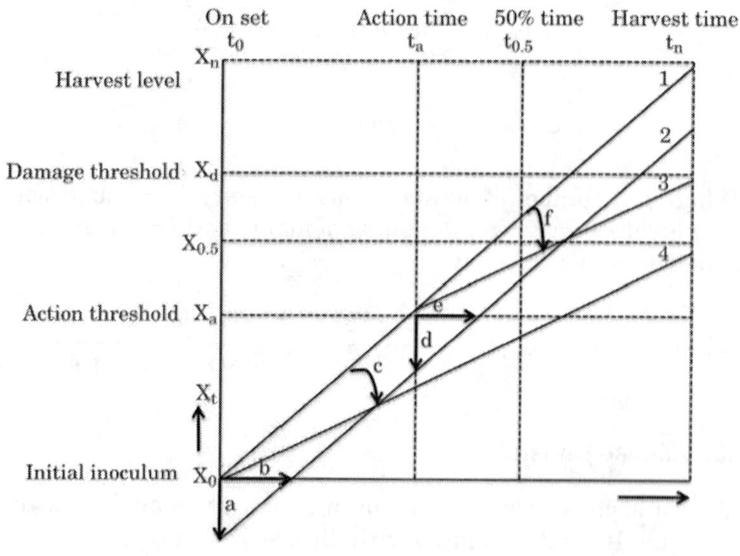

Fig. 8: Model demonstrating the effects of various control action in terms of the equivalence theorem, reduction of r and X_o or X_t (Zadoks and Schein, 1979).

2.11.4.1. *Epidemic threshold*

The disease in which, 50% of the plant and 25% of the leaf shows disease intensity is called epidemic threshold. It is generally utilized during fungicidal application.

2.11.4.2. *Economic damage or action threshold*

It is the highest value of disease severity up to which no economic damage occur. The economic threshold of crop-pathogen system varies with the tolerance level (damage threshold) of crop and this depend on the (a) growth stage of crop when attacked, (b) crop management practices (c) environment and (d) pathogen virulence.

In general it's the density at which control major should be determined to prevent an increasing pathogen population from reaching the economic injury level.

The economic threshold is lower than the economic injury level, to permit sufficient time for the initiation of control measures and for those measures to take effect before the population reaches the economic injury level.

The rate of disease development *r* actually depends on host susceptibility, pathogen virulence and environmental suitability. If amount of initial inoculums level X_0, *r* and damage threshold are know, then we can predicted the time at which the disease will exceed the damage threshold. Particularly, if it's sufficiently before harvest, significant damage may result.

If the damage threshold (X_d) is known, and disease is present, the researcher or practitioner must know when to act-action threshold (X_a). At the right time (t_a) the farmer must apply a control measure that will so reduce *r* that disease will not reach the damage threshold before harvest.

2.11.4.3. *Economic injury level or damage threshold*

It's the lowest population density that will cause economic damage.

Economic damage is the amount of injury which will justify the cost of artificial control measure. In plant disease it has been called as damage threshold.

- The actual determination of the damage threshold is a part of the fine tuning of a local management system and depends not only on biologic information but also on accurate knowledge of the costs of the inputs to the system, the value of outputs, and the requires maintenance level.
- The maximum damage a crop can sustain without yield loss
- Generally used for plant diseases. Since disease pathogens are too small to be easily seen, counting their numbers is impractical, so an estimate is made of the amount of damage caused by them.
- Counting diseased leaf petioles for soybean pod and stem blight
- Estimating the percentage of whole plant infection caused by fungal leaf blights in corn.
- Economic Injury Level (EIL) is often expressed mathematically by the formula:

$$EIL = \frac{C X N}{V x I}$$

- Here C is the unit cost of controlling the pathogen (*e.g.*, \$20/acre); N is the number of pathogen injuring the commodity unit (*e.g.*, 800/acre); V is the unit value of the commodity (*e.g.*, \$ 500/acre) and I is the percentage of the commodity unit inured (*e.g.*, 10% loss).

$$EIL = \frac{20\ X\ 800}{500\ x\ 0.1} = 320$$

2.11.4.4. *Warning threshold*

Warning threshold (X_W) is the level of disease severity that leads to a standby warning. The X_w is lower and earlier than the action threshold and latter is lower and earlier that the damage threshold. As the damage threshold is based on a value judgment, it is subjective, and thus, the other two threshold are also subjective.

Thresholds are useful for (a) for decision making on scheduling of control and control methods and (b) to establish the optimal amount of control which can be used to minimize risk of economic damage and environmental hazards. Different thresholds can be developed from the following factors:

- Amount of physical damage related to various pest densities; how large the pest population can grow before it causes a certain level of damage.
- Monetary value and production costs of the crop at various levels of physical damage
- Monetary loss associated with various levels of physical damage
- Amount of physical damage that can be prevented by the control measure
- Monetary value of the portion the crop that can be saved by the control measure
- Monetary cost of the control measure
- History of the field
- Determining the danger of the pest at different stages of the crops development
- The pest distribution
- How much aesthetic or economic damage can be tolerated
- Establish a treatment level that keeps the pest population small enough so it does not cause an unacceptable level of damage.
- Ultimate destination of the crop, what the standard of the end consumer is.
- Ability to control the pest rapidly and effectively

2.11.5. Yield Loss

The yield loss is the difference between attainable yield and actual yield.

***Attainable yield*:** When crop grown under optimum conditions

***Economic yield*:** Highest net return on expenditure

***Actual yield*:** Obtained under disease management programme.

***Theoretical yield*:** Obtained using crop simulation models.

Yield loss due to disease can be measure at (a) specific growth stage, (b) sequential disease assessment at several stages of a crop growth or (c) by determining the AUDPC *i.e.,* the area between the disease progress curve and the X axis of the graph.

Gashaw *et al.* (2014) have utilized following formula for quantification of yield loss

$$Yield\ Loss\ (\%) = \frac{Yield\ in\ Intensive\ Protected\ Plot - Yield\ in\ Particular\ Treatment\ x\ 100}{Yield\ iin\ Intensive\ Protected\ Plot}$$

Conceptual model for crop loss assessments are presented in Figure 9.

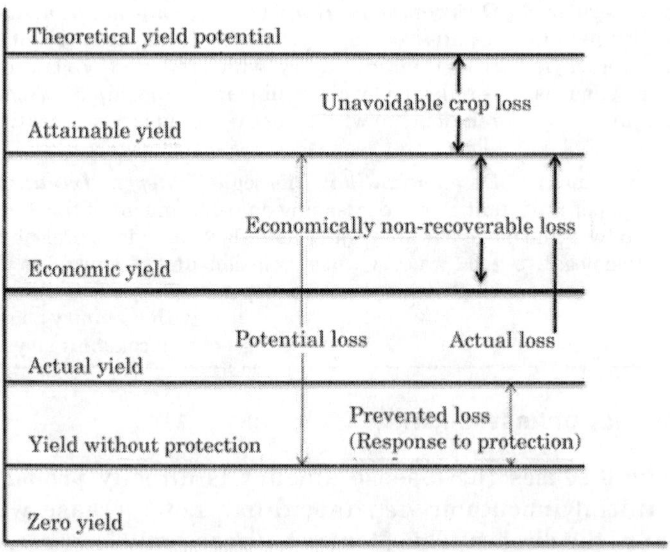

Fig. 9: Crop loss assessment models (adapted from Oerke *et al.*, 1994)

2.11.6. Damage Curve

The relationship between injury and yield is known as the damage curve (Fig. 10).

Damage curve can be divided into several segments that include: tolerance, overcompensation, compensation, linearity, desensitization and inherent impunity.

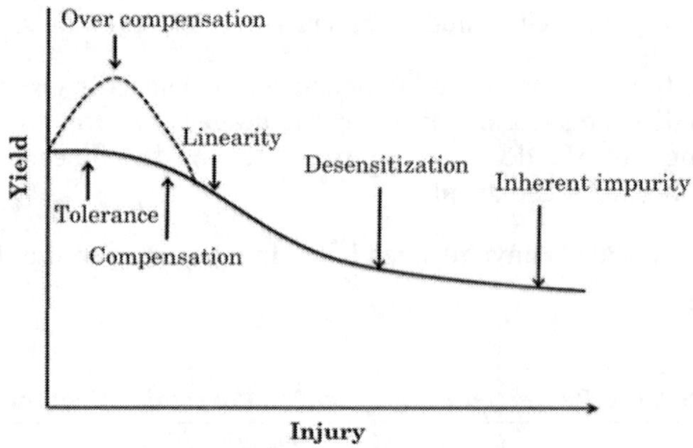

Fig. 10: The damage curve (Peterson and Higley, 2001)

***Tolerance*:** This segment reflects no damage per unit injury. In other words, yield with injury is the same as yield without injury	***Overcompensation*:** The negative damage (yield increase) per unit injury, which is reflected by a curvilinear relationship with a positive slope.	***Compensation*:** An increase in damage per unit injury as reflected by a curvilinear relationship with a negative slop is compensation.
***Linearity*:** The maximum (constant) damage per unit of injury reflected by a linear relationships with a negative slope is linearity	***Desensitization*:** This segment shows decreasing damage per unit injury as reflected by a curvilinear relationship with a negative slope.	***Inherent Impunity*:** This segment of the damage curve indicates no damage unit of injury. The yield with injury is less than the yield with no injury as reflected by a constant slope.

2.11.7. Frequency of Assessment

In monocyclic diseases the disease amount is directly proportion to amount of initial inoculums *i.e.,* inoculums not increase with the advancement of the disease. Thus, for such disease only few assessments are requiring.

While, in polycyclic disease the amount of disease is directly related with amount of inoculums at specific point of time, thus we required several assessment at particular time of event.

In general the disease assessment should be associated with crop growth stage and for cereals it is generally carried out during grain feeling.

2.11.8. Two Pathogen Situations and their Epidemiological Behavior

Two pathogen situations is an interesting and challenging scenario to quantify the different attributes of plant disease epidemiology. Waldir *et al.* (2014) have summarized the interactions between foliar diseases, their concepts and epidemiological approaches.

In general, estimates of disease effects on yield are usually made assuming that each disease acts independently, however, interaction of disease can increase crop damage and complicated to recognize the primary causes of disease and their subsequent management strategies. Such conditions may alter the occurrence and speed of epidemics.

Weber (1996) have advocated that understanding of interspecific interaction between *Septoria nodorum* and *Erysiphe graminis* on wheat are very crucial factor for disease dynamics and their related factors.

- Association or interaction between two pathogens on same host may have strong impacts on crop loss assessment, disease diagnosis, forecasting, modeling and simulation of epidemics and for selection for appropriate management practices.
- Pathogen interactions may also be important because the expected benefit from the control of one pathogen depends on the level of the other pathogens and thus, it indirectly affecting the economics of disease management.
- On the same host presence of two pathogens created a positive, negative or neutral relationships between them, and such conditions may lead the formation of six major types of relationships which are, neutralism (00), Competition (—), Mutualism or proto-cooperation (++, these are based on obligatory or non-obligatory nature of association), commensalism (+ 0), amensalism (- 0) and parasitism (+ -).
- Sporulation by *Puccinia triticina* was reduced substantially by the presence of *Pyrenophora tritici-repentis*; in contrast, the presence of *Puccinia triticina* sometimes increased sporulation of *Pyrenophora tritici-repentis* (Al-Naimi, 2003)
- Examples of such interactions may be find in details in study of Waldir *et al.* (2014).

- In order to compare epidemics of interacting diseases, the area under the disease progress curve have been calculated and analysed using analyses of variance (Savary and Zadoks, 1992).
- Madden *et al.* (1987) applied the classical Lotka-Volterra competition equations to model the disease progress curves of the disease incidence of tobacco etch virus (TEV) and tobacco vein mottling virus (TVMV).
- The model is based on the assumption that the disease progress curves are logistic functions if one disease would be there alone. The competition model is given by the following system of differential equations for the disease severity or incidence of two diseases y_1 and y_2 given as proportions with values between 0 and 1:

  - $\frac{dy_1}{dt} = r_1 y_1 \left(1 - \left[\frac{y_1 + aa_{12} y_2}{K_1}\right]\right)$
  - $\frac{dy_2}{dt} = r_2 y_2 \left(1 - \left[\frac{y_2 + aa_{21} y_1}{K_2}\right]\right)$

  The parameters r_1 and r_2 are the apparent infection rates and K_1 and K_2 the maximum disease levels of both diseases in absence of the other disease. The two parameters aa_{12} and aa_{21} are the coefficients of competition. The coefficient aa_{12} indicates the competitive effect of disease 2 on disease 1. The system of above differential equations can be re-arranged to point out the mutual effects of the diseases on each other Waldir *et al.* (2014):

  - $\frac{dy_1}{dt} = r_1 (1 - aa_{12} Y_2/K_1) x Y_1 (1 - Y_1/[K_1(1 - aa_{12} Y_2/K_1)])$
  - $\frac{dy_2}{dt} = r_2 (1 - aa_{21} Y_1/K_2) x Y_2 (1 - Y_2/[K_2(1 - aa_{21} Y_1/K_2)])$

- In the example of Madden *et al.* (1987) the dynamics of two tobacco virus diseases TEV (*Tobacco etch virus* = 1) and TVMV (*Tobacco vein mottling virus* = 2) occurring at the same time in tobacco fields were modelled. Disease incidence progression was quantified by fitting the Lotka- Volterra equations, resulting in the following parameters values for the data set in 1984 B without insecticide usage: $r_1 = 0.223$/day, $r_2 = 0.261$/day, $K_1 = 0.175$, $K_2 = 1.00$, $\alpha a_{12} = 0.06$ and $\alpha a_{21} = 1.67$. From rearranged equation it can be concluded that the actual rate and the maximum disease level of TEV are reduced by 0.34% when the disease incidence of TVMV increases 1%. The reduction of TVMV is 1.67% per 1% increase of TEV. Due to the interaction modelled as competition, the maximum disease levels of both diseases are reduced.
- Empirical yield-loss equations for a complex of diseases are often obtained via multiple regression analyses with yield or yield loss as the dependent variable. Disease parameters, used directly or after a

transformation, serve as independent variables. For two diseases, the general equation for the total yield loss (*YL*) predicted is given as:

- $YL = b_1 f_1(y_1) + b_2 f_2(y_2) + b_{12} g(y_1 y_2)$

The two disease parameters, y_1 and y=, could be disease severities which can be transformed with the functions f_1 and f_2. The coefficients b_1 and b_2 are the respective damage coefficients for the transformed values of y_1 and y_2. The third term is the interaction term of the model depending on the function g of both disease severities. If the coefficient b_{12} is equal 0, the diseases reduce yield independent from each other. For $b_{12} > 0$ the total yield loss is higher than the sum of the individual yield losses, for $b_{12} < 0$ the effects of $y1$ and $y2$ are less than additive.

2.11.9. Pathogen Spillover

Most epidemiological theory has focused on pathogens that infect a single host species and consequently we have very little information about disease includes multiple host systems.

Pathogens with multiple hosts are common in both plant and animal systems, and understanding their dynamics has become increasingly important for human health, for domestic animal production, and for wildlife and plants of conservation significance.

While some of these generalist pathogens may have host-specific strains or races, many are true generalists. In most multiple host systems, all host species are not created equal.

Species inexorably vary in key epidemiological traits such as resistance, tolerance, and vector preference. As a result, the rates of transmission of a generalist pathogen within and between different host species are commonly highly heterogeneous and asymmetric. In some host populations, epidemic or endemic disease may be primarily driven not by intra-specific pathogen transmission within that population but by transmission from a reservoir species that maintains a relatively high pathogen population.

In such a case, the pathogen typically reaches high prevalence in the reservoir and then spills over into the other host, a process called "the spillover effect" or "pathogen spillover" (Daszak *et al.,* 2000). Power and Mitchell (2004) have defined pathogen spillover as the driving of disease dynamics in one host population by contact with pathogen propagules (regardless of transmission mode) from another host population as a result of high pathogen abundance in this reservoir population.

Power and Mitchell (2004) in their experiment with *Avena fatua* have documented that

- Pathogen spillover resulted in apparent competition among host species. Apparent competition mediated by a shared natural enemy has rarely been demonstrated in field experiments.
- Spillover at relatively small scales can result in pathogen-mediated apparent competition that shapes community structure and community attributes like evenness.
- Spillover at larger scales may be a potent force shaping the population dynamics of non domesticated species

2.11.10. Threshold Theorem

Van der Plank (1963) expressed his threshold theorem as iRc>1, which states that an epidemic will not occur unless the product of infectious period i and basic corrected infection rate Rc exceeds 1.

For consistency with the medical and ecological literature, the theorem has been rewritten as $R_0>1$, where R_0 is the basic reproductive number, *i.e.*, the number of new infected individuals resulting from one introduced infected individual.

This has predominantly utilize in plant epidemiology, which allows for the calculation of not only the risk of pathogen invasion, but also the probability of subsequent persistence in the face of chance events that can lead to extinction at low population densities.

Apart from stochastic forces, the establishment of a pathogen following its successful introduction may be limited by certain demographic features, such as the difficulty to find a compatible mating partner at very low population densities for species with an obligate sexual cycle. This feature leads to an intermediate optimum in the relationship between population growth rate and population density (Allee effect; which represents an intermediate optimum in the relationship between population growth rate and population density, *e.g.*, due to the difficulty to find a compatible mating partner at very low population densities for species with an obligate sexual cycle).

In a deterministic population model of the heterothallic Karnal bunt fungus *Tilletia indica*, inclusion of an Allee effect resulted in a teliospore threshold for establishment about two orders of magnitude higher than in a version of the model without this constraint. This finding has potentially important implications for risk assessments of *T. indica* and other quarantine pathogens with an obligate sexual cycle.

2.12. APPARENT RATE OF INFECTION AND BASIC RATE OF INFECTION

2.12.1. Apparent Infection Rate

It is an estimate of the rate of progress of a disease, based on proportional measures of the extent of infection at different times.

Firstly, a proportional measure of the extent of infection is chosen as the disease extent metric.

For example, the metric might be the proportion of leaf area affected by mildew, or the proportion of plants in a population showing dieback lesions. Measures of disease extent are then taken over time, and a mathematical model is fit.

The model is based upon two assumptions: (a) the progress of the infection is constrained by the amount of tissue that remains to be infected; and (b) if it were not so constrained, the extent of infection would exhibit exponential growth.

There is a single model parameter r, which is the apparent infection rate. It can be calculated analytically using the formula.

$$r = \frac{2.3}{t_2 - t_1}\left(\log\frac{x_2}{1 - x_2} - \log\frac{x_1}{1 - x_1}\right)$$

Or Mbong *et al.*, (2010) proposed its alternate form

$$r = \frac{2.3}{t_2 - t_1}\left[\log 10\left(\frac{x_2\,(1 - x_1)}{x_1\,(1 - x_{12})}\right)\right]$$

According to Nagarajan, (1983), calculating, these values have utility for the evaluation of the host resistance or meteorological factors or effect of fungicide spray etc., on epidemic development. A lower r value denotes a slow rate of growth of disease, which imples that more time will be needed to complete the epidemic.

2.12.2. Basic Infection Rate (R)

R value represents the rate of multiplication of speculating part of the disease symptoms, *i.e.*, it's a measure of sporulation increase and not of the disease. Thus, if include the uninterrupted pustules, sporulating and those that have stopped sproulating, but not yet dead, while *r*- is the apparent infection rate that represent all these, total.

Towards the end of the epidemic though there may be an increase in *r* the R- values falls, and Vanderplank (1964) have provided three possible explanations, which are as follows:

Some parts of the plants are more sensitive to infection than are other and these parts get infected first and progressively only less susceptible are left back.

As an epidemic advances more foliage is lost and this changes the microclimate.

The sporulation ceases as the old lesion exceed the young in number

R can be calculated by following formula

$$R = \frac{1}{x_0(t_2 - t_1)} x \left(1n \frac{1}{1 - x_2} - 1n \frac{1}{1 - x_1}\right)$$

- The importance of *R* is the emphasis it gives for the area under sporulation and hence is directly influenced by the large number of factors that interact with sporulation.
- On the other hand, the apparent infection rate (*r*) is a measure of the rate of increase of the visible infected area. The *r* value is not only governed by *R* but also by the incubation period (*p*) and duration of sporulation (*i*), etc.
- Incubation period is the time difference between the successful establishments of a host-pathogen relationship to the appearance of disease symptoms. Rate of spread of the disease is greatly influenced by *p*, and when the *p* value is less, the pathogen is able to complete the infection cycle quickly enabling rapid disease build-up. The incubation period is not only decided by the environmental factor such as temperature, leaf wetness, light intensity and relative humidity etc.
- Sporulation (*i*) denotes the process of spore formation of the pathogen either sexually or asexually. Sporulatoin various from a few days, to few month and this active period of sporulation varies with host, position of leaf and race of pathogen.
- A variation in climate or weather, gradually affects *r* through different phases of the life cycles (incubation, sporulation, etc.) The changes brought about affect *r*- value only after the incubation period gets completed.
- Mathematically R and apparent infection rate can be write as

$$r = R \frac{x_t - p}{x_t}$$

And, incubation period can be quantified through

$$p = \frac{1}{t_2 - t_1} log_{10} \frac{ax_2(1 - x_1)}{x_1(1 - x_2)}$$

Where, x_1 and x_2 are severity levels at time t_1 and t_2 and $a = y_2/y_1$ (y_2 and y_1 are the amount of susceptible area at time periods t_1 and t_2.

2.12.3. Composite Fitness Index (CFI)

It is proposed by Tooley *et al.* (1986) for quantification of fitness and virulence of *Phytophthora infestans* isolates from sexual and asexual populations and it can be quantified by using following formula

$$CFI = IF \times LA \times SC$$

Where IF is the infection frequency (proportion of inoculated leaflets on which lesion developed); LA is the lesion area (area cm^2 of lesion produced per leaflets), and, SC is the sporulation capacity (number of sporangia produced per square centimetre of lesion).

2.13. REFERENCES

Al-Naimi, F.A. 2003. The competitive interaction between *Pyrenophora tritici-repentis* and *Puccinia triticina*. MS Thesis, Kansas State University. Manhattan KS, USA.

Berger, R.D. 1981. Comparison of the Gompertz and logistic equations to describe plant disease progress. *Phytopathology*, 71: 716–71.

Bjarko, M.E. and Line, R.F. 1988. Heritability and number of genes controlling leaf rust resistance in four cultivars of wheat. *Phytopathology*, 78: 457–461.

Broers, L.H.M., Cuesta Subias, X. and Lopez Atilano, R.M. 1996. Field assessment of quantitative resistance to yellow rust in ten spring bread wheat cultivars. *Euphytica*, 90: 9–16.

Campbell, C.L. and Madden, L.V. 1990. Introduction to Plant Disease Epidemiology. New York: John Wiley and Sons.

Chen, X. and Line, R.F. 1995. Gene action in wheat cultivars for durable, high-temperature, adult-plant resistance and interaction with race-specific, seedling resistance to *Puccinia striiformis*. *Phytopathology*, 85: 567–572.

Das, M.K., Rajaram, S., Mundt, C.C. and Kronstad, W.E. 1992. Inheritance of slow-rusting resistance to leaf rust in wheat. *Crop Science*, 32: 1452–1456.

Daszak, P., Cunningham, A.A. and Hyatt, A.D. 2000. Emerging infectious disease of wildlife: Threat to biodiversity and human health. *Science*, 287: 443–449.

Ferrandino, F.J. and Elmer, W.H. 1992. Reduction in tomato yield due to *Septoria* leaf spot. *Plant Disease*, 76: 208–211.

Forbes, G.A. and Jeger, M.J. 1987. Factors affecting the estimation of disease intensity in simulated plant structures. *Journal of Plant Diseases and Protection*, 94: 113–120.

Forbes, G.A. and Korva, J.T. 1994. The effect of using a Horsfall-Barratt scale on precision and accuracy of visual estimation of potato late blight severity in the field. *Plant Pathology,* 43: 675–682.

Forbes, G., Pérez, W. and Andrade Piedra, J. 2014. Field assessment of resistance in potato to *Phytophthora infestans*. Lima (Peru). International Potato Centre (CIP). p. 35

Gatachew, G., Alemu, T. and Tesfaye, K. 2014. Evaluation of disease incidence and severity and yield loss of finger millet varieties and mycelia growth inhibition of *Pyricularia grisea* isolates using biological antagonists and fungicides *in vitro* condition. *Journal of Applied Biosciences*, 73: 5883–5901.

Haynes, K.G. and Weingartner, D.P. 2004. Use of area under the disease progress curve to assess resistance to late blight in potato germplasm. *American Journal of Potato Research*, 81: 137–141. doi:10.1007/BF02853611.

I.R.R.I. 2002. Standard evaluation system for rice (SES). p. 56.

Jeger, M.J. and Viljanen-Rollinson, S.L.H. 2001. The use of the area under disease progress curve (AUDPC) to assess quantitative disease resistance in crop cultivars. *Theoretical and Applied Genetics*, 102: 32–40.

Madden, L.V., Pirone, T.P. and Raccah, B. 1987. Temporal analysis of two viruses increasing in the same tobacco fields. *Phytopathology*, 77: 974–980.

Madden, L.V. 1980. Quantification of disease progression. *Protection Ecology*, 2: 159–176.

Madden, L.V., Hughes, G. and Van den Bosch, F. 2007. The Study of Plant Disease Epidemics. American Phytopathological Society, St. Paul, MN.

Mbong, G.A., Akem, C.N., Alabi, O., Emechebe, A.M. and Alegbejo, M.D. 2010. Effect of soeing date on the incidence, apparent infection rate and severity of scan of cowpea. *Asian Journal of Agricultural Sciences*, 2(2): 63–68.

Miedaner, T. and Sperling, U. 1995. Effect of leaf rust on yield components of winter rye hybrids and assessment of quantitative resistance. *Journal of Phytopathology*, 143: 725–730.

Mohapatra, N.K. 2002. Studies on durable resistance to blast disease of rice. Dissertation. Orissa: Utkal University.

Mukherjee, A.K., Mohapatra, N.K. and Nayak. 2010. Estimation of area under the disease progress curves in a rice-blast pathosystem from two data points. *European Journal of Plant Pathology*, 1-5.

Mukherjee, A.K., Mohapatra, N.K., Suriya Rao, A.V. and Nayak, P. 2005. Effect of nitrogen fertilization on the expression of slow-blasting resistance in rice. *Journal of Agricultural Science*, 143: 385–393.

Mundt, C.C. 2002. Use of multiline cultivars and cultivar mixtures for disease management. *Annual Review of Phytopathology*, 40: 381–410.

Nutter, F.W., Jr. 1997. Quantifying the temporal dynamics of plant viruses: A review. *Crop Protection*, 16: 603–618.

Oerke, E.C., Dehne, H.W., Schonbeck, F. and Weber, A. 1994. Crop Production and Crop Protection: Estimated Losses in Major Food and Cash Crops. Elsevier, Amsterdam

Papaÿx, J., Touzeau, S. Monod, H. and Lannou C. 2014. Can epidemic control be achieved by altering landscape connectivity in agricultural systems? *Ecological Modeling*, 284: 35–47.

Pasha, A., Babaeian-Jelodar, N., Bagheri, N., Nematzadeh, G. and Khosravi, V. 2013. A field evaluation of resistance of *Pyricularia oryzae* in rice genotype. *International Journal of Agricultural and Crop Sciences*, 5(4): 390–394.

Peterson, R.K.D. and Highley, L.G. 2001. Illuminating black box: the relationships between injury and yield. *In*: Peterson, R.K.D. and Highley L.G. (*Eds.*). Biotic stress and yield loss. *CRC Publication*, pp. 1-12.

Power, A.G. and Mitchell, C.E. 2004. Pathogen spillover in disease epidemics. *The American Naturalist*, 164: 79–89.

Savary, S. and Zadoks, J.C. 1992. Analysis of crop loss in the multiple pathosystem groundnut-rust-late leaf spot. I: Six experiments. *Crop Protection*, 11: 99–109.

Shaner, G. and Finney, R. 1977. The effect of nitrogen fertilization on the expression of slow-mildewing in knox wheat. *Phytopathology*, 67: 1051–1056.

Simko, I. and Hans-Peter, P. 2012. Area under the disease progress stairs: calculation, advantage, and application. *The American Phytopathological Society*, 102(4): 381–389.

Tack, A., Hakala, J., Pet¨aj¨a, T., Kulmala, M. and Laine, A. 2013. Genotype and spatial structure shape pathogen dispersal and disease dynamics at small spatial scales. *Ecology*, 95: 703–714.

Tooley, P.W., Sweigard, J.A. and Fry, W.E. 1986. Fitness and virulence of *Phytophthora infestans* isolates from sexual and asexual population. *The American Phytopathological Society*, 76(11): 1209–1212.

Trapero-Casas, A. and Jimenez-Diaz, R.M. 1985. Fungal wilt and root rot diseases of chickpea in southern Spain. *Phytopathology*, 75: 1146–1151.

Upton, G. and Cook, I. 2008. A Dictionary of Statistics. 2nd ed. United Kingdom: Oxford University Press.

Van der Plank, J.E. 1963. Plant diseases: Epidemics and control. New York: Academic p. 349

Waldir, C.J., Trazilbo, J.P., Miller, S.L. and Bernhard, H. 2014. Interactions between foliar diseases: Concepts and epidemiological approaches. *Tropical Plant Pathology*, 39(1): 1–18.

Weber, G.E. 1996. Modeling interactions between epidemics of *Erysiphe graminis* and *Septoria nodorum* on wheat. *Journal of Plant Diseases and Protection* 103: 364–376.

Wilcoxson, R.D., Skovmand, B. and Atif, A.F. 1975. Evaluation of wheat cultivars for ability to retard development of stem rust. *Annals of Applied Biology*, 80: 275–281.

Zadoks, J.C. and Schein, R.D. 1979. Epidemiology and Plant Disease Management. Oxford University Press, London and New York.

CHAPTER

3 Mathematical Modeling in Botanical Epidemiology

3.1. INTRODUCTION

Epidemiology involves the study of the temporal, spatial, and spatio-temporal dynamics of disease in populations, and the utilization of results of experiments and surveys to describe, understand, compare, and predict epidemics.

Quantitative nature of these epidemics attributes may play an important role to the development and appraisal of efficient control strategies and tactics.

Mathematical modelling of botanical epidemiology is the emerging research theme among the epidemiologist. Statistical tools like regression (linear and multiple), correlation analysis, hypothesis testing, distribution patterns (beta, binomial and the Bayesian approaches) ordination (Principal Component Analysis, Canonical Analysis) and growth analysis models are useful for quantification of spatial and temporal aspects of epidemics.

Recently, compelling arguments have been made for the use of Bayesian decision theory in developing and evaluating real-time disease prediction rules, based on measured disease or weather conditions and either empirical or mechanistic models for disease or control intervention.

Cunniffe *et al.* (2014) have identified the thirteen challenges related with plant diseases modeling and they had categorized these challenges with plant hosts (four), with modeling of the pathogen (five) and with modeling control (four). According to them the challenges related with these three categories are summarized as follows:

Thus, from above comparative analysis it can be emerged that botanical epidemiology and its important attributes are generally judges and predicted with mathematical tools and techniques.

Challenges related with host	***Demerits of present analytical approaches***	***Possible future thrusts***
Linking epidemiological models to crop yield and ecosystem services	Present models concentrating on yield, typically include simple epidemiology with logistic growth dynamics	Models should incorporate sufficient epidemiological realism in order to analyze and predict the effects of disease and host dynamics on yield. *e.g.*, compartmental models proposed by Hall *et al.*, (2007). Ideally models would also account for compensation *via* reduced competition from diseased plants on remaining healthy individuals, and for the effects of pathogens on the full range of ecosystem services (Boyd *et al.*, 2013)
Temporal changes in host availability, from plant organs to populations	Most present models assume constant host population size and their changes in number or density are most often ignored	The default assumption of constant population size should be replaced: models should more routinely include realistic within-season host dynamics and synchronous removal at the end of each season. Model should valued annual, perennial overwintering and alternate host growth also.
Capturing host spatial structure, even when data are limited	Spatial patterns and meta-population dynamics of hosts are not adequately address	Require metrics to quantify how the invisibility of landscapes is conditioned on the interplay between spatial structure and pathogen dispersal, particularly to allow limited control resources to be allocated, and to assess potential evolutionary implications
Beyond a single species: multiple and alternate hosts, spillover and community ecology	Modeling hosts species that perform as alternate or collateral are neglected	Require model that satisfy the dynamics of multi-host pathogen system and host heterogeneity
	Challenges related with pathogen	
Realistic dispersal models, including meteorological and anthropomorphic drivers	For epidemic the aerobiological role is having a prominent importance, through that one can quantify the dispersal pathway's, speed and intrinsic characters associated with pathogen inoculums. Simple models are available for estimation of spore tra-	The challenge is matching the complexity of dispersal to the purpose of the model and, crucially, to the quality of data available for parameterization. While a more detailed treatment of dispersal is attractive, this can only be meaningful if given statistical support *via* fitting, for which available data are typically rather sparse.

Table: (*Contd...*)

Table: (*Contd...*)

Challenges related with host	*Demerits of present analytical approaches*	*Possible future thrusts*
	jectory, however, use of satellite imagery, GIS and associated modeling still in nascent phase	
Network models for human mediated spread	Understanding the spread of disease in these networks could help to identify network characteristics that exacerbate spread and also be used to test sampling and control strategies	Although some insight has already been gained from adaptations of network theoretic models, however, Models must also consider the interaction between natural and artificial spread pathways.
Accounting for time varying infectivity	Models that quantifies the temporal and spatial aspects of host-patho system are needed	Capturing experiential patterns in models is a major challenge because infectivity is so difficult to measure in natural systems, and approaches for scaling from growth chamber measurements to natural conditions are not obvious. Climate change also raises the possibility of inherent non-stationary in all these processes, which poses supplementary challenges in modeling dynamical outcomes and implications of control.
Effect of vector preference on transmission	Models that can provided simulation about vectors preference, transmission mode, incubation time and threshold levels are lacking	Adding vector preference to compartmental models would allow us to understand interactions between host and vector infection status, transmission type and vector preference, together with any evolutionary implications.
Beyond a single species, multiple strains, multiple pathogens and evaluation	Understanding the mechanisms and patterns involved in multiple circulating pathogen are very crucial for disease epidemiology and forcasting. Study of such complex pattern are need to be addressed at greater death	Modeling quantitative disease resistance in populations facing multiple pathogens is a key challenge in predicting disease dynamics and evolutionary responses of pathogens and hosts.
Challenges related with modeling control		
Using Models to Optimize Detection	Surveillance, sample size, frequency of sampling are carried out	Further challenges include optimizing sampling programs in plant trade networks, how to sample to detect

Table: (*Contd...*)

Table: (*Contd...*)

Challenges related with host	***Demerits of present analytical approaches***	***Possible future thrusts***
	through routine procedures, however, there is scope for including the introduction pathways, pathogen dispersal patterns, landscape connectivity and for warning systems	unknown (or unanticipated) pathogens, and the role of passive surveillance and citizen science.
Optimizing dynamic controls in heterogeneous systems	Control procedures generally included various biological, cultural and chemical approaches, however, their accountability and accurate impact on disease management still need refinements	We need to understand how the optimal scale of control depends on the interplay between the epidemiology of the plant-pathogen interaction the implementation of detection and control, and the current state of the epidemic
***Accounting for economics*:** moving optimal control theory to realistic landscapes	At landscape and at different land use levels yield loss assessment models, particularly related with meta-population have greater scope to use algorithms	The challenge now is to integrate and test the inferences from these approaches into models for disease spread in realistic landscapes
Use of models by policy makers and stakeholders	The complexities of disease models are of less concern to stakeholders and policy makers than model validity and usability for practical decision making. Both groups are predominately concerned with avoiding disease, early detection, and control. Regulatory philosophy is to use the latest scientific information that is available	Models that capture the broad characteristics of a host-patho-system and that can be used to predict its dynamics are therefore increasingly attractive to policy makers

In Chapter two of the present book some of the basic quantitative approaches like diseases progress curve, AUDPC, yield loss estimation, basic and apparent infection rates have been addressed and workout. In this chapter some the mathematical approaches that are improved or incorporated in the arena of botanical epidemic will be discussed.

For plant epidemiology mathematical tools uses several variables values as inputs and these variables are considered according to the nature of the problem and the questions to be answered (Van der Plank, 1982). Some of the mathematical models their theories, implication and interpretation are as follows:

3.2. DISEASE PROGRESS CURVES (DPC)

Basic terminology, its attributes and DPC of few diseases are described in chapter two. The curve shows the dynamics of epidemic over time and it provided information about temporal aspect of disease appearance, inoculums amount, and changes in host susceptibility during the growing period, climatic events and the effectiveness of management practices. (Contreras-Medina *et al.*, 2009).

Growth models endow with range of curves that are often similar to DPC and to be considering as most common mathematical tool to describe temporal aspect of disease epidemics (Xu, 2006).

The growth models commonly used are Monomolecular, Exponential, Logistic and Gompertz. In general, growth models that incorporate few variables to describe temporal disease dynamics have a good performance; however, this kind of models sometimes do not satisfy the acquiring process of key characteristics because they frequently ignore relevant variables that affect the epidemic development (Xu, 2006), *e.g.,* host growth, fluctuating environmental condition, length of latent and infectious period, etc.

Both logistic and Gompertz and models have a characteristic sigmoid form and an infection point that means secondary inoculation or plant to plant spread within the crop in comparison to monomolecular model (Fig. 1) which does not have infection point.

3.2.1. Exponential

Also known as the logarithmic, geometric or Malthusian model.

Useful to describe early stages of most polycyclic epidemics. This model suggested that the absolute rate of disease increase is directly proportional to disease intensity (X, *e.g.,* incidence or severity).

It can be write as $X=X_0e^{rt}$, where X_0 is the amount of initial inoculums/ initial disease intensity, e is the exponential and *r* is basic rate of infection and t is time. This model is easy to extrapolate and interpolate (prediction), easy to see changes in initial X and the increase over time and easy to fit.

3.2.2. Monomolecular

Appropriate for modeling monocyclic epidemics.

Also called negative exponential model.

In this model dx/dt is not proportional to x because inoculum produced by infected individual during the current epidemic does not lead to additional infection in the current epidemic. Thus, in other way this model is appropriate foe modeling epidemics where there is no secondary spread within a growing season, meaning that the plant disease has a single cycle during growing season.

However, the dx/dt is proportional to healthy tissue. The quantity of changes in disease (dy/dt) over time from a maximum at the beginning of epidemic and has a negative exponential probability density function.

The disease intensity is on a proportion scale. Thus, disease free (healthy) or level is 1-x

$$dx/dt = r_m (1-x).$$

In this model r_m is the product of amount of inoculums available for infection of disease free plants or units and the probability per unity time.

$$X=X_0 + r_m t$$

3.2.3. Logistic

Appropriate for most polycyclic diseases (having secondary spread within a growing season)

Most widely used for describing plant epidemic.

Particularly this model is most widely used for temporal analyses.

When the rate of infection is constant, the plot of dy/dt versus t increases initially, reaches a maximum (at 50% disease or infection point), and declines to zero.

This curve that is symmetrical about the infection point can be linearized by using logarithmic transformation.

This model having features of both the exponential and monomolecular models.

$$X = 1/1+\exp\{-[1n * (X_0/1-X_0)+r *t]\}$$

3.2.4. Gompertz

Appropriate for polycyclic diseases as an alternative to logistic models.

Gompertz model has an absolute rate curve that reaches a maximum more quickly and declines more gradually than the logistic models.

The curve for dy/dt versus t for Gompertz, unlike logistic model, is not symmetrical around the infection point.

The rate of infection approaches infection point more rapidly and declines more slowly than in the logistic model and the curve is skewed positively *i.e.* a larger portion of the area under curve is to the right of the infection point.

$$X = e^{\ln(X_0)e^{-r*t}}$$

According to Berger (1981), the Gompertz transformation effectively linearized the DPC compared to logistic model and it can also avoid the curvilinearity which is commonly associated with logistically transformed values. He also found that this model more accurately quantified the epidemic rate, projection of future disease severity and determination of initial diseases.

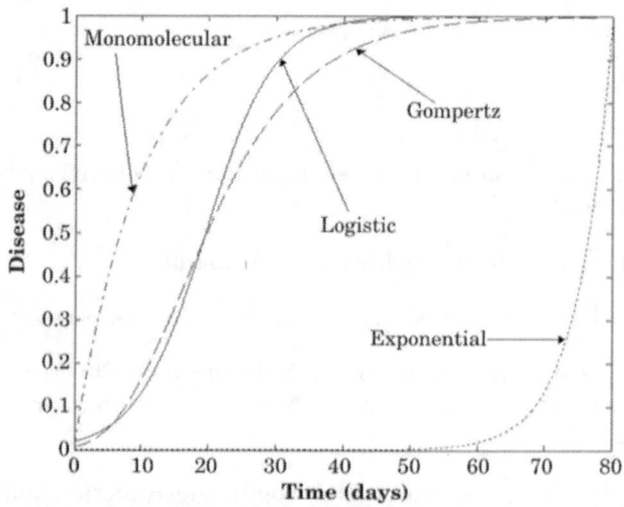

Fig. 1: DPC showing various growth curves.

The proportion of disease increase in a host plant, when plotted against time, gives a sigmoid curve though other types of curves are encountered. It is necessary to linearise such curves in order to evaluate rate reducing resistance for comparison of the epidemics in terms of the rate of disease increase.

This could be achieved by applying either the logistic model or Gompertz model or Richards model or by Weibull model for Gaussian model.

It is not necessary that all the models will not fit well to a specific plant-patho-system nor a specific model will fit in to all the plant-patho-systems or even diseases.

Mohapatra *et al.* (2008) have evaluated the disease progress curve in the rice blast pathogystem and compared with the Logistic and Gompertz models. They had concluded that Gompertz model were more consistent among the studied slow-blasting genotypes over seasons, however, both Gompertz and logistic models fit well in to the rice blast patho-system.

The comparative utility of Gompertz, Exponential, Logistic and Monomolecular models in the analysis of disease progress and disease prediction was studied by Panicker and Ramraj (2010) and the transformations were fitted in the polynomial regression equation to get the absolute rate equation and linearized equation. Expression for the absolute rate of changes in y with (dy/dt) and the integrated linearized forms of these models are presented in Table 1.

Appropriate population growth model for disease progress datasets can be selected by using following steps:

- First stem is to graph y (generally disease or pathogen incidence or severity) versus time t. The shape of disease/pathogen progress curve is very helpful in identifying the best model.
- After viewing these graphs, objective criteria for the acceptance or rejection of a model are provided by the estimated regression parameters and statistics, *i.e.* r^2, coefficient of Variance (CV) and standard deviation (Campbell and Madden, 1990 and Nutter and Parker, 1997)
- Finally, the inspection of residual plots reveals if the model is satisfactory.

Table 1: Various growth models and their mathematical expressions.

Model	***Integrated expression***	***Absolute rate of equation***	***Linearized equation***
Monomolecular	$y=1-(1-y_0)$ *exp* $(-r_M *t)$	$dy/dy\ -r_M * (1-y)$	$Ln[1/(1-y)] = 1n[1(1-y_0)]+r_{m*}t$
Exponential	$y=(y_0)$ *exp $(r*_{E*}\ t)$	$dy/dt = r_E * y$	$1n(y) = 1n *(y_0)+r_E*t$
Logistic	$y = 1/\{1+[(1-y_0)/y_0 *$ exp $*(-r_L *t)$	$dy/dt = r_L\ y\ (1-y)$	$Ln[y/(1-y)] = 1n[y/(1-y_0)]+r_L*t$
Gompertz	$Y=\exp\{1n(y_0)$ *exp $(-r_{G*t)}\}$	$Dy/dt = r_G[-1n(y)]$	$-1n[-in(y)] = 1n[-1n(y_0)] + r_G*t$

In above equations y = disease intensity; t= time; r= rate parameter

(r_M = rate for monomolecular model; r_E = rate for exponential model; r_L = rate for the logistic model; r_G = rate for the Gompertz model) y_0 constant for integration, corresponding to y at t =0.

3.3. LINKED DIFFERENTIAL EQUATIONS (LDE)

This approach use to validate and provide convincing answers related the questions of three important plant disease epidemic attributes *viz.*, host, pathogen and human intervention. Through this approach we can find out the relationships between these three attributes and their relationship with each other as well as with plant disease dynamics.

The LDE models are of the susceptible, exposed, infected and removed (SEIR) types. In this approach, the host population is usually divided into several non-overlapping categories, like as healthy susceptible, lately infected, infectious and removed (post infection).

In the event of host infection, the pathogen moves through the latent stage to become infectious at a rate which is inverse to mean latent period. Infected host lose infectiousness and proceed into the post infection stage at a rate which is the inverse of mean infection period.

LDE equations models are specified for each defined plant category which can express as: $\frac{dP}{dt} = B(P) = D(P)$

In this equation B(P) and D(P) are functions describing the increase and decrease of host population of category P, B(P) and D(P) are jointly determined by host growth function, pathogen attributes, pathogen transmission and disease management.

LDE can provide the answers about (Madden *et al.*, 2000)

- How does the epidemic approach the equilibrium
- What are the conditions necessary for the persistent of epidemic
- System sensitive to model parameters
- What are the values and stability of equilibria and
- How these values affected or related with model parameters.

Zhang *et al.* (2001) used four linked differential equations to illustrate the effect of synergism between plant viruses where each differential equation represents one of the following four categories: healthy host plant, infected host plant with a virus kind A alone, infected with a virus kind B alone, and infected with both viruses A and B, the obtained results were used to understand an epidemic that has been ongoing in Uganda since the late 1980s and the information about the development of Cassava Mosaic Disease (CMD) epidemic in Uganda and neighbouring countries supports model results.

Now a day's computer simulation software's like MATHCAD or MATHEMATICA (Madden, 2006) are available.

A major disadvantage of LDE is that the formulated equation using this tool are mathematically complicated, which is difficult to explore different biological properties of host and pathogen or to know the results of different control strategies on term disease development.

3.4. COMPARTMENTAL MODELS IN BOTANICAL EPIDEMIOLOGY

Epidemiological models of disease spread in plant population typically rely on compartmentalization of infectiousness status or variables like susceptible, exposed (or latent), infectious and removed (or recovered) or subsets or extensions thereof depending on the specific pathogen life-history and host species (Leclerc *et al.*, 2014) and usually a Diseased state is added with the aim to merge infectiousness and pathology status.

Basically in plant pathology, disease expression generally occurs in already infectious host and it does not stop host from being infectious (Motisi *et al.*, 2013). In general host-pathogen system may have some principal state variables (compartmental model or variables) and with these principal variables there are certain epidemiological models proposed by various authors.

These models includes Susceptible-Infectious-Recovered (SIR); Susceptible-Infectious-Detected-Recovered (SIDR); Susceptible-Infectious-Free living inoculums (SIX) and Susceptible-Exposed-Infectious-Recovered (SEIR) while other models includes stochastic SIX, dispersal kernel, meta-population and two pathogen strain (Gilligan, 2008). The graphical illustration of basics of compartmental models/variables is presented in Fig. 2.

However, Leclerc *et al.*, (2014) have reported that for soil borne plant pathogens, incubation period is an important transit state (Fig. 3) and for some host-pathogen system it may be negative exponential. Knowledge of incubation period (*i.e.* time between host infection and expression of disease symptoms) is crucial for understanding epidemiological attributes, behavior and for designing appropriate management practices.

In this connection Leclerc *et al.*, (2014) have realized the importance of incubation period and according them, accurate information about the incubation period distribution of a pathosystem can be critical in assessing

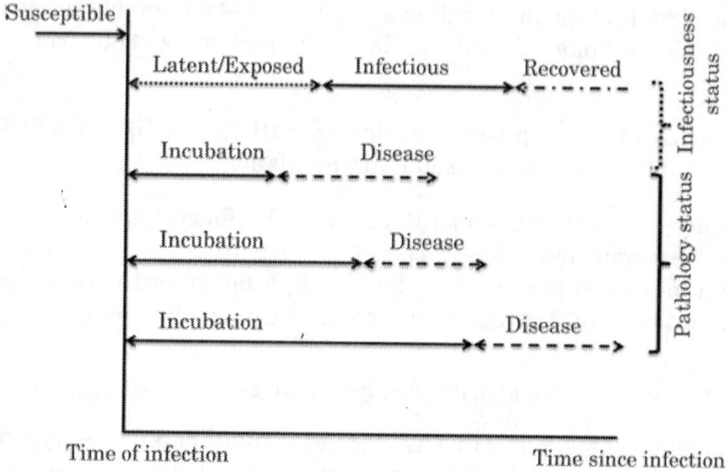

Fig. 2: Compartments of host-pathogen system (Redraw from Leclerc *et al.*, 2014)

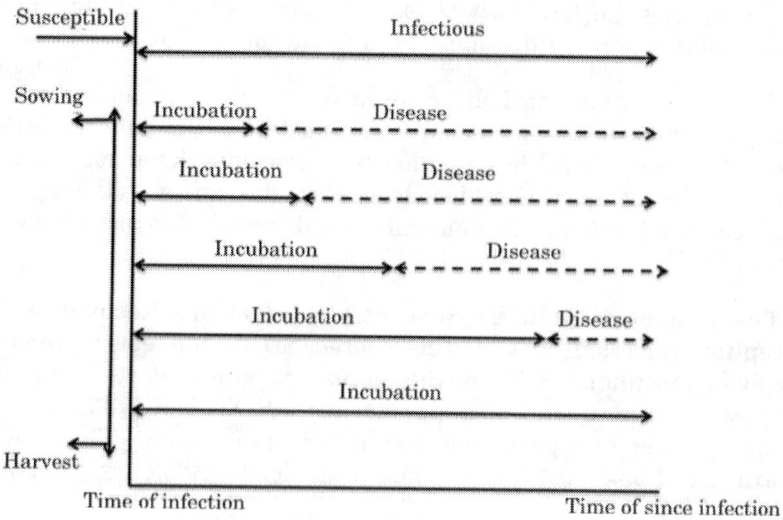

Fig. 3: Compartments of host-pathogen system (in the case of a soil-borne plant disease, the appearance of visible disease symptoms can be delayed when infection occurs late in the crop season and/or the host is mature Redraw from Leclerc *et al.*, 2014).

the true scale of pathogen invasion behind the early disease symptoms in the field and they suggested that reliance on observation of disease symptoms can cause significant delay in detection of soil-borne pathogen epidemics and mislead practitioners and epidemiologist about the timing, extent, and viability of disease control measures for limiting economic loss.

***Susceptible (S)*:** Initially, individual/host is susceptible to infection: no pathogen is present; just a low-level nonspecific immunity within the host.

***Exposed (E)*:** In early stages, the host may or may not exhibit obvious signs of infection and abundance of pathogen may be too low to allow further transmission. In other words, it also represent latently infected.

***Infectious (I)*:** Host encounters infectious individual and becomes infected with a micro-parasite; abundance of the parasite grows with time.

***Recovered*:** The host is either no longer infectious or 'removed' (dead).

***Detected (D)*:** Represents symptomatic and infections, and

Free living inoculums (***X***).

3.4.1. Types Related With Variability's

3.4.1.1. *Demographic stochasticity*

For epidemics, it simply reflects the probabilistic nature of transmission of infection between an infected and a susceptible host under identical environmental conditions.

This leads to a series of stochastic events giving rise to different trajectories for different epidemics, even though the transmission and other parameters remain the same throughout the course of the epidemic.

It is equivalent to saying that if epidemics were initiated in identical fields under the same initial conditions and exposed to identical environmental conditions, the resulting epidemics occurring in each field could still be markedly different.

Analysis of demographic stochasticity allows us to identify the underlying variability of an epidemic, the so-called dynamical landscape.

The enclosure of stochasticity and spatial structure is particularly significant as models featuring these attributes can produce qualitatively very diverse predictions concerning pathogen establishment and diligence than their deterministic mean-field counterparts.

Most importantly, invasion thresholds in stochastic models are higher and the pathogen or strain may be unable to persist following flourishing invasion due to chance events, particularly at low population densities (Gilligan, 2002).

3.4.1.2. *Environmental stochasticity*

This related with sensitivity of plants and micro-organisms to temperature, relative humidity and other environmental driving variables. Here the focus has been more on treating parameters as smoothed deterministic functions of environmental variables.

Pathogenic dispersal can take place by many agencies like abiotic, vertebrates, humans and at various land uses with different magnitude. Based on dispersal distance and it's timing, epidemic may treated as severe, mild or within canopy. Dispersal kernels (***i.e.*, the probability density function of dispersal distances relatively to a point source, constitutes the most basic and synthetic descriptor of the dispersal process; dispersal is anisotropic when propagules disperse differently depending on the direction**), generally questing how far can a disease spread? how far is infection likely to spread? and how might the deployment of control strategies affect the dynamics of spread and the likelihood of disease invasion? (Gilligan, 2008).

Aylor (2003) constructed a stochastic dispersal model to quantifies the rate and extent of seasonal incursions of two aerially dispersed diseases, stem rust of wheat and tobacco blue mould, from southern into northern areas of the US. The model treats the availability of the susceptible crop as a continuum in space, but one that changes over time, sandwiched in a seasonal **'green wave'**. Northward spread determined by crop sowing date. The withdrawing southerly edge is determined by an advancing wave of crop maturity. Both wheat stem rust and tobacco blue mould appear to spread northward on average at about the same rate as the seasonal advance of the green wave of available susceptible host tissue. He concludes that the consistency of the **disease wave** with the green wave underscores two important points.

In principle, dispersal kernels can experimentally be estimated from the real-time tracking of propagules (**Lagrangian methods**) or from the amount and/or diversity of propagules observed at different distances from a source (**Eulerian approach;** Bullock *et al.*, 2006). Lagrangian methods have mostly been applied to animals or animal-dispersed propagules (*e.g.*, seeds, pollen) using mark/recapture or tracking designs. They can barely be applied to wind-dispersed pathogen propagules because of their small size. The Eulerian approach generally utilize for plants (*e.g.*, Bullock *et al.*, 2006; Busi *et al.*, 2008 and Darmency *et al.*, 2009).

- Eulerian approaches require a number of precautions in the experimental design, which can be fastidious when dealing with disease gradients caused by infectious spores.

- First, external contamination must be avoided, which requires performing the experiment in a disease-free area or to use a traceable source of inoculum.
- Second, secondary dispersal events (which occur when inoculum disperse from a plot that has itself been infected by the primary source of inoculum) must be either distinguishable or avoided.
- Third, the sizes of the experimental site and of the inoculum source are both critical and must be large enough to provide measurable disease levels at long distances.
- Fourth, because spore dispersal might be anisotropic (*i.e.*, direction-dependent), trapping should be arranged in several directions
- Fifth, trapping design as well as the sizes and shapes of trap plots need to be carefully conceived, which generally requires a priori knowledge about the dispersal abilities of the species under study (Rieux *et al.*, 2014).

Gilligan (2008) have some basic relevant facts for crop mosaics, heterogeneity and topology of crops in the landscape, and according to him following points affects the epidemic.

- How far and how fast a pathogen spreads through a landscape and, indeed, whether or not it persists and for how long, depends upon the crop mosaic within the landscape.
- Whether or not invasion occurs depends upon the relative magnitude of the modal dispersal distances and the scale of heterogeneity in the landscape.
- For a given pathogen, each field of a susceptible crop is classified as susceptible (*i.e.* healthy), infected or, if appropriate, recovered. Non-host crops are unavailable to the pathogen and are treated as gaps.
- The application of a protectant fungicide temporarily renders a susceptible field unavailable for infection.
- Whether or not a pathogen invades now depends upon the spatial and temporal dynamics of the crop mosaic.
- When spread occurs between adjacent fields, *i.e.* by nearest-neighbour spread, the dispersal kernel then describes the probability distribution for transmission of infection between an infected field and its susceptible neighbour.
- It follows from **percolation theory** that there is a critical probability above which disease spreads and below which it dies out. By increasing the distance between susceptible fields, it is possible to bring the percolation probability for a given lattice below the threshold and so prevent invasion. These results are stochastic: it follows that there

will be some spread when the percolation probability is below the threshold and occasionally invasion may occur but, on average.

- As an epidemic spreads through a heterogeneous landscape, the dynamical contact structure, which measures the exposure of susceptible to infected sites, also changes. This too can slow down or speed up an epidemic and is a property of the topology of the system. So far, we have discussed susceptible and gaps.

Otten *et al.* (2005) showed how to calculate transmission rates from empirical data which, in turn, can be used to predict the effects of changes in the topology and composition of the mixture. By treating fields as units this approach can, in principle, be extended to gain insight into the effects of different levels and densities of partial control at the landscape scale.

Maddan and Van-der Bosch (2002) have graphically illustrated the above variables and according to them total disease (*Y*) can be represent by E (abundance of disease-free plant) +I (latently infected) +I (infectious) and however, all variables depend on time (t).

They is assumed that P_0 units of pathogen inoculum are introduced into a disease-free crop at the start of a season ($t = 0$), considered below as the first season. The introduced inoculum then decays exponentially.

New infections occur based on the contact between *S* and *P*, and the rate depends on the primary infection rate parameter, ν. Infected individuals initially are in the latent state (with a mean duration of 1/σ), and then are in the infectious state (for a mean duration of 1/γ). New infections can occur from infectious plant individuals; the rate of this secondary infection depends on the secondary infection rate parameter, β.

Maddan and Van-der Bosch (2002) have presented the overall model in pseudo-equation form as: (ε)

$$\frac{dS}{dt} = -[Primary\ infections] - [Secondary\ infections]$$

$$\frac{dE}{dt} = +[Primary\ infections] + [Secondary\ infections] \\ - [latent\ Infections\ Becoming\ infectious$$

$$\frac{dI}{dt} = +[latent\ Infections\ Becoming\ infectious] - [Removal\ of\ Infectious]$$

$$\frac{dR}{dt} = +[Removal\ of\ Infectious]$$

$$\frac{dP}{dt} = -[Mortality\ of\ Primary\ Inoculum]$$

3.4.1.3. *Basic quantification of some models*

In botanical epidemiology, models are generally applied (a) to predict invasion of new pathogen or race, (b) for predicting the consequences of control strategies on invasion, and (c) for quantification of persistence and crop loss and to predict the risk of disease.

For modeling of these epidemiological related attributes, knowledge about certain biological and dynamical issues are essential (Gilligan, 2002).

3.4.1.3.1. *Biological issues*

- Dynamics of inoculums.
- Dynamics of primary infection from resident inoculums or immigration.
- Dynamics of secondary infection between infected and susceptible.
- Host dynamics including the rate of production of susceptible tissue.
- Host responses to infection, including the effects of disease on the supply of susceptible tissues.
- Quenching due to changes in host susceptibility or onset of unfavourable conditions
- Vector or antagonist dynamics.
- Genetical structure of the parasite population.
- Definition of scale for the susceptible and infected units, ranging from lesions, leaves, roots, stems, single plants to whole field.

3.4.1.3.2. *Dynamical Issues*

***Non-linearity*:** Small changes that have big effects on epidemics.
***Stochasticity*:** Changes variation within and between epidemics.
***Spatial heterogeneity*:** Local and regional changes that affect spread.
***Temporal heterogeneity*:** Daily and seasonal changes that affects spread.
***More than one scale*:** from micro-through field to regional scale.

3.4.2. Attributes of Some Compartmental Models

1. SIR model with secondary infection, infection period and net birth of susceptible together with criterion for invasions (R_o).
2. SIDR model with secondary infection, cryptic (μ_1^{-1}) and symptomatic (μ_2^{-1}) infection periods with fixed host population size (Dybiec *et al.*, 2004).

3. SIX model with primary β_P and secondary β_S infection, net birth of susceptible and decay of inoculums.
4. Demographic stochastic version of SI model (Gibson *et al.*, 1999).
5. SEIR model with secondary infection, latent (σ^{-1}) and infectious (μ^{-1}) period for fixed host population size.
6. Dispersal kernel (with parameters, θ) for probability that susceptible (s) is infected by secondary infection from all infected host (i) within a neighborhood N_{hood} and by a constant source of primary infection ($\mathring{a}$) (Keeling *et al.*, 2004).
7. Meta-population with transmission between subpopulation in a neighborhood of interaction for a given strength of coupling ($\mathring{a}$) in addition to transmission within subpopulation: models also shows separate density-dependent birth and death processes for susceptible and criterion for invasion with a patch or subpopulation (R_P) and entire meta-population (R_o) (Park *et al.*, 2001).
8. Effect of control agent on transmission rate and infectious period for a single pathogen strain and for two pathogen strains typified by a fungicide-sensitive (I_s) and a fungicide resistant (I_r) strain with finesse costs $\tilde{n}_\beta$, $\tilde{n}_i$ in which the efficiency is a function of fungicide concentration (C)

Conclusively, these compartmental models requires information's about

- *Duration*: Latent and infectious period.
- *Rates*: Contact rates
- *Probabilities*: P(infection/transmission).
- *Demography*: Birth and death rate, and age structure of the population
- *Disease*: Proportion of inapparent infections.

3.4.2.1. *SIR model*

In its most basic form, an SIR model consists of a set of linked differential equations describing the dynamics of susceptible (healthy), infected, and removed (post-infectious) host tissue; commonly, the infected tissue is divided into exposed (latently infected) and infectious compartments, leading to an SEIR model (Madden, 2005).

This model was popularized by Jeger (1982) for use in plant epidemiology.

SIR model with secondary infection, infection period and net birth of susceptible together with criterion for invasions (R_o).

Recent work has added considerable complexity to SIR-type models (Gilligan, 2002), including

- Demographic and environmental stochasticity (Park *et al.*, 2003; Gibson *et al.*, 2004; Otten *et al.*, 2004),
- Seasonal disturbance and multi-year disease dynamics (Madden and van den Bosch, 2002),
- Dynamics of host growth and susceptibility (Gibson *et al.*, 2004),
- Virus vectoring mode and vector performance (Madden *et al.*, 2000; Holt and Colvin, 2001),
- Interactions with bio-control agents (Gibson *et al.*, 2004),
- Spatial structure and meta-population dynamics (Park *et al.*, 2001, 2003), and
- The presence of pesticide-resistant subpopulations (Hall *et al.*, 2004), among others.

The increased complexity of contemporary SIR models adds realism and allows their application to a wider range of problems. Indeed, models are now routinely formulated to accommodate parameters useful in exploring specific management strategies (Jeger, 2000; Gilligan, 2002; Stacey *et al.*, 2004). SIRS, SEIR and SEIRS are the extensions of SIR model. For instance, a linked African cassava mosaic virus–whitefly vector model (Jeger *et al.*, 2004) includes four management-related parameters, namely;

- The rouging rate of infected host plants,
- The insecticide-induced death rate of the vector,
- The virus acquisition and transmission rates of the vector,
- Both of which are determined by the level of host resistance.

Analysis of this model indicated that rouging applied once per month in combination with a modest level of host resistance (specifically one that reduces the product of acquisition rate and transmission rate below 80% of the value of the susceptible host) is sufficient to eradicate the disease, while a combination of rouging and insecticide application is less effective.

This example illustrates that analytical models have come a long way in their capacity to provide specific management recommendations that have traditionally been considered in the realm of more complex simulation models (Scherm *et al.*, 2006).

Basic formulas related with SIR are presented below (Fig. 4)

$$\frac{dS}{dt} = rS\left[1 - \frac{S+I}{k}\right] - \beta SI$$

$$\frac{dI}{dt} = \Box SI - \mu I$$

$$\frac{dR}{dt} = \square I$$

$$R_0 = \frac{\beta k}{\mu}$$

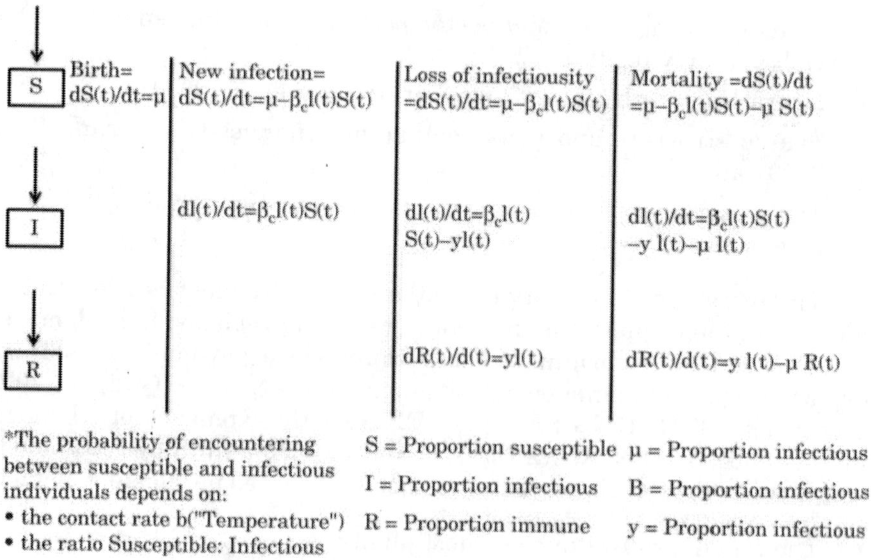

Fig. 4: Dynamic description of SIR model.

Results of SIR model suggests that

- Disease eventually burns itself out, with *I* lead to 0 as t is α
- Not all plants have been infected at the end of the epidemic
- Basic reproductive number (R_0) is the average number of newly infected host if one infected host is introduced into an otherwise susceptible host population.
- In SIR model R_0 is very important and it suggested that
  - $R_0>1$ = disease can invade (but can never persist) and
  - $R_0<1$ = disease cannot invade.
  - R_0 = overall rate of infection *x* infectious period

$$R_0 = \beta_S N\, x \frac{1}{\mu} = \frac{\beta_S N}{\mu}$$

- Thus, conclusively we can say that in SIR model: Invasion no longer inevitable
  - $R_0 >1$ = pathogen invade, causing epidemic

- $R_0<1$ = pathogen does not invade, no epidemic, and
- Not all plants become infected, with some escaping infection.
- *Cryptic infection*: suggesting that increasing μ indefinitely corresponds to immediately noticing and killing infected host before they spread disease and host can be infectious but not yet show symptoms.

Otten *et al.* (2003) have utilized *SI* model (Susceptible-infected) instead of *SIR* for quantification of transmission rates for primary and secondary infection's of a soil borne plant pathogen *Rhizoctonia solani*. From SIR compartmental model they removed R (removed class), since in *R. solani* only S and I classes can be distinguishable. They quantified primary and secondary transmission rates by following equations.

$$\beta_p(t) = \frac{[I_p(t+\Delta t) - I_p(t)]/\Delta t}{[N_{SX}(0,t+\Delta t) + N_{SX}(0,t)]/2}$$

$$\beta_s(t) = \frac{[I_s(t+\Delta t) - I_s(t)]/\Delta t}{[N_{SI}(1,t+\Delta t) + N_{SI}(1,t)]/2}$$

Where β_p and β_s are the rates of transmission for primary and secondary infection, N_{SX} (0,t) is the number of susceptible plants (S) directly in contact with inoculums (X), and N_{SI} (1,t) is the number of pair of nearest neighbors comprising an infected and a susceptible host. The denominators are averaged over time because some plants pass on infection at a rate faster than the scoring interval of one day (Δt)

In SIR compartmental model, a cryptic infection (*i.e.* infectious but yet not symptomatic) can be incorporated *i.e.* SCIR and in this model secondary infection rate β_s, development of symptoms at rate γ and roguing (μ) are the major components and their mathematical representation are as follows:

$$\frac{dS}{dt} = -\beta_s IS - \beta_s CS$$

$$\frac{dC}{dt} = -\beta_s IS + \beta_s CS - \gamma C$$

$$\frac{dI}{dt} = \square C - \mu I$$

$$\frac{dR}{dt} = \mu I$$

and the basic reproductive number (separate components corresponding to cryptic and symptomatic infection)

$$R_0 = R_0^C + R_0^S$$

$$R_0 = \frac{\beta_s N}{\gamma} + \frac{\beta_s N}{\mu}$$

Conclusively

- SCIR model suggests
- However, roguing only affects
- Since >1, roguing will never be successful

3.4.2.2. *SEIR model*

In this model, the population of hosts is divided according to disease status: susceptible (S), exposed (E), infected (I), and removed (R) Fig. 5.

Changes in the number of hosts in each compartment follow coupled nonlinear differential equations, which for the basic SEIR model. In plant pathology the use of SEIR model for botanical epidemiology has been advocated by Gilligna (2002, 2008) and Madden (2006) Sherm *et al.* (2006),

Black / dotted: Movement between classes.

Blue / solid arrows arrow: Level of infectious disease influences the rate at which a susceptible individual moves into the infected class. *Assumption*: account for latent period when pathogen is present in individual (infected) but not transmittable to other susceptible (not infectious).

R_0: Is the basic reproductive ratio. The key threshold is the "basic reproductive number" (sometimes "ratio"), R_0, which is the average number of new infections caused when a single infected host is introduced into a continuously completely susceptible population (compare the progeny/parent ratio of van der Plank). For a population of N hosts, the basic reproductive number equation is $R_0 = \beta N/\mu$.

Timescale of infection: Measured by the infectious period for SIS and SIR infections or by a mixture of exposed and infectious periods in diseases with SEIR dynamics.

Infectious diseases are highly variables in terms of infectious profiles, parameter values, and timescales.

$$\frac{dS}{dt} = -\beta SI$$

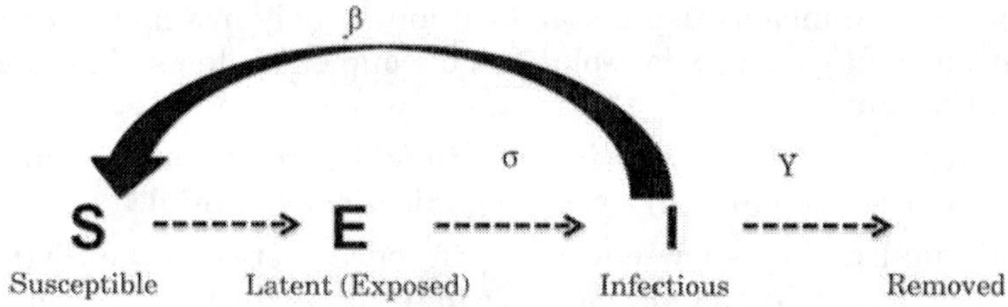

Fig. 5: Attributes of SEIR model.

$$\frac{dE}{dt} = \square SI - \sigma E$$

$$\frac{dI}{dt} = \square E - \mu I$$

$$\frac{dR}{dt} = \square I$$

In equations, β is the per capita rate of infection of susceptible hosts,

γ is the rate at which exposed (*i.e.*, latently infected) hosts become infectious, and

μ is the rate at which infected hosts become removed (*i.e.*, become epidemiologically inert by losing infectivity or dying).

The corresponding mean latent and infectious periods are $1/\gamma$ and 1/î, respectively, and implicit in the formulation is that these periods are distributed exponentially (*i.e.*, that the rate of leaving a class is independent of the time already spent within it).

Although the exponential distributions are selected to ensure that, on average, each host is latently infected or infectious for the correct period, the formulation means that the latent or infectious period of a particular host may be infinitesimally short which, for plant disease, is unrealistic.

Cunniffe *et al.* (2012) have presented the reformulation of SEIR model with addition of some components like multiple exposed (*i.e.*, latently infected) and infected (*i.e.*, infectious) compartments.

They demonstrate how the multiple compartment reformulation

- simplify the analysis of Segarra *et al.* (2001) and characterize the similarities and differences between the SEIR, van der Plank, and Gamma(k) models (in terms of $R0$, f, and r) using only elementary mathematical techniques;

- simulate the models using standard and widely available numerical routines and packages for solutions of coupled systems of differential equations; and
- extend the models very easily; and they illustrate this by adding (a) host demography and (b) free-living infectious material.
- Additionally, they show how the differences between the SEIR, van der Plank, and Gamma(k) models can be of practical significance.
- In particular, they note that the nature of the underlying model used to approximate an epidemic affects the inferred relationship between the initial rate of exponential increase of disease (r) (which is often used to estimate early epidemic spread) and the basic reproductive number (R_0) (frequently used as a measure of ability to invade).
- Moreover, because the final fraction of infected hosts (f) is controlled by $R0$, it follows that fitting an inappropriate model formulation to data from the early part of an epidemic leads to an incorrect assessment of the final severity

Mathematical improvement of SEIR model can be found in Cunniffe *et al.* (2012).

3.4.2.3. *SIDR model*

Stochastic susceptible– asymptomatically infected–detectably infected–removed (SIDR) model on a two-dimensional lattice with long-distance movements.

Dybiec *et al.* (2004, 2005) examined a series of control strategies of different sizes centered around a symptomatic site

$$\frac{dS}{dt} = \beta S\,(I + D)$$

$$\frac{dI}{dt} = \beta S\,(I + D) - \omega_1 I$$

$$\frac{dD}{dt} = \omega_1 I - \omega_2 I$$

$$\frac{dD}{dt} = \omega_2 D$$

3.4.2.4. *SIX*

$$\frac{dS}{dt} = rS\left[1 - \frac{S + I}{k}\right] - (\beta_p X + \beta_S I)S$$

$$\frac{dI}{dt} = (\beta_p X + \beta_S I)S$$

$$\frac{dX}{dt} = gI - hX$$

3.4.2.5. *Stochastic Six*

Deterministic to stochastic models

There are many strong arguments for moving away from deterministic to stochastic model. Real processes are stochastic. At the very least, adding stochasticity to a model gives it a great deal of added flexibility so the model better fits real data.

A stochastic framework can allow for important parameters of a model to be identified even when they are unidentifiable in a deterministic setting.

Can understand distributions associated with characteristics of a process. Often of great interest. Simulation of such stochastic models may be easier (though expensive).

Rigorous procedures may be available for inference

$$\Pr[I(t + dt) - I(t) + 1] - (\beta_p X + \beta_p I(t))Sdt$$

3.4.2.6. *Dispersal kernel*

In this model type rate of infection is depend on distance (Fig. 6).

How far the infection spreads within and between fields is defined by the dispersal kernel

The conventional question concerning dispersal kernels is, how far can a disease spread? By this is usually meant, what is the furthest distance over which transmission of inoculum can give rise to infection and disease? This simple question, however, is perplexed by very low probabilities of very long-distance dispersal.

More open-minded questions are, how far is infection likely to spread? and how might the deployment of control strategies affect the dynamics of spread and the likelihood of disease invasion?

The tail of the distribution for dispersal kernels is certainly important. Much thought and experimental ingenuity has been given to the distinction between exponential and thick-tailed distributions.

Exponential kernels result in expanding waves about an initial focus, tending to a constant velocity of expansion after an initial period of build-up. Thick-tailed distributions give rise to distinct daughter foci and an entirely different spatial dynamic for epidemic spread with dispersive spread in which the velocity of spread increases with time. Owing to the complications of controlling experimental conditions, most experiments to identify dispersal gradients have been confined to single foci spreading through single homogeneous field plots.

Scaling-up to predict what happens at the larger scales is still in its infancy for spread through heterogeneous landscapes. It is a major problem for emerging epidemics of new, rare or mutated forms of pathogens.

The challenge of estimating parameters for epidemics with multiple sites of initial infection has, with some notable exceptions, received little attention.

A simple analysis based upon ideas from percolation, derived in statistical physics illustrates an important relationship between dispersal kernels and invasion. When spread occurs between adjacent fields, *i.e.* by nearest-neighbour spread, the dispersal kernel then describes the probability distribution for transmission of infection between an infected field and its susceptible neighbour.

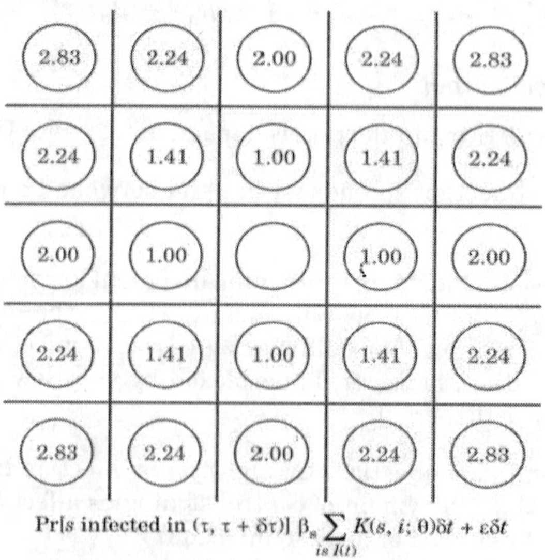

$$\Pr[s \text{ infected in } (\tau, \tau + \delta\tau)]\ \beta_s \sum_{i s\, I(t)} K(s, i; \theta)\delta t + \varepsilon\delta t$$

Fig. 6: Illustration of Dispersal Kernel.

Dispersal kernel (with parameters, θ) for probability that susceptible (s) is infected by secondary infection from all infected host (i) within a neighborhood N_{hood} and by a constant source of primary infection (ω) (Keeling *et al.*, 2004).

3.4.2.7. *Control single pathogen strain*

$$\frac{dI}{dt} e_{\beta}(C)\beta SI - \frac{\mu}{e_{\mu}(C)} I, e_i = \frac{1}{1+a_1 C}$$

3.4.2.8. *Two pathogen strains*

$$\frac{dI}{dt} = e_{\beta s}(C)\square SI_s - \frac{\mu}{e_{\mu s}(C)} I_s$$

$$\frac{dI}{dt} = e_{\beta,r}(C)\rho_{\beta}\square SI_r - \frac{\mu}{e_{\mu,s}(C)} I_r$$

$$1 - e_{\beta}(\bar{C})e_{\mu}(\bar{C}) < \frac{R_{0,r}}{R_{0,s}}$$

3.4.2.9. *Metapopulation*

$$\frac{dS_i}{dt} - (b_0 - b_1 N_j)S_j - (d_0 + d_1 N_j)S_j - \lambda_j S_j$$

$$\frac{dI_j}{dt} = \lambda_j S_j - (\mu + d_0 + d_i N_j)I_j$$

$$\lambda_j = \beta I_j + \varepsilon\beta \sum_{k \varepsilon N_{hood}} I_k$$

$$R_p = \frac{\beta_k}{\mu + d_0 + d_1 k}$$

$$R_0 = R_p(1 + z(N_{hood})\varepsilon)$$

Theoretical information's on compartmental models like Control Single Pathogen Strain, Two Pathogen Strains and about metapopulation can be found in exhaustive review work carried out by Gilligan *et al.* (2002).

3.5. COMPARTMENTAL MODEL FOR VECTOR BORNE PLANT EPIDEMICS

In general it's a very well established fact that insect vectors like aphids, leafhoppers, plant hopper, mealworm have close relations with many kinds of plant diseases Gilligan (2002 and 2008), particularly related with plant virus disease like Cucumber mosaic virus, Broad bean wilt virus, Beet curly top virus, and Maize streak virus (Liang *et al.*, 2001). Similar to soil borne plant epidemics {as described earlier in this chapter with example of work of Otten *et al.*, 2003}, vector-borne plant diseases also needs appropriate modification or modeling for compartmental epidemic models.

In most recent work, Shi *et al.* (2014) analyze the dynamics of vector-borne plant disease model. They have constructed the model and obtain the basic reproduction number (R_0) by using the next generation matrix model. They have also point out the local stability of the equilibria by the **Jacobian matrix method, Lyapunov function and the global stability of the disease-free equilibrium and endemic equilibrium.** Their most significant findings suggested that;

- If $R_0 < 1$, then the disease-free equilibrium is globally asymptotically stable, which means that the plant disease will disappear eventually.
- If $R_0 > 1$, then the endemic equilibrium is globally asymptotically stable, this indicates that the plant disease will persist for all time.

Following important assumptions are essential to construct the model for vector borne plant epidemic (Shi *et al* 2014).

For modeling the compartmental models related with insect vector population, the total population is divided into two categories, X and Y (these categories denotes the densities of the susceptible vector and infective vector at time t, respectively).

For the plant host population, the total population is divided into three categories, S (susceptible), I (infective), and R (recovered) at time t

Thus, total number of plants $K = S+I+R$ is a positive constant with an assumption that the number of plants in one area is fixed is logical (total number of plant can be always keep fixed by adding a new plant when a plant has died and for this birth rate of susceptible plant host is $f^*(S, I) = \mu K + dI$.

The susceptible plants can be infected not only by the infected insect vector but also by the infected plants.

A susceptible vector can be infected only by an infected plant host and after it is infected, it will hold the virus or vector for the rest of its life and further not vertical infection being considered.

The replenishment rate of insect vectors is a positive constant, and all the new born vectors are susceptible.

With following the principles of compartmental model, SIR models for vector born system can be formulated (Shi *et al.*, 2014).

$$S = f\ (S,I) - \mu S - \left(\frac{\beta_p\, Y}{1 + \alpha_p I} + \frac{\beta_s\, I}{1 + \alpha_s I}\right) x\, S$$

$$I = \left(\frac{\beta_p\, Y}{1 + \alpha_p I} + \frac{\beta_s\, I}{1 + \alpha_s I}\right) x\, S - (d + \mu + \gamma) x I$$

$$R = \Box I - \mu R$$

$$X = \wedge - \frac{\beta_1\, IX}{1 + \alpha_1 I} - m\, X$$

$$Y = \frac{\beta_1\, IX}{1 + \alpha_1 I} - m\, Y$$

- In these model S = number of susceptible plant hosts; I = number of infected plant hosts; R = number of recovered plant hosts; K = sum the total plant hosts; X = density of the susceptible insect vectors; Y = density of the infected insect vector; N = sum of the total insect vectors density; β_1 = infection ration between infected hosts and susceptible host plants; β_P = biting rate of infected vector on the susceptible host plants; β_S = infection incidence between infected and susceptible hosts; α_1 = determine the level at which the force of infection saturates (01); α_P = determines the level at which the force of infection saturates (0.2); α_S = determines the level at which the force of infections saturates (0.2); γ = the conversion rate of infected hosts to recovered hosts (0-025); μ = natural death rate of plant host; ^ = birth or immigration of insect vectors; m = natural death of insect vectors and d = disease-induced mortality of infected hosts.

Different sections and theorem formulated by Shi *et al.* (2014), for vector born plant epidemics it can be concluded that

- If $R_0 < 1$, then the disease free equilibrium is globally asymptotically stable and the unique endemic equilibrium is globally asymptotically stable provided that $R_0 > 1$. Thus, the basic reproduction number R_0 plays an important role in determining the persistence or dying out of the disease.
- Total number of host plant, birth rate of the vector and incidence rate of the disease can positively affect the value of R_0
- However, the death rate of t he host plant, the death rate of the vector and the disease-induced death rate can negatively affect the value of R_0, but the saturation rate of the incidence has no relation to the value of R_0.

3.6. SOME IMPORTANT CONSIDERATION'S FOR PATHOGEN DEVELOPMENT

In the first analytical model of Van der Plank (1963, 1982) disease development was proportional to the sum of infectious (sporulating) and accessible tissue with a constant infection rate (multiplication factor of disease per unit), a constant latent period and with random dispersion of inoculum between infected and susceptible hosts.

There have been several endeavors to combine host and pathogen growth since in order to account for the effect of changing environments on disease progress. From the pathogen point of view, temperature and leaf age are generally documented as the driving force for pathogen expansion (growth and sporulation of lesions), whilst wind factors (direction, speed) directed inoculum (spore) dispersal (Rabbinge and Bastiaans, 1989).

For modeling the plant epidemiology and their associated attributes, the dynamics of the pathogen population on leaves can be categorize into three key processes: (a) infection, (b) colony or mycelium growth, and (c) sporulation and dispersion. The time between infection and sporulation was described as the latent period, and the duration of sporulation as the infectious period.

3.6.1. Temperature for Spore Infection and Mycelium Growth

Quantitative analysis of thermal time calculations is a prerequisite parameter for modelling the phases of host-pathogen system. For several plant pathogens, various ontogenetic phases have been reported to be dependent on thermal time accumulation, for example ascospore maturation and release, germination, latent period, and sporulation .

Thermal time is usually calculated using air temperature. While air temperature may be appropriate to many studies of systems such as plant growth and plant pathogens, soil temperature may be more suitable for studies of soil-borne and root-infecting pathogens.

Variation in development of ontogenetic phases may well be observed if there are temperature differentials within the host. For example, upper canopy leaves that are subject to high levels of direct radiation may be warmer than lower canopy leaves that receive diffuse radiation. Therefore due consideration should be given to any possible variations between the actual temperature measured and that in which the pathogen resides. Not all processes are suited for modelling by thermal time.

Ontogenetic phases of pathogen development may also differ in their responses to temperature, so modelling using thermal time must reflect this. For example, Xu (1999) found that when the latent period for

Podosphaera (= *Sphaerotheca*) *pannosa* (powdery mildew) in roses was divided into two stages, incubation period and post incubation period, the rate of fungal development had a nonlinear relationship with temperature for incubation, but a linear relationship post-incubation.

The standard formula provided by Analytis (1980) can be use to describe the effect of temperature on the rate of spore infection and on the rate of mycelium growth. *F* was a function of a normalized temperature T_n

$$F\,(T_n) = \frac{(m+n)^{(m+n)}}{n^n m^m} T_n^n\,(1-T_n)^m, with\, T_n = \frac{T\,(t) - T_{min}}{T_{max} - T_{min}}$$

Where, *T(t)* is temperature at time *t* and T_{max} and T_{min} were the cardinal temperatures for either mycelia growth or infection (T_{maxg} and T_{ming} respectively), with m and n (mg and ng, respectively) shape parameters of the curve. *F* followed a gamma function with temperature.

3.6.2. Infection

Infection by pathogen can be quantified through equation having experimental data relating percentage, I of spore capable of infecting a leaf of age (A), 5I the decay rate of leaf susceptibility, I_0 the maximum infection rate at optimum temperature, and $F(T_n)$ the gamma function with change in temperature (T, obtain from above equation).

$$I = I_0 F\,(T_n) e^{-\tau * A}$$

For an example $I_0F(T_n)$ is considered almost zero at T<10°C and T>30°C for powdery mildew disease of grapevine and it had a maximum at T= 22°C (Sall, 1980 and Calonnec *et al.*, 2008). Further, again for an example the comparative optimum temperature for the various activities for *Pythium* and *Phytophthora* are as follows:

Events infastance	***Pythium***	***Phytophthora***
Most destructive	24–30°C	24–30°C
Oospores direct germinate through germ tube	28°C	21°C
Oospores indirect germinate through zoospore	17°C	12°C
Sporulation		9–26°C
Mycelial growth		16–18°C

3.6.3. Colony Growth

Colony diameter (D), over time *t,* can be quantified through following equation:

$$D(t+1) = D(t) + K_{max}\ x\ F(T_n)\ x\ \frac{r\ x\ e^{r\,(t^*-t)}}{[1+\ e^{r\,(t^*-t)}]^2}$$

With, K_{max} the maximum colony diameter, r the rate of growth, t^* the time to $K_{max}/2$, and $F(T)_n$ the rate of mycelium growth with temperature. Further, it should also be consider that the maximum diameter of t he colony also decrease as a function of leaf age *i.e.* $K_{max} = (A\text{-}B)\ x\ e^{(-C*age)} + B$, here A, B and c are the scaling parameters.

3.6.4. Latent Period

The latent period represents the time between initial infection and the appearance of new spores. The expected latent period (ρ{T}, at temperature T is

$$\rho(T) = \frac{\rho_{min}}{F(T_n)}$$

Where, ρ_{min} is the minimum time before the appearance of the first spores. The latent period is considered completed when

$$[1/\rho(T)] = 1$$

3.6.5. Sporulation

The quantity of spores released from a lesion depend on wind speed. The fraction, Q_s, of spores dispersed from the sporulating leaf at a given wind speed, u (m x s^{-1}, takes the following form with parameters α and b.

$$Q_s = \frac{e^{\alpha\,u+b}}{1+e^{\alpha\,u+b}}$$

The quantity of spores, Q, produced during a given period is a function of the colony size, s, thus: $Q = \beta å^{(5\ \text{-}s)}$ here, 5 and β are scaling parameters, and the quantity of spores release from a colony of a given size at time *t* is $Q_r = Q_s$ [Q(t) +R(t-1)] with R(t-1) the quantity of infectious spores not dispersed at time t-1.

3.6.6. Inoculum Dispersal

Spore dispersal generally described as the motion and short range dispersal of a large sample of particles released from each source (colony). At each time step, a quantity, Q_r of spores release is depend on wind intensity and size of colony. The spore dispersed within a cone of angle θ with respect to an axis given by the wind direction.

However, spore density can be decrease through two ways, (1) as an exponential dilution with axial distance to the source, *d* (with *cid* a parameters of decay with the distance) and (2) linearly according the radial position, r(θ), within cone. The quantity of spores deposit on the leaf Q_c depends upon its distance from source, (d, r(θ)), and can be proportional to surface area (Fig. 7).

3.6.7. Lesion Expansion

In many pathosystems, lesions in infected leaves continue to grow radially or linearly after their initial appearance, until much or the entire host unit is symptomatic.

$$LEXP = 0.012 \ x \ DLA \ x \ \frac{TLA - DLA}{TLA}$$

$$DLA = A_{LAT\ (Symptoms)} + A_{INF} + A_{POS}$$

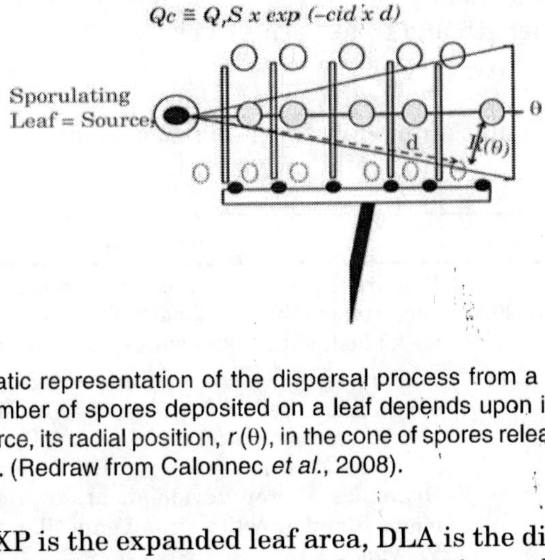

Fig. 7: Schematic representation of the dispersal process from a sporulating source. The number of spores deposited on a leaf depends upon its distance, *d*, from the source, its radial position, *r*(θ), in the cone of spores released, and its surface area, *S*. (Redraw from Calonnec *et al.*, 2008).

Where, LEXP is the expanded leaf area, DLA is the diseased leaf area showing symptoms, and TLA is the total leaf area in a cohort, A_{LAT} is the area in latent stage but shows symptoms, A_{INF} is the area in infectious stage and A_{POS} is the area in post infectious stage.

3.6.8. Proportion of Disease Severity (PDS)

$$PDS = \left[1 - \left(1 - \frac{TLA}{NDLA}\right)\right] \ X \ \frac{DLA}{TLA} + \left(1 - \frac{TLA}{NDLA}\right)$$

Here, NDLA is non-disease total leaf area

3.7. SIMULATION IN BOTANICAL EPIDEMIOLOGY

Simulation is basically referred to the execution of a model and this has practical importance in plant disease epidemiology, particularly related with

- Exploration of hypothesis
- Integrated the epidemiological knowledge with experimental results and
- It enable connecting epidemiology with other fields of science
- It helps to identify the factors that determine the behavior of pathosystem
- Thus, simulating modeling is a scientific approach as it enables us to explore the possible futures.

A very basic simulation requires

- Analytical and numerical integration
- Dimensions: Dimensions can be represented between brackets like L], [T] stands for length and time, respectively.
- Time constant

Example of few simulation models for plant disease epidemiology are as follows

Name	***Attributes***
STELLA®RICEPEST (Willocquet *et al.* (1998, 2000, 2002, 2004)	It address pest association and damage related with bacterial leaf blight, sheath blight, brown spot, tungro, neck blast, sheath rot, white head, weeds
WHEATPEST (Willocquet *et al.*, 2008)	It address pest association and damage related with powdery mildew, yellow rust, brown rust, *Septoria nodorum* blotch, *Septoria tritici* blotch, take all of wheat, eyespot of wheat, *Fusarium* stem rot, BYDV and weeds
GENECROP	Dynamics of crop development, accumulation of crop biomass and growth in number of tillers
GENEPEST	Varying Pest
EPIDEMRES	Model structure and equations, and run the model with varying values of components of resistance
EPIRICE (Savary *et al.*, 2012)	Brown spot, leaf blast, bacterial blight, sheath blight, and rice tungro disease
CLIMEX (http://www.csu.edu.au)	The CLIMEX meteorological database contains 2031 locations worldwide including 229 locations in Australia. This quantifies rice blast, leaf blast, collar rot, neck rot and panicle blast, node infection
PROCULTURE (Jarroudi *et al.*, 2009)	*Septoria* leaf blotch on Wheat
EPIWHEAT (Savary *et al.*, 2015)	Brown rust (leaf rust; *Puccinia triticina*) and Septoria tritici blotch (*Zymoseptoria tritici*)

Computer Program for Diseases Assessments

There are certain computerized disease assessment program designed by **Forrest W. Nutter, Jr.** (Associate Professor, Department of Plant Pathology, Iowa State University, Ames; **http://www.public.iastate.edu/~fwn/**) to help people working in the agricultural sciences improve their ability to estimate disease proportion.

Disease. Pro Program	Disease. Pro Program simulates four foliar diseases of peanut: peanut rust, early leaf spot, late leaf spot, and mixed leaf spots.
Barley. Pro Program	The program simulates four foliar diseases of barley: barley rust, spot blotch, and scald
Alfalfa. Pro Program	The program simulates five foliar diseases of Alfalfa: alfalfa rust, spring black stem and leaf spot, summer black stem and leaf spot, stemphylium leaf spot, and bacterial leaf spot.
Corn. Pro Program	The program simulates nine foliar diseases of corn: common rust, southern rust, northern leaf blight, northern leaf spot, southern leaf blight (race O), gray leaf spot, eyespot, and Stewart's disease.

In these programs user, will be asked to select the number of leaves that wish to evaluate (1 to 50), the size of lesions that will appear on a simulated peanut leaf (small, medium, large, or random mixture), and whether or not you want the actual disease proportion displayed after user have keyed in his estimate of the disease proportion (immediate feedback). Finally, the program allows to evaluate users rating performance by (1) producing a graph of the error of estimated disease proportions versus the actual proportions as calculated by the program, (2) producing a table and score report of these data, (3) producing a graph of the estimated severity versus the actual severity as well as calculating and displaying the linear regression line and equation, and (4) outputting the data to a file or inputting data from a file

3.8. HYPOTHESIS AND ITS COMPONENTS

In mathematical representation of epidemic development, overall modeling can be divided into three steps: (a) model development, (b) model analysis and (c) hypothesis testing. When developing a model, biological characteristics of the host-pathogen system are expressed in mathematical relationships.

In model analysis, epidemic dynamics are investigated in relation to the parameters of interest like variables which may be either related with pathogen, host or with environment. Finally in hypothesis testing, the results from model analysis are used to test or verify whether or under what combination the hypothesis is valid. Depending on the hypothesis to be tested the number of these categories used varies greatly. For example when modeling the effect of induced resistance an extra category, healthy resistant may also be required.

Esker *et al.* (2007) in his work on an application of space time analysis to improve the epidemiological understating of the papaya-papaya yellow crinkle patho-system have hypothesized that there would be a contagion process, where TBB- and SPLL-V4 (tomato big bud and sweet potato little leaf V4 SPLL-V4)-infected papaya would be aggregated and not random. To test this hypothesis, they have used a point pattern spatial analysis using Monte Carlo simulation. Their results support the hypothesis that papaya infection by both phytoplasma strains appears to be the result of a contagion process, providing support for the contention that insect vectors are the most likely mechanism for acquisition, dispersal, and transmission.

Batista *et al.* (2008) have hypothesized that epidemiological patterns of Citrus Sudden Death in Brazil and citrus tristeza in Cuba are indistinguishable.

In botanical epidemiology, certain hypothesis concerns the area of the system and its boundaries. The system under consideration may consist of a 1 m^2 crop area surrounded by similar systems. One would obviously have to increase this size when considering most perennials or semi-perennial crops. The assumption also implies that these boundaries allow fluxes of propagules to enter and exit the system in a steady-state. What is obscure by 'similar crop areas' surrounding selected system is that the amount of disease is the same in surrounding areas.

For certain experimental area, microclimatic conditions may be more or less similar, thus, such system under consideration is kept as simple as possible; the limited size of the system, the variability of the host population size over time, the consequences the host population size may have on microclimate or disease spread, for instance, are disregarded for the sake of simplicity.

The hypothesis of a limited system surrounded by similar systems with which it is in a dynamic equilibrium (*e.g.*, flows of propagules, heat, water vapor) is often referred as the 'mean field' hypothesis.

Whichever approach is chosen, whether analytical or numerical, the model considered here is based on the major assumption that all healthy sites are equally accessible for infection, or that, conversely, propagules all have equal probabilities to reach and (possibly) infect healthy sites. This is what may be called a mean field hypothesis.

Canopies are, in the real world, heterogeneous; leaves or fruits, or plant organs-sites, in general are not all equally exposed to incoming inoculum; tissues vary in their susceptibility; gradients of propagule dispersal vary widely across pathosystems, and these gradients often depend on several, not one, dispersal mechanisms; or again, the microclimate in a canopy (say, in an apple tree, but in a wheat or rice field, too) is bound to show spatial variability. All these elements, which may explain why an epidemic occurs, or why it does not, are left aside at this stage.

Area hypothesis is logistic. In many cases, this assumption is not appropriate. This is especially the case if the sites under consideration were root-sites: root growth, in many crops, is very rapid in the beginning of the growing season, and stops long before the reproductive stage (in annual crops) is over (*e.g.*, Gardner *et al.*, 1985).

The aerial tissues of a crop canopy also have an asymmetric rate of growth, which rapidly increases in the beginning of the growing season, and progressively decreases later. As a result, a flush of healthy sites would become rapidly available to possible infection, with strong differences in the shape of disease progress. There are a number of equations to account for such asymmetrical growth, for the host or for the pathogen (*e.g.*, Kranz, J., 1976; Berger, 1981; Madden *et al.*, 2007).

Alternatively we can also hypothesize that researcher assumed that only healthy sites (HSites) contribute to crop growth. This hypothesis can in turn be partitioned into two questions:

- Are all diseased sites truly not contributing to crop growth, and
- Are all healthy sites equally contributing to crop growth, creating a differential in disease occurring in some healthy sites rather than others.

The first question may, to some extent, be related to the biology of the considered pathogen. If one is dealing with a necrotroph (Cooke and Whipps, 1980), then infected sites become very rapidly dysfunctional (*e.g.*, Savary *et al.*, 1990; Bastiaans, 1993; Lopes and Berger, 2001), and the underlying hypothesis seems to hold.

If one is dealing with a biotroph, then infected sites still continue to be functional, but their photosynthetic activity is only directed to the maintenance, growth, and reproduction of the pathogen (*e.g.*, Savary *et al.*, 1990; Lopes and Berger, 2001). In the case of some host-pathogen interactions, the biotrophic pathogen in fact enhances the photosynthetic activity of infected tissues. Thus, one may consider a possible underestimate of the effect of infection on the population of sites, at least among the infected ones, and possibly, among the healthy sites. The latter remark brings us back again to the underlying hypothesis of evenly distributed disease in plant tissues, in this case, in terms of crop growth – disease progress relationships.

3.8.1. Null Hypothesis and Type I and Type II Errors

A hypothesis deals with a possible solution or answer derived from a thought process using evidence provided in the established body of knowledge in the area of research. There are four criteria to be satisfied by a good hypothesis (a)States an expected relationship among variables, (b) Provides justification for the hypothesis from literature, (c) Is testable under the constraints of the study and (d) Is clear and brief.

3.8.1.1 *Element of hypothesis*

- *Null hypothesis*: Proposes that there is no relationship among the identified variables - if the study is an experiment, then a statistical test of significance will measure support for the null hypothesis. A null hypothesis is non directional since there is no relationship (Ho)
- *Alternative hypothesis*: Remaining possible outcomes other than the null hypothesis. Alternative hypothesis usually have a direction stated in the relationship of variables (*H*).
- *Test statistic*: Quantity based on sample data and null hypothesis used to test between null and alternative hypothesis
- *Rejection region*: Values of the test statistic for which we reject the null in favor of the alternative hypothesis

<table>
<tr><th>Type I Error/α Error</th><th>Type II Error/β Error</th></tr>
<tr><td>The probability of making a error as level of significance i.e., consider as 0.05 (5%). For computing the sample size its specification in term of Zα is required. The quantity of Zα is a value from the standard normal distribution corresponding to α. Sample size is inversely proportional to type I error.
Alpha is traditionally set to 5%, but if the risks involved in the 'exposed' group is high, then it can be set to 10%. Conversely, if the subject is not critical it could be set to 1%.
• Alpha = 10% means that with a little evidence the difference will be claimed significant.
• Alpha = 1% means that you need a very strong evidence to claim significance</td><td>For computing the sample size its specification in terms of Zα is required. The quantity of Zα is a value from the standard normal distribution corresponding to α. A type II error is frequently due to small sample sizes and the exact probability of type II error is generally unknown.
• Beta is defined as 1 - power. The power of a test is a function of the difference between the groups.
• If the difference is Zero, the Power = Alpha.
• If the difference is very large, the Power = 1.
• That is why we must specify Beta (1-power) to be small when the difference in prevalence is small.
• If a study finds no difference, the power plays a major role in defining the maximum size that the true difference in the population may have.
• A power of 50% is a coin-flip chance to identify the true association as statistically significant.</td></tr>
</table>

3.8.1.2. *Hypothesis testing*

For experiments, once we know what kind of data we have, we should consider the desired confidence level of the statistical test. This confidence is expressed as α; it gives one the probability of making a Type I error (Table 2) which occurs when one rejects a true null hypothesis.

Typically that level for α is set at 0.05, meaning that we are 95% confident ($1-\alpha= 0.95$) that we will not make a Type I error, *i.e.* 95% confident that we will not reject a true null hypothesis.

For many commonly used statistical tests, the p-value is the probability that the test statistic calculated from the observed data occurred by chance, given that the null hypothesis is true.

- If $p < \alpha$ we reject the null hypothesis;
- If $p \geq \alpha$ we do not reject the null hypothesis.

A Type II error, expressed as the probability β occurs when one fails to reject a false null hypothesis.

Unlike α, the value of β is determined by properties of the experimental design and data, as well as how different results need to be from those stipulated under the null hypothesis to make one believe the alternative hypothesis is true.

Table 2: Elements of Null Hypothesis

	The Null Hypothesis (H_0) is	
Statistical result	***True***	***False***
Reject the null hypothesis	*Type I error*: A value = probability of falsely rejecting H_0	Probability of correctly rejection H ($1-\beta$) = power
Accept the null hypothesis	Probability of correctly rejection H ($1-\beta$)	*Type II error*: B value = probability of falsely rejecting H_0

3.8.2. Practical Exercise for Different Levels

For example a researcher wants to prove that that new pathogen dispersal is random one and not following regular pattern:

A. ***Null hypothesis*** – dispersal is random then regular

$$H_0 : \mu_{New} - \mu_{Std} \leq 0 \qquad (\mu_{New} - \mu_{Std} = 0)$$

B. ***Alternative hypothesis*** – dispersal is regular

$$H_A : \mu_{New} - \mu_{Std} > 0$$

C. ***Experimental sample data***

$$\bar{y}_{New} \quad \bar{y}_{Std}$$

$$s_{New} \quad s_{Std}$$

$$n_{New} \quad n_{Std}$$

D. In large samples, the difference in two sample means is approximately normally distributed.

$$\bar{Y}_1 - \bar{Y}_2 \sim N\left(\mu_1 - \mu_2, \sqrt{\frac{\sigma_1^2}{n_1} + \frac{\sigma_2^2}{n_2}}\right)$$

E. Under the null hypothesis, $\mu_1 - \mu_2 = 0$ and

F. σ_1^2 and σ_1^2 are unknown and estimated by s_1^2 and s_2^2

Type I error: Concluding that the new fungicide is better than the standard (H_A) when in fact it is no better (H_0). Ineffective drug is deemed better. Traditionally a = P(Type I error) = 0.05

Type II error: Failing to conclude that the new fungicide is better (H_A) when in fact it is. Effective drug is deemed to be no better. Traditionally a clinically important difference (D) is assigned and sample sizes chosen so that: $b = P(\text{Type II error} \mid \mu_1 - \mu_2 = \Delta) \leq .20$

G. ***Test statistic*:** Difference between the Sample means, scaled to number of standard deviations (standard errors) from the null difference of 0 for the Population means:

$$T.S.: z_{obs} = \frac{\bar{y}_1 - \bar{y}_2}{\sqrt{\frac{s_1^2}{n_1} + \frac{s_2^2}{n_2}}}$$

H. ***Rejection region*:** Set of values of the test statistic that are consistent with H_A, such that the probability it falls in this region when H_0 is true is a (we will always set a=0.05).

I. **P-value:** Measure of the strength of evidence the sample data provides against the null hypothesis: P(Evidence This strong or stronger against $H_0 \mid H_0$ is true).

$$P-val: p = P(Z \geq z_{obs})$$

J. ***Large-sample test H_0:*** $\mu_1-\mu_2=0$ *vs.* $H_0:\mu_1-\mu_2>0$ H_0: $\mu_1-\mu_2=0$ (No difference in population means H_A: $\mu_1-\mu_2 > 0$ (Population Mean 1 > Pop Mean 2)

$$\bullet\ T.S.: z_{obs} = \frac{\bar{y}_1 - \bar{y}_2}{\sqrt{\frac{s_1^2}{n_1} + \frac{s_2^2}{n_2}}}$$

$$\bullet\ R.R.: z_{obs} \geq z_\alpha$$

$$\bullet\ P-value: P(Z \geq z_{obs})$$

Conclusion: Reject H_0 if test statistic falls in rejection region, or equivalently the *P*-value is £ *a*.

For example
$$\bar{y}_1 = 10.1 \quad s_1 = 3.6 \quad n_1 = 33$$
$$\bar{y}_2 = 7.7 \quad s_2 = 3.4 \quad n_2 = 35$$

$$\bullet\ H_0: \mu_1 - \mu_2 = 0$$

$$\bullet\ H_A: \mu_1 - \mu_2 > 0$$

$$\bullet\ T.S.: z_{obs} = \frac{10.1 - 7.7}{\sqrt{\frac{(3.6)^2}{33} + \frac{(3.4)^2}{35}}} = \frac{2.4}{0.85} = 2.82$$

$$\bullet\ R.R.: z_{obs} \geq z_\alpha = z_{.05} = 1.645$$

$$\bullet\ P-val: P(Z \geq 2.82) = .0024$$

***Conclusion*:** Pathogen dispersal is random and not regular one (since 2.82 > 1.645 and *P*-value < 0.05).

3.9. DATA CATEGORIZATION FOR SAMPLE SIZE DETERMINATION

For determination of sample size we can categorize our data in to continuous and into categorical data. Continuous (or quantitative) is data where the values can change continuously, and you cannot count the number of different values. Examples include weight, price, profits, counts, etc.

Basically, anything you can measure or count is quantitative. Categorical data, in contrast, is for those aspects of your data where you make a distinction between different groups, and where you typically can list a small number of categories. For both types of data we require three basic types of information's. These are as follows.

Alpha level	*Acceptable margin of error*	*Variance estimation*
• The alpha level used in determining sample size in most research inventories is either 0.05 or 0.01 (Bartlett *et al.*, 2001). In sample size formula the alpha level is incorporated into the formula by utilizing the t value for the alpha level selected (*e.g.*, t value for alpha level of 0.05 is 1.96 for sample size above 120) Researchers should ensure they use the correct t value when their research involves smaller population, *e.g.*, t value for alpha of 0.05 and a population of 60 is 2.00. • In general, an alpha level of 0.05 is acceptable in most research (Mathur, 2013). An alpha level of 0.10 or lower may be used if the researcher is more interested in identifying marginal relationships, differences or other statistical phenomena as a precursor to further studies. An alpha level of 0.01 may be used in those cases where decision based on the research are critical and error may cause substantial financial or personal harm *e.g.*, major programmatic changes	The general rule relative to acceptable margin of error in survey works is as follows: • For categorical data, 5% margin of error is acceptable. • For Continuous data, 3% margin of error is acceptable. • For example, a 3% margin of error would result in the researcher being confident that the true mean of a seven point scale is within ±0.21 (0.03 times seven points on the scale)	This is the most critical component of sample size estimation. The researcher does not have direct control over variance and must incorporate variable estimation into research design. There are four ways to estimating population for sample size determination: • Take sample in two steps, and use the results of the first step to determine how how many additional responses are needed to attain an appropriate sample size based on the variance observed in the first step data. • Use pilot study results. • Use data from previous studies of the same or a similar population and • Estimate or guess the structure of the population assisted by some logical mathematical results. A researcher typically needs to estimate the variance of scaled and categorical variable. To estimate the variance of scaled variable, one most determine the inclusive range of the scale, and then divide by the number of standard deviation that would include all possible values in the range and then square this number. For example, if a researchers used a seven-point scale and given that six standard deviation (three to each side of the mean) would capture 98% of all responses, and the calculation would be as follows:

Power is the probability that the null hypothesis is rejected, if a specific alternative hypothesis is true. ß represents Type II error, the probability of not rejecting the null hypothesis when the given alternative is true.

$$1 - \beta = \text{power}.$$

The power of a study should be minimally 80% and often, studies are designed to have 90-95% power to detect a particular treatment or impact. Following factors affects power

- α is the level of significance, the probability of a Type I error. This is usually 5% or 1%, meaning the investigator is willing to accept this level of risk of declaring the null hypothesis false when it is actually true.
- The **effect size** is the deviation from the null that the investigator wishes to be able to detect. The effect size should be clinically meaningful. It may be based on the results of prior or pilot studies. For example, a study might be powered to be able to detect a relative risk of 2 or greater.
- Sometimes a standardized effect size is given, *i.e.*, the effect size divided by the standard deviation. This is a unitless value. If power is calculated in this manner, the standardized effect size is usually between 0.1 and 0.5, with 0.5 meaning H_1 is 0.5 standard deviations away from H_o.
- Variability may be expressed in terms of a standard deviation, or an appropriate measure of variability for the statistic. If the hypothesis are concerned with a population proportion, the value of the proportion and the sample size are used to calculate the variability. The investigator will need an estimate of the variability in order to calculate power. Reasonable estimates may be obtained from historical data, pilot study data or a literature search.
- A study may have multiple sources of variation, each accounted for in the analysis. For example, a repeated measures design will need to account for both within-subject and between-subject variability.

The **baseline** incidence rate is related to the effect size. If it is hypothesized that a rate has increased or decreased, the baseline rate and the effect size must both be known to calculate the power for detecting such a change. With knowledge of the above factors, power of a statistical test can be calculated for a given sample size. Alternatively, the required sample size for a given power can be calculated.

Power is directly related to effect size, sample size and significance level. An *increase in the effect size, the sample size or the significance level will produce increased statistical power*, all other factors being equal. Power is inversely related to variability. *Decreasing variability* will increase the power of a study. If the power of a study is relatively high and a statistically significant effect is not observed, this implies the effect, if any, is small. Epidemiologic studies can be population-based or non-population-based, such as case-control studies.

Population-based studies (cohort or cross-sectional studies	*Case-control studies (e.g., calculating an odds ratios)*
Differences in proportions (*e.g.*, attributable risk)	Unmatched study designs
Ratios (*e.g.*, relative risks, relative rates, prevalence ratios)	Multiple controls/case Matched study designs

3.9.1. Example of Sample Size Determination for Continuous and Categorical Data.

Sample Size for Continuous Data

$$\underline{n} = \frac{(\underline{t})^2 x (\underline{s})^2}{(\underline{d})^2}$$

- Assume that a researcher want to know the appropriate sampling population out of 1,679 total population and has set the alpha level a priori at 0.05, plans to use a seven point scale, has set the level of acceptable error at 3% and has estimated the standard deviation of the scale as 1.16.

$$\underline{n} = \frac{(1.96)^2 x (1.16)^2}{(7 x 0.03)^2} = 118$$

- Where 1.96 is the selected alpha level at 0.025 (the alpha level of 0.05 indicates the level of risk the researchers is willing to take that true margin of error may exceed the acceptable margin of error). 1.16 is the standard deviation at described above. 0.21 is the acceptable margin of error (number of points on primary scale x acceptable margin of error). Therefore, for a population of 1,679, the required sample size is 118. However, since this sample size exceeds 5 % of the population (1,679 x 0.05 = 84). Therefore following correction formula can be applied

$$\underline{n} = \frac{n}{(1 + \frac{n}{Population})}$$

Sample Size for Categorical Data

$$\underline{n} = \frac{(\underline{t})^2 x (p)(q)}{(\underline{d})^2}$$

- Assume a researcher has set the alpha level a priori at 0.05 plans to use a proportional variable, has set the level of acceptable error at 5%, and has estimated the standard deviation of the scale as 0.5.

$$\underline{n} = \frac{(1.96)^2 (0.5) x (0.5)}{(0.05)^2} = 384$$

- However, since this sample size exceeds 5 % of the population (1,679 x 0.05 = 84). Therefore following correction formula can be applied

$$\underline{n} = \frac{n}{(1 + \frac{n}{Population})}$$

$$\underline{n} = \frac{(384)}{1 + 384/_{1679}} = 313$$

- In this example it was anticipated that a response rate of 65 % would be achieved based on prior researcher experience, therefore actual sample size would be 313/0.65 = 482.

Sample Size for Continuous Data	***Sample Size for Categorical Data***
$n = \frac{(118)}{1+118/1679} = 111$ • In this example it was anticipated that a response rate of 65 % would be achieved based on prior researcher experience, therefore actual sample size would be 111/0.65 = 171	

3.9.2. Sample Size Determination Table

Population size	***Sample types***					
	Continuous data (Margin of error = 0.03)			***Categorical data (Margin of error = 0.05)***		
	Alpha=0.10 t=1.65	***Alpha=0.5 t=1.96***	***Alpha=0.01 t=2.58***	***P=0.50 t=1.65***	***P=0.50 t=1.96***	***P=0.50 t=2.58***
100	46	55	68	74	80	87
200	59	75	102	116	132	154
300	65	85	123	143	169	207
400	69	92	137	162	196	250
500	72	96	147	176	218	286
600	73	100	155	187	235	316
700	75	102	161	196	249	341
800	76	104	166	203	260	363
900	76	105	170	209	270	382
1000	77	106	173	213	278	399
1500	79	110	183	230	306	461
2000	83	112	189	239	323	499
4000	83	119	198	254	351	570
6000	83	119	209	259	362	598
8000	83	119	209	262	367	613
10000	83	119	209	264	370	623

The margins of error used in the table were 0.03 for continuous data and 0.05 for categorical data. Researchers may use this table if the margin of error shown is appropriate for their study; however, the appropriate sample size must be calculated if these error rates are not appropriate (Mathur, 2013).

3.9.3. Sample Size in Plant Pathology Context

Sample size should optimize the balance between the number of samples and reliability of the data. Although faster and easier to perform, small sample sizes can provide unreliable data, whereas a very high number of samples may contribute little to reliability when compared with the confidence level and precision of an intermediate number of samples (Campbell and Madden, 1990). There are at least three methods to estimate sample size, which depend on the operational definition of the reliability and on the costs of sample collection:

- Reliability is defined by the coefficient of variation of the mean or standard error;
- Equations of probability define reliability;
- Components of variance and cost functions are used to optimize sample number, considering that each sample type has an associated cost (Campbell and Madden, 1990).

Considering that the reliability of disease estimate is directly related to sample size and spatial heterogeneity of disease, all above methods can be associated to distributions, which represent different models of spatial distribution of the disease in the field (Perry, 1994).

Michereff *et al.*, (2011) have quantified appropriate sample size for quantification of *Cercospora* leaf spot on sweet pepper based on a random pattern of diseased plants. For their analysis the workers have selected the fields with 3.5 km as a minimum distance between fields. The trench system was employed in all fields, with flood irrigation and a spacing of 1.0 × 0.80 m between rows and plants, respectively. On each field, a 0.5 ha area was chosen, consisting of 50 rows and 125 plants/row, for a total of 6,250 plants. Two diagonals (X) were established in each area and 25 plants were evaluated along each diagonal, with the first plant located approximately 5 m from the beginning of the diagonal and the remaining plants spaced at distances of approximately 5 m. With the aid of the empiric scale, which includes levels from 0 to 50% of damaged leaf area (Michereff *et al.*, 2006), the leaf damage by *Cercospora* leaf spot was estimated using 50 selected plants, with further selection of 3, 6, 9 and 12 leaves/plant. In the three-leaf sample, one leaf from each third of the plant (lower, middle and upper) was evaluated. In the 6-, 9- and 12-leaf samples, two, three and four leaves were evaluated in each third of the plant, respectively. Ideal sample size (n) were quantified for each crop areas based on the coefficient of variation of the mean (CV_X) and random pattern of diseased plants, considering sample of 3,6,9 and 12 leaves/plant and for this they have use following equation:

$N{=}S^2 / (X^2\text{-}CV^2{}_X)$, with a pre-established acceptable errors of 5, 10, and 20% ($CV_{X=}$ 0.05, 0.1 and 0.2). Mean ideal sample size was calculated when 3,6,9 and 12 leaves/plant were assessed. Further, they have quantified mean sample size within each category of acceptable error and were compared with the Krusal-Wallis test (P=0.05).

They have concluded a sample of 3 leaves/plant with 10 and 20% rate of acceptable error, the mean ideal sample size was 103 and 26 plants, respectively, for each 0.5 ha of cultivated field.

Gent *et al.* (2006) have developed sampling curves to predict the number of sampling units that necessary to estimate the incidence of disease cone with varying level of precision (Maddan *et al.*, 1995 and Madden and Hughes, 1999). According to them the number of sampling units (N) needed to estimate the p with a coefficient of determination, C, of 0.2, 0.3 and 0.4 can be calculated by as follows

$$N = ap^{b-2}\frac{(1-p)^b}{C^2}$$

Where C = standard error $C = [\hat{p}]/\hat{p}$; $a = A_p n^{-b}$ and A_p and b are parameters from the binary power law which can be quantified as follow

$$1n\,(V_{obs}) = 1n(A_p) + b\;1n\,[\hat{p}(1-\hat{p})]/\bar{n}$$

Where 1n (A_p) and b are the intercept and slope parameters of a straight line, respectively. When $Ap = 1$ and $b = 1$, indicates a random pattern of diseased cones that can be represented by the binomial distribution. When $Ap > 1$ and $b = 1$, disease incidence has an aggregated pattern, but aggregation is not dependent on p. Values of $b > 1$ indicate that aggregation is systematically related to p. When $b > 1$, the beta-binomial parameter θ can be expressed in terms of A_p and b of the binary power law according to the formula:

$$\theta = [a\text{-}f\,(p)/n]/[f(p)\text{-}a]$$

Where $f(p) = 9p(1\text{-}p)]^{1\text{-}b}$ and $a = A_p n^{-b}$.

3.10. PARAMETRIC AND NON-PARAMETRIC TESTS

In the literal meaning of the terms, a **parametric statistical test** is one that makes assumptions about the parameters (defining properties) of the population distribution(s) from which one's data are drawn, while a **non-parametric test** is one that makes no such assumptions.

If the information about the population is completely known by means of its parameters then statistical test is called parametric test and examples includes t-test, f-test, z-test and ANOVA.

However, if there is no knowledge about the population or parameters, but still it is required to test the hypothesis of the population, then it is called non-parametric test and example includes Whitney, rank sum test and Krusal-Wallis test.

A comparative analysis between these two is presented in Table 3 and the flow diagram in Fig. 8 gives a simple decision tree to choose the right test.

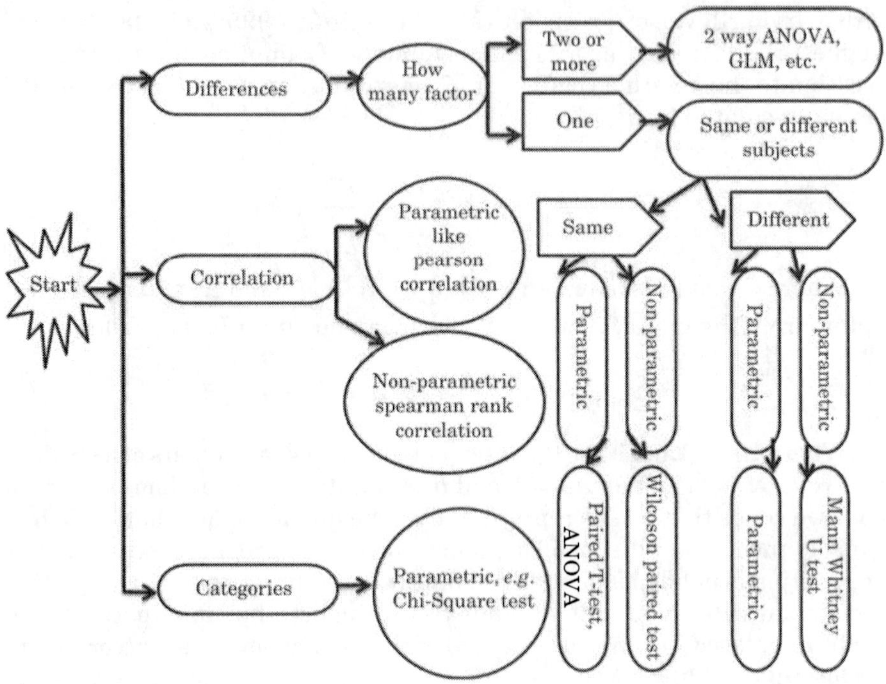

Fig. 8: Comparative illustration for strategy for selection of parametric and non-parametric approaches

Table 3: Attributes of parametric and non-parametric tests

Attributes	***Parametric test***	***Non-parametric test***
Test statistic	Based on the distribution	Is arbitrary
Applicable for	Parametric tests are applicable only by variable	It is applied both variables and attributes
Data type	No parametric test exist for nominal scale data	Non parametric test exist for nominal and ordinal scale data
Strength	Parametric test is powerful, if its exist	It's not so powerful like parametric test
Distribution	Normal	Any
Variance	Homogenous	Any
Example tests		
Correlation	Pearson	Spearman
Independent data	t-test for independent samples	Mann-Whitney U test
Independent data more than 2 groups	One way ANOVA	Kruskal Wallis ANOVA
Two repeated measures, 2 groups	Matched pair t-test	Wilcoxon paired test
Two repeated measures, >2 groups	Repeated ANOVA	Friedman ANOVA

Different types of parametric tests like t- test, F-test, Z-test and ANOVA can be find in any statistical book. Thus, in latter part of this chapter, different types of non-parametric tests along with their working examples will only be discussed.

3.10.1. Non-Parametric Test

Nonparametric statistics (also called "distribution free statistics") are those that can describe some attribute of a population, test hypothesis about that attribute, its relationship with some other attribute, or differences on that attribute across populations, across time or across related constructs, that require no assumptions about the form of the population data distribution (s) nor require interval level measurement.

Advances of non-parametric test

- They are simple and easy to understand
- It will not involve complicated sampling theory
- No assumption is made regarding the parent population
- This method is only available for nominal scale data
- These methods are easy applicable for attribute dates
- A non-parametric test does not require that the data be samples from any particular distribution (*i.e.* they are distribution-free). In other words Non-parametric tests are valid for both non-Normally distributed data and normally distributed data.

Disadvantages of non-parametric test

- It can be applied only for nominal or ordinal scale
- For any problem, if any parametric test exist it is highly powerful
- Lesser efficient compare to parametric test

3.10.1.1. *Conditions for using nonparametric test and a caution to consider*

Data are not normally distributed	r, Z, t, F and related statistics are rather "robust" to many violations of these assumptions
Data are not measured on an interval scale	Most psychological data are measured "somewhere between" ordinal and interval levels of measurement. The good news is that the "regular stats" are pretty robust to this influence, since the rank order information is the most influential (especially for correlation-type analyses).
Sample size is too small for "regular stats"	Do we really want to make important decisions based on a sample that is so small that we change the statistical

(Contd...)

(Contd...)

	models we use? Remember the importance of sample size to stability.
Robustness of parametric statistics to most violated assumptions	Difficult to know if the violations or a particular data set are "enough" to produce bias in the parametric statistics. One approach is to show convergence between parametric and nonparametric analyses of the data.
Poorer power/sensitivity of nonpar statistics (make Type II errors)	Parametric stats are only more powerful when the assumptions upon which they are based are well-met. If assumptions are violated then nonparametric statistics are more powerful.
Mostly limited to uni- and bivariate analyses	Most research questions are bivariate. If the bivariate results of parametric and nonparametric analyses converge, then there may be increased confidence in the parametric multivariate results.
Not an integrated family of models, like GLM	There are only 2 families — tests based on summed ranks and tests using ×2 (including tests of medians), most of which converge to Z-tests in their "large sample" versions.
H_0:s not parallel with those of of parametric tests	This argument applies best to comparisons of "groups" using quantitative DVs. For these types of data, although the null is that the distributions are equivalent (rather than that the centers are similarly positioned to H_0: for t-test and ANOVA), if the spread and symmetry of the distributions are similar (as is often the case and the assumption of t-test and ANOVA), then the centers (medians instead of means) are what is being compared by the significance tests. In other words, the H0:s are similar when the two sets of analyses make the same assumptions.

In plant pathology, these tests could be used to measure the degree to which two or more pathogens, or two or more diseases, are associated (Trumper and Gorla, 1997; Turechek and Madden, 2000; Petybridge and Turechek, 2003, Turechek, 2004; Carisse *et al.*, 2008; Henne *et al.*, 2010; Peterson *et al.*, 2014 and Martini *et al.*, 2015).

Associations can be characterized based on two underlying properties: covariation and occurrence. Covariation is a measure of how one disease's intensity (incidence or severity) increases or decreases in response to a change in intensity of another disease. Occurrence measures the degree to which two (or more) diseases occupy the same habitat (*e.g.*, leaf, plant, field, region, etc.). Turechek and Stevenson (1998) used Kendall's sample rank correlation coefficient to measure the degree of association between components of partial resistance to pecan scab caused by *Cladosporium caryigenum*.

The two most commonly used nonparametric or so-called "rank correlation coefficients" are Kendall's tau (τ) and Spearman's rho (ρ). The most common question surrounding the use of these two correlation coefficients is "which of these should researcher use?" Spearman's ρ and Kendall's τ are, in general, measuring the same property but imply different interpretations: ρ can be thought of as the regular Pearson's product moment correlation coefficient applied to ranks. That is, ρ represents the proportion of variability that can be attributed to association between diseases. Kendall's τ represents a probability; it is the difference between the probabilities that two variables are in the same order versus that they are in different order. In general, the two statistics will seldom lead to different conclusions when applied to the same set of data.

Spearman's ρ is more sensitive to outliers than Kendall's τ, but less so than Pearson's coefficient. Kendall's τ can also be used to derive partial correlation coefficients, whereas Spearman's ρ cannot (Sprebt and Smeeton, 2001). In the following part of this chapter non-parametric test like Kendall's tau, the Mann-Whitney U test, the Wilcoxon signed-ranks test, the sign test, association and their related indices and The jackknife will be discussed in detail.

Before applying any method, researchers must know the variable types; in general variables may be nominal, ranked, circular and ratios (McDonald, 2008). Comparative knowledge of these variables types are as follows

Variables types	***Description***
Nominal Variables	These variables, also called "attribute variables" or "categorical variables," classify observations into a small number of categories. Examples of nominal variables include sex (the possible values are male or female), genotype (values are *AA, Aa,* or *aa*). Nominal variables are often used to divide individuals up into classes, so that other variables may be compared among the classes. Nominal variables are often summarized as proportions or percentages.
Ranked Variables	Ranked variables, also called ordinal variables, are those for which the individual observations can be put in order from smallest to largest, even though the exact values are unknown.
Circular Variables	These have the property that the highest value and the lowest value are right next to each other; often, the zero point is completely arbitrary. The most common circular variables in biology are time of day, time of year.
Ratios Variables	Biological variables are ratios of two measurement variables (denominator and numerator)

3.10.1.2. *Kruskal–Wallis test*

The Kruskal–Wallis test is most commonly used when there is one nominal variable and one measurement variable, and the measurement variable does not meet the normality assumption of an ANOVA.

Kruskal-Wallis test can be considered as a backup method for ANOVA where the independent variable is categorical but the dependent variables are not normally distributed. On the other hand, Kruskal-Wallis test can also be considered an alternative method for Mann-Whitney test where it is a nonparametric test but the independent variable could have more than two categories. In summary, Kruskal-Wallis test is preferred when:

- The independent variable has three or more categories and randomly selected;
- For each sample (that determined by each categories of the independent variable), there is a least 5 observations. If the sample size is fewer than 5
- There is no requirement that the populations have a normal distribution or any other particular distribution.

It is the non-parametric analogue of a one-way ANOVA. A one-way ANOVA may yield inaccurate estimates of the P-value when the data are very far from normally distributed.

The Kruskal–Wallis test does not make assumptions about normality.

Like most non-parametric tests, it is performed on ranked data, so the measurement observations are converted to their ranks in the overall data set: the smallest value gets a rank of 1, the next smallest gets a rank of 2, and so on.

If the original data set actually consists of one nominal variable and one ranked variable, you cannot do an ANOVA and must use the Kruskal–Wallis test

In plant disease epidemiology this method was testified by Zivanov *et al.* (2014) and Martini *et al.* (2015). Metz *et al.* (2012) use this method to test differences between the two field screening and the variance for disease component (*i.e.*, lesion size and overall plant disease severity) for *Stenotaphrum secundatum* grass infected with *Magnaporthe oryzae*

With this test the two forms of hypothesis may be

- ***Null hypothesis***: Assumes that the samples are from identical populations.

- ***Alternative hypothesis*:** Assumes that the samples come from different population

Procedure of this test includes following steps

1. Arrange the data of both samples in a single series in ascending order.
2. Assign rank to them in ascending order and in the case of a repeated value, or a tie, assign the ranks to them by averaging their rank position.
3. Then sum up the different ranks for each of the different groups and
4. To calculate the value by apply the following formula

$$H = \left[\frac{12}{N\ (N+1)}\ x\ \Sigma\frac{Tc^2}{n_c}\right] - 3\ x(N+1)$$

Where, H = Kruskal-Wallis Test Statistic; N = total number of observation in all samples and T is the sum of the rank assigned and n_c is the number of participant in each group

3.10.1.2.1. *Working example of Kruskal-Wallis test*

In an artificial experiment a researcher want to quantify does the leaf wetness period increase the disease incidence and severity of apple scab caused by Venturia inaeqalis. *The appropriate test here is Kruskal-Wallis test.*

Researcher has three separate variables, each of whom gives single score on a rating scale. Ratings are examples of an ordinal scale of measurement, and so the data are not suitable for a parametric test.

	Leaf wetness period	***Disease incidence***	***Disease severity***
	23	22	59
	36	27	66
	51	39	38
	49	29	49
	58	46	56
	37	48	60
	29	49	56
	44	65	62
Mean Rating	39.63	40.63	55.75
Standard Deviation	**12.85**	**14.23**	**55.75**

Rank all the scores, ignoring which group they belong to. Ranking can be calculated as: the lowest scores get the lowest rank and if two or more scores are the same they are tied

	Leaf wetness period	*Disease incidence*	*Disease severity*
	23 (2)	22 (1)	59 (20)
	36 (3)	27 (4)	66 (24)
	51 (16)	39 (9)	38 (8)
	49 (14)	29 (5.5)	49 (14)
	58 (19)	46 (11)	56 (17.5)
	37 (7)	48 (12)	60 (21)
	29 (5.5)	49 (14)	56 (17.5)
	44 (10)	65 (23)	62 (22)
Mean Rank	(9.56)	(9.94)	(18.00)
Standard Deviation	(6.25)	(6.84)	(5.09)
Sum of Rank (T_C)	**76.5**	**79.5**	**144**

22 is the lowest score, therefore it get rank 1

23 is the next lowest score this get rank 2

26 is the next lowest score this get rank 3

27 is the next lowest score this get rank 4

There are two instances of 29 and this tie, so the next two ranks are 5 to 6 the average of 5+6/2 is 5.5. Thus, both 29 gets a rank of 5.5.

Similar case for 38 and 39 and accordingly they get 8 and 9 respectively, while 44, 46 and 48 gets 10,11 and 12 accordingly. However there are three case of 49 so this is tie again, and in the above cited example 49 recorded at three instance (13+14+15/3=14). The similar procedure follows for entire cases.

Calculate the sum of rank (T_{-C}) for each; Leaf Wetness Period T_C = 76.5; Disease incidence T_C = 79.5 and Disease severity T_C =144.

Now find the H with the equation mentioned above and in that equation N is the total number of participants (all groups combined and in this example N is 24), nc is the number of observation in each groups *viz.*, nc 1=8; nc2 =8 and nc3 =8.

$$H = \left[\frac{12}{24 \; x \; (24+1)} \; x \; \sum \frac{Tc^2}{n_c}\right] - \; 3 \; x \; (24+1)$$

$$\sum \frac{Tc^2}{n_c} = \frac{76.5^2}{8} + \frac{79.5^2}{8} + \frac{144^2}{8} = 731.53 + 790.03 + 2592 = 41113.56$$

$$H = \left[\frac{12}{600} x\ 4113.56\right] - 75; H = [002\ x\ 4113.56] - 75; H = \mathbf{7.27}$$

- Degree of freedom is the number of groups minus one and in this example we have 3 groups, thus degree of freedom will be 2.
- Now assess the significance of *H* depends on the number of participants and the number of groups and if we have three groups, with five or fewer participant in each group, then we need to use the small sample size table. But if we have more than five participants per group, then treat *H* as Chi-Square. *H* is statistically significant if it is equal to or larger than the critical value of Chi-Square with particular degree of freedom.
- In this example we have eight variables per parameter and so treat *H* as Chi-Square. With 2 degree of freedom the Chi-Square value is 5.99 at p=0.05 level. Obtained value of 7.27 is even larger than this. Thus, this test revealed that leaf wetness period significantly affects the disease severity and disease incidence. (H =(2) =7.27, $p<0.05$).

3.10.1.3. *Mann–Whitney U test*

The Mann–Whitney U-test (also known as the Mann–Whitney–Wilcoxon test, the Wilcoxon rank-sum test; or Wilcoxon two-sample test) is limited to nominal variables with only two values; it is the non-parametric analogue to Student's t-test.

It is a non-parametric counterpart of the *t*-test for independent samples.

It uses a different test statistic (*U* instead of the *H* of the Kruskal–Wallis test), but the P-value is mathematically identical to that of a Kruskal–Wallis test.

The Mann-Whitney U-test is useful when researchers want to know if two independent sets of data show a significant overall difference in the degree of the variable they are interested in, but they cannot use the *z* or *t* test because the hypothesis relating to level of measurement, sample size, normality or equality of variance are not suitable.

The test assumes only an ordinal level of measurement, since it is based on the ranking of scores.

This test is used to calculate the probability of finding any given difference in rank sums under the null hypothesis that the samples were collected from population with the same distribution.

Null hypothesis: There is no difference in scores of the two groups (*i.e.* the sum of ranks for group 1 is no different than the sum of ranks for group 2).

Alternative hypothesis: There is a difference between the scores of the two groups (*i.e.* the sum of ranks for group 1 is significantly different from the sum of ranks for group 2).

Basic formula for this test is as follows:

$$U_1 = N_1N_2 + \frac{N_1 x (N_1 + 1)}{2} - R_1$$

$$U_2 = N_1N_2 + \frac{N_2 x (N_2 + 1)}{2} - R_2 \ or \ U_2 = N_1N_2 - U_1$$

Where U_1 is the smaller group and U_2 is the larger group and similarly R_1 and R_2 and the rank score in all class and N_1 and N_2 are the number of values in each group. This situation is particularly refers to the condition when we have two dissimilar group, but in case of similar group size that N_1 and N_2 are similar in each group then U_1=U_2.

The ranking and tie criteria will be similar as describe in Kruskal–Wallis test.

Group N_2 (9 values)	*Rank*	*Group N_1 (8 values)*	*Rank*
7	9	7	9
8	13.5	4	1
6	5.5	5	3
9	16	6	5.5
10	17	8	13.5
7	9	5	3
7	9	5	3
8	13.5	7	9
8	13.5		
			R_1 = 47

Now by using the above formula the U_1 will be = (8 x 9) + 8 (8+1)/2-47 =61 and U_1 will be = (8 x 9)-61 =72-61=11.

Since U_2 is the smaller of two values, we have U=U_2 =11. The significance of the result can be check through tables (append at last of this chapter) at both 0.05 and 0.01 levels. In particular this example we have N_1 =8 and N_2 = 9 and with table at 5 per cent level the critical value is 15.

In this example, since the calculated value is less than this, therefore, we can reject the null hypothesis, and claim that there is a significant difference between the two sets of rating.

Interestingly, for N_1 and N_2 values about 20, the test statistic U conforms to an approximately normal distribution, however for large samples research must have calculate a Z value by using following equation

$$Z = \frac{\frac{U - N_1 \, x \, N_2}{2}}{\sqrt{\frac{N_1 \, x \, N_2 \, (N_1 + N_2 + 1)}{12}}}$$

If we have 30 and 35 scores we got a U value of 618 and Z value will be

$$Z = \frac{\frac{618 - 30 \, x \, 35}{2}}{\sqrt{\frac{30 \, x \, 35 \, (30 + 35 + 1)}{12}}} = 1.22$$

The test is non-directional and the at the 5 per cent level, the critical value of z is 1.96 and because the calculated value is smaller than this, we cannot reject the null hypothesis.

Tables related with critical values of the Mann-Whitney U test for two tailed and one tail testing are append at the end of this chapter.

3.10.1.4. *Wilcoxon signed-ranks test*

As mentioned earlier the Mann-Whitney U test can be used as a non-parametric alternative of the t test for independent samples. However, the non parametric equivalent of the t test for correlated samples is the Wilcoxon Signed Ranks test.

This test relay on the fact that a researcher can assume that he can rank differences between paired observation.

For this test the data should consist of a number of pairs of scores, each related with a single subject, or from a pair of matched or related subjects.

Under the null hypothesis that there is no difference in the distribution of the populations from which the samples are drawn.

Conclusively the Wilcoxon signed ranks test assesses the significance of disparity and the basis of this test is very simple, and similar to that of the Mann-Whitney test for independent samples.

For quantification of this test we have to find out the difference between each pair of scores, subtracting consistently (second from first, or *vice versa*) and recording the signs.

Ranking tie methods are similar with Mann-Whitney test. Each rank is now given the sign of the difference it corresponds to.

The sum of the positive rank is found and also that of the negative ranks. And, the smaller of these two sums is the test statistic *W*.

Disease intensity at filed 1	*Disease intensity at filed 2*	*1–2*	*Rank*
8	10	–2	–4.5
7	6	+1	+2
4	4	0	–
2	5	–3	–7.5
4	7	–3	–7.5
10	11	–1	–2
17	15	+2	+4.5
3	6	–3	–7.5
2	3	–1	–2
11	14	–3	–7.5

Conclusively the value of *W* must be smaller than or equal to the critical value if the null hypothesis is to be rejected.

For a working example we want to quantify this test for disease intensity of Green ear diseases of Bajara at two different fields.

For above cited example first we have to give rank and sign of the difference it corresponds to. The pair with zero difference is dropped from the analysis, and the N, the number of pair, is decreased accordingly. In this particular example the sum of the positive ranks is 6.5, while negative ranks is 38.5.

By rule we take the smaller of these, 6.5 as value for *W*. and from tables the critical value of *W* for $N = 9$ and at 5 per cent significance level in a non-directional test is 5. Since calculated value of *W* is larger than tabular value, we cannot reject the null hypothesis and must conclude that no significant difference has been observed.

For larger sample size (N>20) the distribution of *W* is almost normal, and a *z*-score can be quantified through following formula

$$Z = \frac{W - \frac{Nx\,(N+1)}{4}}{\sqrt{\frac{N\,x\,(N+1)x\,(2N+1)}{24}}}$$

3.10.1.5. *Kendall's test of concordance*

This test requires bivariate observations like (X_1, Y_1), *i.e.* one observation on each of n subjects and assumes that the n (X,Y) pairs are mutually independent and identically distributed according to some continuous bivariate distribution.

In this test the null hypothesis that X and Y variables are independent.

The test statistic, *K*, can be calculated as $K = K'\text{-}K''$, where K' = is the number of concordant pairs while, K'' is the number of discordant pairs.

X, Y pair is said to be concordant if $(X_i\text{-}X_j)\ (Y_i\text{-}Y_j)>0$, and discordant if $(X_i\text{-}X_j)\ (Y_i\text{-}Y_j)<0$.

This approach to calculating *K* is applicable when no ties exist among X' and no ties exist the *Y*.

Like most other correlation coefficient, this test (5I) also assumes a value between +1 and -1, the correlation coefficient can be calculated using 5I = 2*K*/[*n (n-1)*].

Working example, assume that disease severity level of potato scab in different location are scored as follows, according to the value we have calculated concordant and discordant values.

Location	*Disease severity score*	*Concordance value*	*Discordant value*
1	2	10 (how many values are higher than 2) (how man values are lower than 2)	1
2	1	10	0
3	4	8	1
4	3	8 (we have 8 values higher than 3, see only those values should count which are located underneath the 3)	0
5	6	6	1
6	5	6	0
7	8	4	1
8	7	4	0
9	10	2	1
10	9	2	0
11	12	0	1
12	11		
	Total	60	6

So Kendall's tau is = sum of concordant pair - sum of discordant pair/ sum of concordant pair + sum of discordant pair. Thus, 60-6/60+6 = 54/66 = 5I = 0.818.

Statistical significance can be quantified through *Z* value.

$$Z = \frac{3\ x\ tau\ x\ \sqrt{n\ (n-1)}}{\sqrt{2\ (2n+5)}}$$

$$Z = \frac{3\ x\ \ 0.818\ x\ \sqrt{12\ (12-1)}}{\sqrt{2\ (2x12+5)}}$$

$$Z = \frac{3\ x\ \ 0.818\ x\ 11.489}{7.616} = 3.70$$

3.10.1.6. *Test of association*

Study of species interactions is a useful criteria to identify the impacts of various biotic and abiotic factors on distribution, abundance and interactions among species. Further, depending on whether or not two species select or avoid the same habitat, have some mutual attraction or repulsion, or have no interaction whatsoever, a certain pattern of inter-specific association results. This association may be positive, negative or absent.

In plant pathology particularly botanical epidemiology, this concept can be utilized as an important quantitative tool for a condition where two or more than two pathogen exists on crop, field or within experimental area. Nature of their interaction (+, - or 0) provide a healthy handout about prediction of any outbreak, multiple losses, and management approaches (chemical, cultural or biological).

Spatial associations of diseases at the field level may result from environmental (*e.g.*, climatic) and D or biological conditions affecting both diseases over large scales. Dissociations at the fruit level, for example, could be due to antibiosis or competition of the pathogens for infection sites (Nelson and Campbell, 1992).

For an example a complex of more than 60 fungal species (Dý´az Arias *et al.*, 2010) has been associated with sooty blotch and flyspeck disease complex (SBFS) which blemishes the cuticle of apple fruit (Williamson and Sutton, 2000). Several species within the SBFS complex appear to be cosmopolitan, such as *Schizothyrium pomi*, usually associated with the flyspeck-type of symptom on fruit, and *Peltaster fructicola*, *Leptodontium elatius* and *Geastrumia polystigmatis* associated with the punctate, fuliginous, and ramose-type symptoms, respectively (Johnson *et al.*, 1997).

Scott *et al.* (2003), analyzing spatial associations of oilseed rape mildew between lagged time periods (t vs. t-1) suggested that newly infected sites contributed less to the epidemics than expansion of previously infected sites, probably the result of localized conidial inoculum from secondary sources. For bull's eye rot, there was evidence of spatial association of rotted fruit between years, which suggests that inoculum sources overwinter in survival sites such as cankers within the apple trees themselves, and from where it can be dispersed, especially the conidia. Such association was not detected for bitter rot, and this may be related to other factors influencing epidemics, such as varying environmental conditions or inoculum sources that are relatively distant from the infected trees (Turechek and Madden, 2000).

Significant spatial associations ($P < 0.0001$) were detected between total bull's eye rot incidence (as well as for visible rot) between years and between visible rotted and latently infected fruit at harvest in the two seasons by Spolti *et al.* (2012). The strong evidence for the spatial association found between visibly rotted and latently infected fruit at harvest in the same year, for both rots, is indicative of the contribution of both primary and secondary inoculums sources to the epidemics. Spolti *et al.* (2012) have further suggested reported that when spatial associations between different diseases were measured, there was a significant spatial association between bitter rot and bull's eye rot. Similarities in means of spore dispersal, overwintering sites, and environmental requirements for the establishment and development of these rots may explain the positive spatial association between these diseases. Both fruit rot diseases have rain splash as their principal means of inoculum dispersal and similar requirements of temperature and wetness for infection (Noe and Starkey, 1982 and Henrý´quez *et al.*, 2008).

The study of species association involves two distinct components. The first is a statistical test of the hypothesis that two species are associated or not at some predetermined probability level. The second is a measure of degree of strength of the association

For quantification of species association, following variables are needed (for *e.g.* two pathogen species A and B)

- a = The number of sampling units where both species occur
- b = The number of sampling units where species A occurs but not B
- c = The number of sampling where species B occurs, but not A

- d = The number of sampling units where neither A nor B are found
- N = The total number of sampling units (N = a+b+c+d)

With using these parameters, the following matrix can be visualized with following 2 x 2 contingency table

Species A		***Species B***		
		Present	***Absent***	
	Present	a	b	m=a+b
	Absent	c	d	n=c+d
		r=a+c	s=b+d	N =a+b+c+d

With using above mentioned parameters many indices can be calculated. There are at least 25 different ecological indices alone for characterizing the property of occurrence (Ludwig and Reynolds, 1988). Some of most frequently used indices are Jaccard, Ochiai and Dice indices. These indices are equal to 0 at "no association" and 1 at "maximum association". Computation of these three indices are as follows

Ochiai Index (OI)	***Dice Index (DI)***	***Jaccard Index (J)***
This index is based on the geometric mean of a/m and a/r	This index is based on the harmonic mean of a/m and a/r	This index is the proportion of the number of sampling units where both species occur to the total number of sampling units where at least one of the species is found
$OI = \frac{a}{\sqrt{a+b} x \sqrt{a+c}}$	$DI = \frac{2a}{2a+b+c}$	$JI = \frac{a}{2a+b+c}$

One reason for choosing the Jaccard index over other indices is that the Jaccard index does not consider disease-free sampling units ("double zero") in its calculation as an indication of association. Other indices, such as the Ochiai and Dice indices also do not use double zero sampling units in their calculation (Ludwig and Reynolds, 1988). However, the Jaccard is simpler to interpret. The Jaccard index simply represents the probability of encountering a sampling unit with both species (diseases) in the population of sampling units with either disease. Other indices, such as the Dice and Ochiai, use alternative weightings of the proportion of sampling units occupied by both species relative to sampling units occupied by only a single species to measure association, making it difficult to interpret these indices biologically.

In selecting the Jaccard index, however, one surrenders the opportunity to perform a simple, parametric test of the hypothesis of independence in lieu of, presumably, a more meaningful measure of association. Counting double-zero sampling units would allow the analyst to apply a simple-to-calculate chi-square test designed for two-way tables to test the null hypothesis of independence because the sampling distribution of this statistic is known (Turechek, 2004).

For multi-species association analysis, the following steps need to quantify

- ***Sate hypothesis*:** The null hypothesis is that there is no association among the S species. This is true under two situations (a) the species are independent, and (b) positive and negative association between species. The alternate hypothesis is that there is a net positive or negative association among species.
- ***Compute test statistic*:** For this parameters first compute the sample variance for the occurrences of the S species in the sample as

$$\sigma_T^2 = \sum_{i=1}^{S} p_i (1 - p_i)$$

Where $p_i = n_i/N$

- Now estimate the variance in total species number as

$$S_T^2 = \frac{1}{N}\sum_{j=1}^{N} (T_j - t)^2$$

Where t is the mean number of species per sample

- Now calculate the variance ratio, which serves as the index of overall species association. The expected value under the null hypothesis of independence is VR>1 = suggests that overall, the species exhibit a positive association while if VR<1 a negative net association

$$VR = \frac{S_T^2}{\sigma_T^2}$$

- *W* statistic is computed through multiplication of variance ratio and number of species. This statistics is used to test whether deviations from one are significant or not by using *Chi-Square* test.

Working example with multi-pathogen (12) at 11 different sampling sites. For this example we are utilize a binary data that presence (1) and absence (0) of a particular pathogen at different experimental sites.

	Experimental site													Total number of pathogen/(30) total number of sampling site (II)	
	1	2	3	4	5	6	7	8	9	10	11		n1		
Pathogen identified									= SUM(B5:M5)					Test static	2.72
Pathogen 1	1	1	0	0	0	1	0	0	1	0	1		5	0.24793	
Pathogen 2	1	1	0	0	0	0	0	0	0	0	1		3	0.19835	
Pathogen 3	0	1	0	0	0	0	0	0	0	0	0		1	0.08264	
Pathogen 4	0	0	0	0	0	0	0	1	0	0	0		1	0.08264	
Pathogen 5	0	0	1	1	0	0	1	1	0	0	0		4	0.2314	= (N5/11)*(1-N5/11)
Pathogen 6	1	1	0	0	1	0	0	0	0	1	1		5	0.24793	
Pathogen 7	0	0	0	0	0	1	0	0	1	0	0		2	0.14876	
Pathogen 8	0	0	1	1	1	0	1	0	0	0	1		5	0.24793	
Pathogen 9	0	0	0	0	0	0	0	1	0	0	0		1	0.08264	
Pathogen 10	0	0	0	0	1	0	0	0	0	0	0		1	0.08264	
Pathogen 11	0	0	0	0	0	0	0	1	0	0	0		1	0.08264	
Pathogen 12	0	0	0	0	0	0	0	0	0	1	0		1	0.08264	
Total number of pathogen	= (B16-2.72)^2 3	4	2	2	3	2	2	4	2	2	4		30	1.81818	Test static
	0.784	1.64	0.5	0.52	0.08	0.5	0.5	1.6	0.5	0.5	1.6		8.18	= SUM(O4:O15)	
= 1/11	0.09091	0.74	= 0.09091/8.18												
Species affinity	0.74				= Species affinity/sum of										
Varicne ratio	0.40663	negative ass			test static *i.e.* 0.74/1.818										
W	4.4726	3.9	1												
=(11)*(0.4066)	df = 10														

Since VR<1 there is possibility of negative association

In above example under the null hypothesis of no association, there is a 90% probability that *W* should lies between the limit 1.14<*W*>11.07. Thus, we accept the null hypothesis of no association

In an another example we have five sampling site with six pathogen and the various indices are as follows

Pathogen	***Sampling site***				
	1	***2***	***3***	***4***	***5***
Pathogen 1	1	1	1	0	0
Pathogen 2	0	1	1	1	0
Pathogen 3	1	1	1	0	0
Pathogen 4	1	0	1	1	1
Pathogen 5	0	0	0	0	1
Pathogen 6	0	1	0	1	1
Variance Ratio 0.1875					
W 0.93					

Pair	*Association type*	*Ochichi*	*Dice*	*Jaccard*	
1	2	+	0.667	0.667	0.5
1	3	+	1	1	1
1	4	-	0.577	0.571	0.4
1	5	-	0	0	0
1	6	-	0.33	0.33	0.33
2	3	+	0.667	0.667	0.5
2	4	-	0.577	0.571	0.4
2	5	-	0	0	0
2	6	+	0.667	0.667	0.5
3	4	-	0.577	0.571	0.4
3	5	-	0	0	0
3	6	-	0.333	0.333	0.2
4	5	+	0.5	0.4	0.25
4	6	-	0.577	0.571	0.4
5	6	+	0.577	0.5	0.33

3.11. THE REGRESSION AND BAYESIAN APPROACHES

In general, Null Hypothesis Testing (NHT) suffers from following drwabacks

- the dependence of the test results on sample size: how large sample should be followed to increase the chances of a statistically significant result,
- the fact that null hypothesis is always false,
- false to reject the null hypothesis is not the same as evidence supporting the null hypothesis,
- the arbitrary traditional P value for significance (typically P = 0.01 and 0.05) and
- the tendency to resort to data dredging (collecting and analysis data on more and more variable to try to come-up with adequate explanation), when attempting to revise rejected hypothesis.

3.11.1 Regression Analysis

To overcome above cited problems Moerman (1996) had developed a linear regression model including residual analysis which assessed the importance of a various attributes of regression analysis. The regression analysis seeks to quantify the degree to which severity or incidence of a plant disease has an unexpectedly high or low with various independent variables like (climate, number of day or planting, host factors which may include any physiological, anatomical or genotypic factors etc.

Residual analysis can be carried out with simple Microsoft Excel methods 'LINEST'.

In botanical epidemiology, uses of this approach have been explained and documented by (Butt *et al.*, 1990; Godoy *et al.*, 2003 and Mila *et al.*, 2004).

3.11.2 Bayesian Approaches

Bayes was a seventeenth century mathematician who introduces a set of rules for what is known as conditional probability. Bayesian method explicitly recognize and combing few components of knowledge. Prior knowledge and new data are combined using a model to product posterior knowledge. These four components may be reported as

$$Prior + Data \xrightarrow{Model} Posterior$$

The Bayesian approach sees parameters such as mean, standard deviations and regression coefficient as random or unknown variable. Rather than erecting hypothesis to be tested and accepted or rejected. An important aspect of Bayesian is that for each hypothesis, they initially incorporate prior information through what are termed prior probability based on existing knowledge and previous research.

The Bayesian methods ask "*What is the probability of the hypothesis being tree given the observed data*". Thus this method then takes the prior probability for a given hypothesis. They are then compared using various statistics. In practice, Bayesian method differs from likelihood method by weighting the likelihood value by the prior probabilities to obtain posterior probabilities.

- Bayes theorem allows us to connect the prior and posterior odds for disease when using a predictive model or a diagnosis test with known predictive accuracy. The posterior odd is found to be the product of the prior odds and the likelihood ratio of the predictor. The likelihood of the predictor depends on the predictive accuracy of the model or diagnosis test. One argument for adopting a Bayesian approach to evaluating decision tools is that it allows users to operate the tool at a threshold which suits their attitude to risk. This type of approach is currently very much in vogue with funders and policy-makers who wish to promote end-user involvement in research. However, the Bayesian approach is inherently probabilistic so it confronts the user with uncertainty.

- Arguably the difference between Bayesian and frequentist approaches is more important in parameter estimation than in prediction. The important difference is that in the Bayesian framework parameter estimates are based on a combination of prior information about parameter values and the likelihood estimated from the data. In the frequentist (*i.e.* classical) approach only information from the likelihood is used. One advantage for the Bayesian approach (which may also be a weakness) is that it can incorporate expert knowledge and, consequently, makes the estimation possible even where empirical data are scarce.
- Yuen and Hughes (2002) have documented that rule based approaches for disease prediction can be evaluated through many different way and out of them probability of disease occurrence is one of them. They further summarized that Bayes's Theorem can be a useful tool to examine how a disease forecast affects the probability of occurrence without knowing the risk preferences of the targeted decision makers.

This approach had been studies for various host-pathogen systems like *Sclerotinia sclerotiorum* stem rot of oil seed rape, *Pseudocercosporella herpotrichoides* Rye by Yuen and Hughes (2002), Rhizoctonia-radish system by Gibson *et al.* (2006), Madden and Paul (2009) and Mila and Ngugi (2011).

3.11.2.1 *Bayes rule for conditional probability*

Bayes rules are based on conditional probability. Consider two events: event C and event D. Pr(C) = the probability of event C occurring; Pr (D) = the probability of event D occurring, and Pr (C and D) = the probability of both event occurring together. Thus conditional probability theory tell us that Pr(C and D) = Pr (D) X Pr (C | D)

Which in words is: the probability of event C and D both occurring is equal to the probability of event C occurring given that event D has occurred multiplied by the probability of event D occurring (independent of event C). The | symbol means given the truth or occurrence of.' The above can be rearranged to give Pr (C | D) = Pr (C and D) /Pr (D).

For example *Phytophthora* race 1 present in a sample with probability of 0.03 and *Phytophthora* race 2 with probability 0.35. Therefore, we can calculate the conditional probability that race 1 is present given that race 2

Pr (race 1 | race 2) = Pr (race 1 and 2)/ Pr (race 2)

= 0.03/0.35 = 0.086

3.11.2.2 *Bayes' rule for a finite Number of hypothesIs*

Conditional probability states that for two events C and DL

$$Pr\ (C \text{ and } D) = Pr\ (D) \times Pr\ (C \mid D).$$

C and D are simply labels for events (outcome) that can be swapped arbitrarily, so that the following is also true:

$$Pr\ (D \text{ and } C) = Pr\ (C) \times Pr\ (D \mid C).$$

These two equivalent expressions for Pr (C and D) can be set equal to each other:

$$Pr\ (D) \times Pr\ (C \mid D) = Pr\ (C) \times Pr\ (D \mid C)$$

It is then straight forward to obtain

$$Pr\ (C \mid D) = Pr\ (C) \times Pr\ (D \mid C) / Pr\ (D).$$

Let us assume that event C is that a particular hypothesis is true, and event D is the occurrence of data. Then, the posterior probability that *Phytophthora* is absent given the data (Pr (H_a ´ D)) is:

$$Pr\ (H_a \mid D) = Pr\ (H_a) \times (pr\ (D \mid H_a) / Pr\ (D).$$

The various components of the equation are the prior probability that he *Phytophthora* is absent (Pr(H_a)), the probability of obtaining the data given that it is absent (Pr (D | H_a), which is the likelihood, and the probability of obtaining the data independent of the hypothesis being considered (Pr (D)).

The probability of obtaining the data (the *Phytophthora* was not detected given H_a is true (the Plant is absent) was provided when using the likelihood based methods:

$$Pr\ (D \mid H_a) = 1.0$$

Similarly, given the presence of the race

$$Pr\ (D \mid H_p) = 0.2$$

The value of Pr (D) is the same regardless of the hypothesis being considered (H_P the plant is present or H_a the plant is absent), so it simply acts as a scaling constant. Therefore, Pr (H_a | D) is proportional to Pr (H_a) X Pr (D | H_a) and Pr (H_p | D) is proportional to Pr (H_p) X Pr (D | H_p), with both expressing having the same constant of proportionality (1/Pr (D)).

Pr (D) is calculated as the sum of the values Pr (H) X Pr (D | H) under all hypothesis. Even prior probabilities are equal

(Pr (H_a) = Pr (H_p) = 0.5: Pr (D) = [Pr (H_a) X Pr (D | H_a)] + [(Pr (H_p) X Pr (D | H_p)]

= (0.5 X 1)+ (0.5 X 0.2) = 0.6.

Therefore, the posterior probabilities are 5/6 (0.5/0.6) for the absence of the plant and 1/6 (0.1/0.6) for the presence of the plant. So, for a finite number of hypothesis, Bayes' rule states that the probability of the hypothesis given the data is calculated using the prior probabilities of the different hypothesis (Pr (H_j)) and the probability of obtaining the data given the hypothesis (Pr (D | H_j)):

$$\Pr(H_j \mid D) = \frac{\Pr(H_i) X \Pr(D \mid H_i)}{\Sigma_j \Pr(H_j) X \Pr(D \mid H_j)}$$

3.11.2.3. *Estimating a mean for a normal model analytically*

When the data and prior have normal distribution, the posterior distribution also has a normal distribution, the mean and variance of which depends, not surprisingly, on the mean and variance of the prior.

The posterior distribution also depends on the sample size, mean and variance of the data. The mean and variance of the posterior can be calculated from the following formula.

$$\mu_{post} = \frac{\frac{\mu_{post}}{\sigma^2_{Prior}} + \frac{\mu_{data}}{\sigma^2_{data}}}{\frac{1}{\sigma^2_{Prior}} + \frac{n}{\sigma^2_{data}}}, and$$

$$\sigma^2_{Prior} = \frac{\sigma^2_{Prior} X \sigma^2_{data}/n}{\sigma^2_{data}/\sigma^2_{Prior}},$$

where n is equal to the sample size, σ^2_{prior}, σ^2_{data}, σ^2_{post} are the prior, data and posterior and μ_{prior}, μ_{data}, μ_{post} are the mean of distribution.

These formulas provide useful insight into Bayesian statistics. The mean of the posterior is a weighted average of the means of the prior and data. The weights are the precisions of the prior ($1/\sigma^2_{prior}$) and the data (n/σ^2_{data}).

The Influence of the data and prior on the posterior mean depends on which is more informative. When there are no data (n = 0), the mean of the posterior is equal to the mean of prior.

When the variance of prior is very large, $1/\sigma^2_{prior}$ approaches zero and the mean of the posterior will be close to the mean of the data. The prior is said to be uninformative when the posterior influences exclusively by the data. This is achieved by using a prior with a large variance. The variance of the posterior has similar properties, but these are most obvious when its formula is re-arranged to be expressed as inverse of the variance:

$$\frac{1}{\sigma^2_{Post}} = \frac{1}{\sigma^2_{data}} + \frac{1}{\sigma^2_{Prior}}$$

The inverse of the variance measures precision. Large values for the precision mean the variance is small.

The quantity of n/σ^2_{data} is the inverse of the standard error squared, and it measures precision in an ordinary frequents analysis. Therefore, the precision of the posteriors is simply equal to the precision based on the data (the inverse of the standard error square) plus the precision of the prior.

The precision of an estimate is increased by the using prior information. Disease severity at cotton plant at ten different locations 42, 43, 58, 70, 47, 51, 85, 63, 58, 46. These having a mean of 56.3 and variance of 184.9. Given the prior has a mean and variance 53 and 25, the posterior distribution for the mean of uses in new system has the following mean and variance:

$$\mu_{post} = \frac{\frac{53}{25} + 56.3\, X \frac{10}{184.9}}{\frac{1}{25} + \frac{10}{184.9}} = 54.9$$

$$\sigma^2_{post} = \frac{25\, x \frac{184.9}{10}}{\frac{184.9}{10} + 25} = 10.6$$

The standard deviation of the posterior is 3.26 (0.6). Therefore, there is an approximate 95% chances that the mean of uses of species is between 48.5 to 61.3 (The mean of the posterior plus or minus 1.96 times of the standard deviation of the posterior

3.11.2.4. *General Bayesian method*

Bayesian inference derives the posterior probability as a consequence of two antecedents,

- Prior probability and a

- Likelihood function derived from a probability model for the data to be observed.
- Bayesian inference computed the posterior probability according to Bayes rules:

$$P\ (H|E) = \frac{P\ (H|E)\ x\ P\ (H)}{P\ (E)}$$

H Stands for E		***P (H)***	***P (H%E)***	***P(E%H)***	***P (E)***
Any hypothesis whose probability may be affected by data often there are competing hypotheses, from which one chooses the most probable	The evidence E corresponds to new data that were not used in computing the prior probability	The prior probability is the probability of H before E is observed. This indicates one's previous estimate of the probability that a hypothesis is true, before gaining the current evidence.	The posterior probability is the probability of H given E *i.e.* After E is observed. This tells us what we want to know: The probability of a hypothesis given the observed evidence	The probability of observing E given H is also known as likelihood. It indicates the compatibility of the evidence with the given hypothesis	Called marginal likelihood or model evidence. This factor is same for all possible hypothesis being considered. This means that this factor does not enter into determining the relative probabilities of different hypotheses.

3.11.2.5. *Integration of Bayesian method in botanical epidemiology*

To correlate and prove the usefulness of this method with affiliation of disease severity with planting date, this statistical tool can be summarized in following manner,

- Suppose a researcher wants to evaluate whether a given pathogen or its race j, j = 1..... J of totally groups departs from the expected proportion θ with planting date, *i.e.* whether the probability distribution of θ_j departs from the probability distribution of θ.
- The null hypothesis (H_{0}.) is: disease severity (*j*) has a proportion of planting date (θ_j) equal to the overall proportion of disease (θ).
- Prior to sampling it can be assume that θ_j as a random variable, which can take any value between [0, 1] with a certain probability that can be call as prior probability. This will indicate the uncertainty, which can be reduced but not eliminated whatever sample size.
- Since prior to sampling and analyzing the data the researcher cannot predict and believe that pathogen disease severity have a higher or lower probability to planting date.

- It can be assumed that θ_j is uniformly distributed between 0 and 1 and this can be denotedxas $\pi(\theta_j) = 1$.
- After sampling, disease severity rank n_j and the number of days since planting x_j for all J sites, researcher can update uniform prior assumption obtaining the posterior probability of $\theta_{j,}$ $\pi(\theta_j)$, x_j, $n_{j.}$

In order to calculate $\pi(\theta_j |, x_j, n_j)$, in each species in a class j can be selected independently from the selection of another. Consequently it can be assumed that the number of used species is a random variable, and it can be write as

$$\Pr(X_j = x_j|\Theta_j = \binom{n_j}{x_j} \Theta_j^{xj} (1-\Theta_j)^{n_j - x_j}$$

Pr $(X_j = x_j - \Theta_j)$ as a function of θ_j that can use to obtain the likelihood function $L(\theta_j)\, \alpha \theta_j^{xj}(1-\theta_j)^{nj-xj}$, which indicates the values of θ_j more likely to expected conditionally on the observed data $_{xj}$, and $_{nj}$. Further the taxonomic group is identified as incompatible with H_0 (random selection of plant species used for medicines) if

$$P_j = \Pr(X_j \geq x_j|\Theta_j = \hat{\theta} = \sum_{xj}^{nj} \binom{n_j}{x_j} \hat{\theta}^{xj} (1-\hat{\theta})^{n_j - x_j}$$

The value of pj can be used as a measure of evidence against the hypothesis H_0. The most important consideration for this approach that it does not consider the uncertainty around θ and neither evaluates the probability of H_0 to be true conditionally on the sampled data only. Instead it considers hypothetical and never observed, values of X_j larger than the observed ones, violating the principle that only observed data should be considered for inference.

3.12. ORDINATION ANALYSIS

Statistical classification and ordination techniques are commonly used to study multivariate ecological relationships (*e.g.*, species distributions in relation to environmental gradients) but in a few cases they have also been used to measure and visualize patterns of inter informant similarity/ dissimilarity in regards to their ecological knowledge and practices.

To carry out these types of analyses, a distance or similarity matrix is set up showing the pairwise distance (*e.g.*, Euclidean distance, percentage dissimilarity) or similarity (*e.g.*, correlation, covariance) between all pairs of sample objects (in this case, the objects refer to informants). In classification, the analysis seeks to partition the set of heterogenous objects into relatively homogenous groups based on the calculation of class centers, density, variance, number of members, and distinctiveness of delimitation.

Ordination is a collective term for multivariate techniques which adapt a multi-dimensional swarm of data points in such a way that when it is projected onto a two dimensional space any intrinsic pattern the data may possess becomes apparent upon visual inspection.

Basically, ordination serves to summarize community data (such as species abundance data) by producing a low-dimensional ordination space in which similar species and samples are plotted close together, and dissimilar species and samples are placed far apart.

Generally, ordination techniques are used to describe relationships between species composition patterns and the underlying environmental gradients which influence these patterns.

The basic objective of ordination is to reduce multidimensional data sets to a lower number of dimensions to facilitate analysis.

It does this by screening out redundant variables and retaining those characteristics of the data set that contribute most to its variance (*i.e.* compressing many measurements into a fewer number). The statistical procedures for most types of ordination are computation-intensive, especially for large data sets, and hence one of the biggest limitations with this technique is the need for computers.

3.12.1. Eigen Analysis-based Indirect Gradient Analysis What they are

An introduction to eigen analysis is beyond the scope of this article. However, in the context of ordination there are several points worth making. For eigen analysis-based methods:

- An eigen analysis is performed on a square, symmetric matrix derived from the data matrix
- There is a unique solution to the eigen analysis, no matter the order of data.
- Each ordination axis is an eigen vector, and is associated with an eigen value. The coordinates for the ith sample along a given axis is the ith element of the axis' eigen vector.
- Axes are ranked by their eigen values. Thus, the first axis has the highest eigen value, the second axis has the second highest eigen value, etc.
- Eigen values have mathematical meaning that can aid in interpretation. In principal components analysis, eigen values are 'variance extracted'.

In methods related to correspondence analysis, eigen values are 'inertia extracted', or equivalently, correlation coefficients.

- Axes are orthogonal to each other.

There are a potentially large number of axes (usually, the number of samples minus one, or the number of species minus one, whichever is less) so there is no need to specify the dimensionality in advance. However, the number of dimensions worth interpreting is usually very low.

Species and samples are ordinated simultaneously, and can hence both be represented on the same ordination diagram (if this is done, it is termed a biplot).

3.12.2. Principal Components Analysis

The simplest and oldest eigen analysis-based method is Principal Components Analysis (PCA). It is used for many purposes, but i will only discuss its applicability as an ordination method here. Geometrically, PCA is a rigid rotation of the original data matrix, and can be defined as a projection of samples onto a new set of axes, such that the maximum variance is projected or "extracted" along the first axis, the maximum variation uncorrelated with axis 1 is projected on the second axis, the maximum variation uncorrelated with the first and second axis is projected on the third axis, etc.

One of the biggest differences between PCA and PCoA is that the variables (*i.e.* species) representing the original axes are projected as biplot arrows. The eigen values represent the variance extracted by each axis, and are often conveniently expressed as a percentage of the sum of all eigen values (*i.e.* total variance). In most applications of PCA (*e.g.,* as a factor analysis technique), variables are often measured in different units. For example, PCA of taxonomic data may include measures of size, shape, color, age, numbers, and chemical concentrations. For such data, the data must be standardized to zero mean and unit variance (the typical default for most computer programs).

3.12.3. Principal Coordinates Analysis

Principal coordinates analysis (PCoA) is similar to Polar ordination in that it attempts to represent the distances between samples.

In particular, it maximizes the linear correlation between the distances in the distance matrix, and the distances in a space of low dimension (typically, 2 or 3 axes are selected). PCoA is perhaps best understood geometrically.

The distance between two items can be faithfully represented by one dimension (a line). The distances between three items are faithfully represented by 2 dimensions (a plane): that is, the items will form the vertices of a triangle, a planar object. Distances between four objects define a tetrahedron (a 3-dimensional object).

To generalize, distances between N objects can be faithfully represented in N-1 dimensions. Unfortunately, it is difficult for the human mind to grasp more than 3 dimensions simultaneously, so we need to project such multidimensional objects onto lower dimensional space.

The PCoA algorithm is analogous to rotating the multidimensional object such that the distances (lines) in the 'shadow' are maximally correlated with the distances (connections) in the object.

Although PCoA is based on a distance matrix, the solution can be found by eigen analysis. When the distance metric is Euclidean, PCoA is equivalent to Principal Components Analysis (thus the bryophyte analysis for PCoA will not be presented here, but rather in the PCA section).

3.12.4. Nonmetric Multidimensional Scaling

PCoA suffers from a number of flaws, in particular the arch effect. These flaws stem, in part, from the fact that PCoA maximizes a linear correlation. Non-metric Multidimensional Scaling (NMDS) rectifies this by maximizing the rank order correlation. The algorithm in brief outline, proceeds as follows:

- The user selects the number of dimensions (N) for the solution, and chooses an appropriate distance metric.
- The distance matrix is calculated.
- An initial configuration of samples in N dimensions is selected. This configuration can be random, though the chances of reaching the correct solution are enhanced if the configuration is derived from another ordination method.
- A measure of 'stress' (mismatch between the rank order of distances in the data, and the rank order of distances in the ordination) is calculated
- The samples are moved slightly in a direction that decreases the stress
- 4 and 5 are repeated until 'stress' appears to reach a minimum. The final configuration of points may be rotated if desired.

NMDS arranges multidimensional objects in a low-dimensional space suitable for visualization in two- or three-dimensional graphic displays. The objects are reproduced in the graph to reflect the observed distances among them with the least amount of stress (*i.e.,* deviation from real distance).

Besides providing the researcher with a visual display of the clustering and dispersal of the objects in relation to each other, the objective here is to detect meaningful underlying dimensions that allow the researcher to explain observed similarities or dissimilarities between the objects. NMDS has several advantages over linear ordination: (a) it makes no assumptions of linearity, (b) it can be used with ordinal-scale data, and (c) the underlying dimensions of contrast come from the respondents' own judgments or answers.

3.12.5. Detrended Correspondence Analysis

Detrended Correspondence Analysis (DCA) eliminates the arch effect by detrending. There are two basic approaches to detrending: by polynomials and by segments.

Detrending by polynomials is the more elegant of the two: a regression is performed in which the second axis is a polynomial function of the first axis, after which the second axis is replaced by the residuals from this regression.

Similar procedures are followed for the third and higher axes. Unfortunately, results of detrending by polynomials can be unsatisfactory and hence detrending by segments is preferred.

To detrend the second axis by segments, the first axis is divided up into segments, and the samples within each segment are centered to have a zero mean for the second axis. The procedure is repeated for different 'starting points' of the segments. Although results in some cases are sensitive to the number of segments.

3.12.6. Contrast between DCA and NMDS

DCA and NMDS are the two most popular methods for indirect gradient analysis. The reason they have remained side-by-side for so long is because, in part, they have different strengths and weaknesses.

While the choice between the two is not always straightforward, it is worthwhile outlining a few of the key differences. Some of the issues are relatively minor: for example, computation time is rarely an important consideration, except for the hugest data sets.

Some issues are not entirely resolved: the degree to which noise affects NMDS, and the degree to which NMDS finds local rather than global options still need to be determined (in the case of the bryophyte data, it took several iterations, with different optimization criteria, before the solution in was reached – so blind acceptance of the first solution is not recommended). Since NMDS is a distance-based method, all information about species identities is hidden once the distance matrix is created. For many, this is the biggest disadvantage of NMDS.

3.12.7. Correspondence Analysis

Correspondence Analysis (CA) is also known as reciprocal averaging, because one algorithm for finding the solution involves the repeated averaging of sample scores and species scores (citations). Instead of maximizing 'variance explained', CA maximizes the correspondence between species scores and sample scores. First Axis species scores and sample scores are assigned such that the weighted correlation between the two is maximized where the 'weight' is the abundance of the species.

The eigen value of the CA axis is equivalent to the correlation coefficient between species scores and sample scores. If species scores are standardized to zero mean and unit variance, the eigen values also represent the variance in the sample scores (but not, as is often misunderstood, the variance in species abundance). In the context of CA, we term this variance the inertia of an axis.

3.12.8. Canonical Correspondence Analysis

Canonical Correspondence Analysis is the marriage between CA and multiple regression.

Like CCA, CA maximizes the correlation between species scores and sample scores

However, in CCA the sample scores are constrained to be linear combinations of explanatory variables. Because of the 'constraint', eigen values in CCA will be lower than in CA.

3.12.9. Redundancy Analysis

As mentioned previously, most of the discussion of CCA pertains to Redundancy Analysis (RDA). However, note that RDA is a linear method. Some of the special properties of RDA include:

Since it is a linear method, species as well as environmental variables are represented by arrows. In most cases, it is best to represent the two sets of arrows in two figures for ease of display.

CCA focuses more on species composition, *i.e.,* relative abundance. Thus, if you have a gradient along which all species are positively correlated, RDA will detect such a gradient while CCA will not.

With RDA, it is possible to use 'species' that are measured in different units. If so, the data must be centered and standardized. But in general, as an ordination technique, the species should not be standardized.

RDA can useful when gradients are short. In particular, RDA may be the method of choice in a short-term experimental study. In such cases, the treatments are the explanatory variables (and are usually dummy variables). The sample ID or block might be a covariable in a partial RDA, if one wishes to factor out local effects.

'Variance explained' is actually a variance explained, and not merely inertia. Thus, variance partitioning, and interpretation of eigen values, are more straightforward than for CCA.

PCA was applied to various epidemic variables in order to select the best subset of variables with the highest explanatory capacity to describe observed 60 disease progress curves of papaya ringspot (papaya ringspot virus type P [PRSV-P]) incidence on papaya (Mora-Aguilera *et al.*, 1996).

Standardised area under disease progress curve, shape parameter of the Weibull distribution function, and time between transplant date and first symptoms were selected as the most important variables and represented 83.5% of the overall variance. These selected variables were then used in cluster analysis.

In analysing the effects of sowing date on *Fusarium* wilt of chickpea, PCA was applied to a set of epidemic variables, including final disease intensity, AUDPC and parameters from the fitted Richards model over 108 epidemics (Navas-Cortes *et al.*, 1998).

Three components, accounting for c. 98% of the total variance, were retained and they provided plausible epidemiological interpretations: representing a temporal positional factor, the AUDPC/final disease and the uniqueness of the estimated value for the point of inflection, respectively.

PCA scores for the retained components can also be used as independent variables in regression analysis. Scholosser *et al.* (2000) applied the canonical correlation analysis to characterize the relationship between plant morphological characters and disease variables in rice blast.

3.13. REFERENCES

Analytis, S, 1980. Obtaining of sub-models for modeling the entire life cycle of a pathogen. *Journal of Plant Diseases and Protection*, 87: 371–82.

Aylor, D.E. 2003. Spread of plant disease on a continental scale: Role of aerial dispersal of pathogens. *Ecology*, 84: 1989–1997.

Bastiaans L. 1993. Effects of leaf blast on growth and production of a rice crop. 1. Determining the mechanism of yield reduction. *Netherlands Journal of Plant Pathology*, 99: 323–334.

Batista, L., Bassanezi, R.B. and Laranjeira, F.F. 2008. Comparative epidemiology of *citrus tristeza* in Cuba and citrus sudden death in Brazil. *Tropical Plant Pathology*, 33(5): 348–355.

Berger, R.D. 1981. Comparison of the Gompertz and logistic equations to describe plant disease progress. *Phytopathology*, 71: 716–719.

Boyd, I.L., Freer-Smith, P.H., Gilligan, C.A. and Godfray, H.C.J. 2013. The consequence of tree pests and diseases for ecosystem services. *Science*, 342: 57–73.

Bullock, J.M., Shea, K. and Sltarpaas, O. 2006. Measuring plant dispersal: An introduction to field methods and experimental design. *Plant Ecology*, 186: 217–234.

Busi, R., Yu, Q., Barrett-Lennard, R. and Powles, 2008. Long distance pollen mediated flow of herbicide resistance genes in *Lolium rigidum*. *Theoretical and Applied Genetics*, 117: 1281–1290.

Butt, D.J. and Royle, D.J. 1990. Multiple regression analysis in the epidemiology of plant diseases. *In*: Kranz, J. (*Ed.*) Epidemics of Plant Diseases. Berlin. Springer. pp. 143–180.

Calonnec, A., Cartolaro, P., Naulin, J.M., Bailey, D. and Langlais, M. 2008. A host-pathogen simulaton model: Powdery mildew of grapevine. *Plant Pathology*, 57: 493–508.

Campbell C.L. and Madden, L.V. 1990. Introduction to Plant Disease Epidemiology. John Wiley and Sons, New York, NY, USA.

Carisse, O., Savary, S. and Willocquet, L. 2008. Spatiotemporal relationships between disease development and airborne inoculum in unmanaged and managed Botrytis leaf blight epidemics. *Phytopathology*, 98: 38–44.

Contreras-Medina, L.M., Torres-Pacheco, I., Guevara-Gonzalez, R.G., Romero-Troncoso, R.J., Terol-Villalobos, I.R. and Osornio-Rios, R.A. 2009. Mathematical modeling tendencies in plant pathology. *African Journal of Biotechnology*, 8(25): 7399–7408.

Cooke, R.C. and Whipps, J.M. 1980. The evolution of modes of nutrition in fungi parasitic on terrestrial plants. *Biological Reviews*, 55: 341–360.

Cunnifee, N.J., Stutt, R.O.J.H., Bosch, F.V. and Gillignan, C.A. 2012. Time dependent infectivity and flexible latent and infectious periods in compartmental models of plant disease. *The American Phytopathology*, 102(4): 365–380.

Cunniffe, N.I., Koskella, B., Metcalf, C.J.E., Parnell, S., Gottwald, T.R. and Gilligan, C.A. 2014. Thirteen challenges in modeling plant diseases. *Epidemics, 10: 6-10*

Darmency, H., Klein, E.K., De, Garanbe, T.G., Gouyon, P.H., Richard-Molard, M., *et al.* 2009 Pollen dispersal in sugar beet production fields. *Theoretical and Applied Genetics,* 118: 1083–1092.

Dý´az Arias, M.M., Batzer, J.C. and Harrington, T.C. 2010. Diversity and biogeography of sooty blotch and flyspeck fungi on apple in the Eastern and Midwestern United States. *Phytopathology*, 100: 345–355.

Dybiec, B., Kleczkowski, A. and Gilligan, C.A. 2004. Controlling disease spread on networks with incomplete knowledge. *Physiology Review Series E,* 70: 066145. (doi:10.1103/ PhysRevE.70.066145) p 5.

Dybiec, B., Kleczkowski, A. and Gilligan, C.A. 2005. Optimizing control of disease spread on networks. *Acta Physica Polonica B*, 36: 1509–1526.

Esker, P.D., Gibb, K.S., Dixon, P.M., and Nutter, Jr., F.W. 2007. An application of space-time analysis to improve the epidemiological understanding of the papaya-papaya yellow crinkle pathosystem. Online. Plant Health Progress, doi:10.1094/ PHP-2007-0726-02-RS.

Gardner, F.P., Pearce, R.B. and Mitchell, R.L. 1985. Physiology of Crop Plants. Iowa University Press, Ames.

Gent, D.H., Mahaffee, W.F. and Turechek, W.W. 2006. Spatial heterogeneity of the incidence of powdery mildew on hop cones. *Plant Disease*, 90: 1433–1440.

Gibson, G.J., Kleczkowski, A. and Gilligan, C.A. 2004. Bayesian analysis of botanical epidemics using stochastic compartmental models. *Proceedings of the National Academy of Sciences*, 101: 12120–12124.

Gibson, G.J., Gilligan, C.A. and Kleczkowski, A. 1999 Predicting variability in biological control of a plant pathogen system using stochastic models. *Proceeding of Royale Society B*, 266: 1743–1753.

Gibson, G.J., Otten, W., Filipe, J.A.N., Cook, A., Marion, G. and Gilligan, C.A. 2006. Bayesian estimation for percolation models of disease spread in plant populations. *Statistical Computation*, 16: 391–402.

Gilligan, C.A. 2002. An epidemiological framework for disease management. *In*: Advances in Botanical Research. Callow, J.A. (*Ed.*). Academic Press, San Diego, CA. pp. 1-64

Gilligan, C.A. 2008. Sustainable agriculture and plant disease: An epidemiological perspective. *Philosophical Transaction Royale Society of London B*, 363: 741–759.

Goodoy, C.V., Amorim, L., Filho, A.M., Silva, H.P., Silva, W.J. and Berger, D. 2003. Temporal progress of southern rust in maize under different environmental conditions. *Fitopathologica Brasilica*, 28(3): 273–278.

Hall, R.J., Gubbins, S. and Gilligan, C.A. 2004 Invasion of drug and pesticide resistance is determined by a trade-off between treatment efficacy and relative fitness. *Bulletin of Mathematical Biology*, 66: 825–840.

Hall, R.J., Gubbins, S. and Gilligan, C.A. 2007. Evaluating the performance of chemical control in the presence of resistant pathogens. *Bulletin Mathematical Biology*, 69: 525–537.

Henne, D.C., Workneh, F., Wen, A., Price, J.A., Pasche, J.S., Gudmestad, N.C. and Rush, C.M. 2010. Characterization and epidemiological significance of potato plants grown from seed tubers affected by Zebra Chip disease. *Plant Disease*, 94: 659–665.

Henrý´quez, J.L., Sugar, D., Spotts, R.A. 2008. Effects of environmental factors and cultural practices on bull's eye rot of pear. *Plant Disease*, 92: 421–424.

Holt, J. and Colvin, J. 2001. Observation and theory of whitefly borne virus disease epidemics. *In*: Jeger, M.J. and Spence, N.J. (*eds*.). Biotic Interactions in Plant–Pathogen Associations. CABI Publishing, Wallingford, UK pp 331–343.

Jarroudi, E.I., Delfosse, M., Maraite, P.,Hoffmann, H. and Tychon, B. 2009. Assessing the accuracy of simulation model for Septoria leaf blotch disease progress on winter wheat. *Plant Disease,* 93: 983–992.

Jeger, M.J., Holt, J., Van den Bosch, F. and Madden, L.V. 2004. Epidemiology of insect-transmitted plant viruses: Modelling disease dynamics and control interventions. *Physiological Entomology*, 29: 291–304.

Johnson, E.M., Sutton, T.B. and Hodges, C.S. 1997. Etiology of apple sooty blotch disease in North Carolina. *Phytopathology*, 87: 88–95.

Keeling, M., Brooks, S.P. and Gilligan, C.A. 2004 Using conservation of pattern to approximate spatial parameters from a single snapshot. *Proceeding of National Academy of Science*, 101: 9155–9160.

Kranz, J. 1976. Effect of some plant diseases on the loss of leaves. *Zeitschrift fur Pflanzenkrankheiten und Pflanzenschutz*, 72: 1373–1377.

Leclerc, M., Dore, T., Gilligan, C.A., Lucas, P. and Filipe, J.A.N. 2014. Estimating the delay between host infection and disease (incubation period) and assessing its significance to the epidemiology of plant diseases. PLos ONE 9 (1): e86568. doi:10.1371/journal.pone.0086568.

Liang, X., Ru, R., Wu, Y. and Peng, X. 2001. Research progress of vector borne plant disease. *Biological Engineering Progress*, 21: 11–17.

Lopes, D.B. and Berger, R.D. 2001. The effects of rust and anthracnose on the photosynthesis competence of diseased bean leaves. *Phytopathology*, 91: 212–220.

Ludwig, J.A. and Reynolds, J.F. 1988. Statistical Ecology. John Wiley and Sons, New York.

Maddan, L.V. and Bosch, F.V.D. 2002. A population-dynamics approach to assess the threat of plant pathogens as biological weapons against annual crops. *BioScience*, 2(1): 65–74.

Madden, L.V. 2006. Botanical epidemiology: Some key advances and its continuing role in disease management. *European Journal of Plant Pathology*, 115: 2–23.

Madden, L.V. and Van den Bosch, F. 2002. A population dynamics approach to assess the threat of plant pathogens as biological weapons against annual crops. *Bioscience*, 52: 65–74.

Madden, L.V., Jeger, M.J. and Van den Bosch, F. 2000. A theoretical assessment of the effects of vector–virus transmission mechanism on plant virus disease epidemics. *Phytopathology*, 90: 576–594.

Madden, L.V. and Hughes, G. 1999. Sampling for plant disease incidence. *Phytopathology*, 89: 1088–1113.

Madden, L.V. and Paul, P.A. 2009. Assessing heterogeneity in the relationship between wheat yield and *Fusarium* head blight intensity using random-coefficient mixed models. *Phytopathology*, 99: 850–860.

Madden, L.V., Hughes, G., and Ellis, M.A. 1995. Spatial heterogeneity of the incidence of grape downy mildew. *Phytopathology,* 85:269–275.

Madden, L.V., Jeger, M.J. and Van den Bosch, F. 2000. A theoretical assessment of the effects of vector-virus transmission mechanism on plant virus disease epidemics. *Phytopathology*, 90: 576–594.

Madden, L.V., Hughes, G. and Van den Bosch, F. 2007. The Study of Plant Disease Epidemics. The American Phytopathology Press. St Paul, Minnesota, USA.

Martini, X., Hoffmann, M., Coy, M.R., Stelinski, L.L., and Pelz-Stelinski, K.S. 2015. Infection of an Insect Vector with a Bacterial Plant Pathogen Increases Its Propensity for Dispersal. PLoS ONE 10(6): e0129373. doi:10.1371/journal.pone.0129373

Mathur, M. 2013. Attributes of Plant Spatial Analysis. Today and Tomorrow Printer and Publishers, New Delhi. p. 365.

McDonald, J.H. 2008. Handbook of Biological Statistics. Sparky House Publishing, Baltimore, Maryland. p. 291.

Mendgen, K. 1981. Nutrient uptake in rust fungi. *Phytopathology*, 71: 963–964.

Metz, A.S.P., Chandra, A., Genovesi, A.D. and Campos, C.F. 2012. Comparsion of screening methods to evaluate the response of St. Augustine grass to *Magnaporthe oryzae*. *Plant Pathology*, 61: 255–261.

Michereff, S.J., Noronha, M.A., Andrade, D.E.G.T., Oliveira, E.P., Xavier Filha, M.S. and Moreira, P.A.A., 2006. Elaboração e validação de escala diagramática para a cercosporiose do pimentão. *Summa Phytopathologica*, 32: 260–266.

Michereff, S.J., Martins, R.B., Noronha, M.A. and Machado, L.P. 2011. Sample size for quantification of Cercospora leaf spot in sweet pepper. *Journal of Plant Pathology*, 93(1): 183–186.

Mila, A.L. and Ngugi, H.K. 2011. A Bayesian approach to meta-analysis of plant pathology studies. *Phytopathology*, 101: 42–51.

Mila, A.L., Carriquiry, A.L. and Yang, X.B. 2004. Logistic regression modeling of prevalence of soybean Sclerotinia stem rot in the North Central region of the United States. *Phytopathology*, 94: 102–110.

Moerman, D.E. 1996. An analysis of the food plants and durg plants of native North America. *Journal of Ethnopharmacology*, 52(1): 1–22.

Mohapatra, N.K., Mukherjee, A.K., Suriya, A.V. and Nayak, P. 2008. Disease progress curves in the rice blast pathosystem compared with the logistic and Gompertz model. *ARPN Journal of Agricultural and Biological Science*, 3(1): 28–37.

Mora-Aguilera, G., Nieto-Angel, D., Campbell, C.L., Teliz, D. and Garcia, E. 1996. Multivariate comparison of papaya ringspot epidemics. *Phytopathology*, 86: 70–78.

Motisi, N., Poggi, S., Filipe, J.A.N., Lucas, P. and Dore, T. 2013. Epidemiological analysis of effects of biofumigation for biological conrol of roor rot in sugar beet. *Plant Pathology*, 62: 69–78.

Navas-Cortes, J.A., Hau, B. and Jimenez-Diaz, R.M. 1998. Effect of sowing date, host cultivar, and race of *Fusarium oxysporum* f. sp. *ciceris* on development of *fusarium* wilt of chickpea. *Phytopathology*, 88: 1338–1346.

Nelson, S.C. and Campbell, C.L. 1992. Incidence and patterns of association of pathogens in a leaf spot disease complex on white clover in the Piedmont region of North Carolina. *Phytopathology*, 82: 1013–1021.

Noe, J.P. and Starkey, T.E. 1982. Relationships of apple fruit maturity and inoculum concentration to infections by *Glomerella cingulata*. *Plant Disease*, 66: 379–381.

Nutter, F.W., Jr. and Parker, S.K. 1997. Fitting Disease Progress Curves Using EPIDEMIOL. Excercises in Plant Disease Epidemiology", Francl, L. J. and Neher, D. A. (*Eds*.). APS Press, St. Paul, MN, pp. 24–88.

Otten, W., Bailey, D.J. and Gilligan, C.A. 2004. Empirical evidence of spatial thresholds to control invasion of fungal parasites and saprotrophs. *New Phytologist*, 163: 125–132.

Otten, W., Filipe, J.A.N., Bailey, D.J. and Gilligan, C.A. 2003. Quantification and analysis of transmission rates for soilborne epidemics. *Ecology*, 84(12): 3232–3239.

Otten,W., Filipe, J.A.N. and Gilligan, C.A. 2005. Dynamics of contact structure and disease transmission in mixed species populations: An analysis of damping-off epidemics. *Ecology*, 86: 1948–1957.

Panicker, S. and Ramraj, B. 2010. Studies on the epidemiology and control of Ascochyta blight of peas (*Pisum sativum* L) caused by *Ascochyta pinodes*. *Archives of Phytopathology and Plant Protection*, 43(1): 51–58,

Park, A.W., Gubbins, S. and Gilligan, C.A. 2003. Extinction times for closed epidemics: The effects of host spatial structure. *Ecology Letters*, 5: 747–755.

Perry, J.N. 1994. Sampling and applied statistics for pests and diseases. *Aspects of Applied Biology*, 37: 1–14.

Peterson, E., Hansen, E. and Hulbert, J. 2014. Source or sink? The role of soil and water borne inoculum in the dispersal of *Phytophthora ramorum* in Oregon tanoak forests. *Forest Ecology and Management*, 322: 48–57.

Pethybridge, S.J. and Turechek, W.W. 2003. Analysis of the association among three viruses infecting hop in Australia. *Plant Pathology*, 52: 158–167.

Rabbinge, R. and Bastiaans, L. 1989. Combination models, crop growth and pests and diseases. *In*: Rabbinge, R, Ward, S.A. and Van Laar, H.H., (*eds.*) Simulation and Systems Management in Crop Protection. Wageningen, Netherlands: PUDOC: *Simulation Monographs,* 32: 217–39.

Rieux, A., Soubeyrand, S., Bonnot, F., Klein, E.K., Ngando, J.E., Mehl, A., Ravigne, V., Carlier, J. and Bellaire, L.L. 2014. Long distance wind dispersal of spores in a fungal plant pathogen: Estimation of anisotropic dispersal kernels from an extensive field experiment. *Plos One*, 9(8): 1–13.

Sall, M. 1980. Epidemiology of grape powdery mildew: A model. *Phytopathology*, 70: 338–342.

Savary, S., De Jong, P.D., Rabbinge, R. and Zadoks, J.C. 1990. Dynamic simulation of groundnut rust, a preliminary model. *Agricultural Systems*, 32: 113–141.

Savary, S., Nelson, A., Willocquet, L., Pangga, I. and Aunario, J. 2012. Modeling and mapping potential epidemics of rice disease globally. *Crop Protection*, 34: 6-17.

Savary, S., Stetkiewicz, S., Brun, F., and Willocquet, L. 2015. Modelling and mapping potential epidemics of wheat disease examples on leaf rust and *septoria tritici* blotch using EPIWHEAT. *European Journal of Plant Pathology*, 142(4): 771–790.

Schlosser, I., Kranz, J. and Bonman, J.M. 2000. Characterization of plant type and epidemiological development in the pathosystem "upland rice/rice blast" (*Pyricularia grisea*) by means of multivariate statistical methods. *Zeitschrift Fur Pflanzenkrankheiten Und Pflanzenschutz-Journal of Plant Diseases and Protection*, 107: 12–32.

Scott, J.B., Hay, F.S., Wilson, C.R., Cotterill, P.J. and Fist, A.J. 2003. Spatiotemporal analysis of epiphytotics of downy mildew of oilseed poppy in Tasmania, *Australia Phytopathology*, 93: 752–7.

Segarra, J., Jeger, M.J. and Van den Bosch, F. 2001. Epidemic dynamics and patterns of plant disease. *Phytopathology*, 91: 1001–1010.

Sherm, H., Ngugi, H.K. and Ojiambo, P.S. 2006. Trends in theoretical plant epidemiology. *European Journal of Plant Pathology*, 115: 61–73

Shi, R., Zhao, H. and Tang, S. 2014. Global dynamic analysis of a vector-borne plant disease model. *Advances in Difference Equations*, 59: 1–16.

Spolti, P., Valdebenito-Sanhueza, R.M., Laranjeira, F.F. and Del Ponte, E.M. 2012. Comparative spatial analysis of the sooty blotch/flyspeck disease complex, bull's eye and bitter rots of apples. *Plant Pathology*, 61: 271–280.

Sprent, P. and Smeeton, N.C. 2001. Applied Nonparametric Statistical Methods. 3rd ed. Chapman and Hall/CRC Texts in Statistical Sciences, Boca Raton, FL.

Stacey, A.J., Truscott, J.E., Asher, M.J.C. and Gilligan, C.A. 2004. A model for the invasion and spread of rhizomania in the United Kingdom: Implications for disease control strategies. *Phytopathology*, 94: 209–215.

Trumper, E.V. and Gorla, D.E. 1997. The sensitivity of the ordinary runs test for evaluating the spatial pattern of infected plants. *Ecologia Austral*, 7: 79–86.

Turechek, W.W. 2004. Nonparametric tests in plant disease epidemiology: Characterizing disease associations. *Phytopathology,* 94: 1018–1021.

Turechek, W.W. and Madden, L.V. 2000. Analysis of the association between the incidence of two spatially aggregated foliar diseases of strawberry. *Phytopathology,* 90: 157–170.

Turechek, W.W. and Stevenson, K.L. 1998. Effects of host resistance, temperature, leaf wetness, and leaf age on infection and lesion development of pecan scab. *Phytopathology*, 88: 1294–1301.

Van der plank, J.E. 1982. Host-pathogen interaction in plant disease. Academic Press, New York, USA.

Venkatasubbaiah, P., Sutton, T.B. and Chilton, W.S. 1995. The structure and biological properties of secondary metabolites produced by Peltaster fructicola, a fungus associated with apple sooty blotch disease. *Plant Disease*, 79: 1157–1160.

Willocquet, L., Elazegui, F.A., Castilla, N., Fernandez, L., Fischer, K.S., Peng, S., Teng, P.S., Srivastava, R.K., Singh, H.M., Zhu, D., and Savary, S. 2004. Research priorities for rice pest management in tropical Asia: A simulation analysis of yield losses and management efficiencies. *Phytopathology*, 94: 672–682.

Willocquet, L., Lebreton, L., Sarniguet, A. and Lucas, P. 2008. Quantification of within-season focal spread of wheat take-all in relation to pathogen genotype and host spatial distribution. *Plant Pathology*, 57: 906–915.

Willocquet, L., Savary, S., Fernandez, L., Elazegui, F.A. and Teng, P.S. 1998. Simulation of yield losses caused by rice diseases, insects, and weeds in tropical Asia. IRRI Discussion Paper Series no 34. IRRI, Los Baños, Philippines.

Willocquet, L., Savary, S., Fernandez, L., Elazegui, F.A. and Teng, P.S. 2000. Development and evaluation of a multiple-pest, production situation specific, simulation model of rice yield losses in tropical Asia. *Ecological Modelling*, 131: 133–159.

Xu, X. 2006. Modeling and interpreting disease progress in time. *In*: The epidemiology of plant disease. Cooke B.M., Gareth, Jones D. and Kaye, B. (*eds.*). Springer, Dordrecht, The Netherlands, pp. 215–238.

Xu, X.M. 1999. Effects of temperature on the latent period of the rose powdery mildew pathogen, Sphaerotheca pannosa. *Plant Pathology*, 48: 662–667.

Yuen, J.E. and Hughes, G. 2002. Bayesian analysis of plant disease prediction. *Plant Pathology*, 51: 407–412.

Zhang, X.S., Holt, J. and Colvin J. 2001. Synergism between plant virus: A mathematical analysis of the epidemiological implications. *Plant Pathology*, 50: 732–746.

Zivanov, D., Jevtic, R., Tancic, S., Vasiljevic, S. and Masirevic, S. 2014. Control of winter forage pea diseases by pea-oat intercropping under filed conditions. *Pestic Phyttomed*, 29(2): 131–136.

CHAPTER

4 Spatial Patterns Analysis in Botanical Epidemiology

4.1. WHAT IS SPATIAL PATTERN'S?

There are many ways to describe plant communities and the different plant species of which they are comprised. One approach is to study the spatial patterns of plants; that is, the physical arrangement or distribution of species within the community.

Spatial patterns are often described in terms of the two-dimensional location of individual plants (*e.g.*, latitude and longitude), analogous to the (X, Y) coordinates of the plant. Spatial patterns are useful descriptors of horizontal structure in plant communities.

However, their chief value lies in the fact that they can provide useful information on processes acting on a given community (Hubbel *et al.,* 2001; Wiegand and Moloney, 2004). Their importance may be realized directly as in the provision of biomass or indirectly through modification of microclimate (Mathur and Sundaramoorthy, 2008). Further Darbemamiet *et al.,* (2011) have demonstrated that uses of dispersion indices seem to be convenient decision making methods for management programs because of their easy calculation procedure and simple result.

Though their interpretation does not necessarily indicate cause-effect relationships (a given spatial pattern may reflect two or more process), it nonetheless enables us to restrict the spectrum of possible acting processes and thus allows us to frame more specific questions which can be addressed with more direct observation or experiments. This approach is particularly useful in little known regions (Lopez *et al.,* 2010). Moreover, spatial pattern analysis has been employed in different ecosystem and with many species belonging to all life forms.

Uses of attributes of spatial pattern analysis in epidemiology not only related with analysis and describing the spatial patterns of disease, but also to identify and specify the physical and biological mechanism that account for the evaluation of spatial pattern over time.

Spatial pattern of plant pathogen are seldom truly random or regular in nature. Diseased plants and pathogen are often found to be aggregated Johnosn *et al.* (1988). The pattern of disease is not static; powdery mildew on wheat leaves changed from an aggregated to random pattern as the epidemic progressed, while citrus canker changed from a non-aggregated pattern in early stage of disease development to an aggregated pattern after secondary spread had occurred (Rouse *et al.*, 1981).

An aggregated, rather than random, pattern can be expected for many polycyclic plant diseases. Many spatial statistical methods have been developed to quantify and model spatial patterns of forest diseases (Reich and Lundquist, 2005). Typically, the locations of unaffected, diseased and dead hosts are analyzed for spatial pattern; usually, such populations of hosts are represented by various spatial point data derived through field sampling or mapping from remotely sensed imagery.

As such, spatial point pattern analysis has been intensively investigated to reveal the scale, extent, and dynamics of mortality patterns and test potential hypotheses related to spatial mechanisms of disease spread.

For example, Batista and Maguire (1998) modeled the spatial structure of host mortality as a "thinning" process on the existing live hosts in tropical stands. Cole and Syms (1999) applied various spatial analysis techniques to differentiate between climate-induced and pathogen-induced mass mortalities of the kelp Ecklonia radiata in north-eastern New Zealand.

Kelly and Meentemeyer (2002) quantified the scale and extent of spatial clustering of oak mortality in a selected portion of a hardwood forest associated with Sudden Oak Death forest disease using Ripley's K-function in California.

Nearest Neighborhood and Ripley's K function ware quantified by Liu *et al.* (2007) for Oak host mortality caused by *Phytophthora ramorum*. Rouse *et al.* (1988) have utilized many spatial indices (Morisits's index of dispersion, index of clumping, negative binomial distribution and variance to mean ratio) to quantify the spatial pattern of downy mildew in hop yards, they used semi-variogram distance method to measure the degree of correlation between population counts from spatially contiguous hills at successively larger distance. Further, they reported that the distance over which the semi-variogram continues to increase the range of influence of diseased hill on nearby hill and thus, this tool could indicate the spread of infections and direction of spread.

In plant pathology, frequency distribution, indices of aggregation (variance to mean ration, Lloyd's index of patchiness, Morisita's, Taylor's Power law, Iwao's patchiness regression), quadrate variance, spatial autocorrelation, runs, doublets and geostatistics are utilized by various workers like Alby *et al.* (1983); Noe and Cambell (1985); Alston and Schmitt (1987); Hughes and Madden, (1993); Hughes *et al.* (1995); Tanne *et al.* (1996); Xiao *et al.* (1997); Hughes and Maddan, (1998); Wallace and Hawking (1994); Reich and Lundquist (2005); Liu *et al.* (2007); Gent *et al.* (2008); Jones *et al.* (2011) Spolti *et al.* (2012) and Kamu *et al.* (2015).

Description and quantification of spatial patterns of plant disease are integral to botanical epidemiology, providing insight into biological processes associated with pathogen dispersal and disease development, and allowing for the development of testable hypotheses to explain observed patterns (Gent *et al.*, 2008). Quantification of spatial pattern is prerequisite to designing statistically sound sampling methods (Hughes, 1999; Madden and Hughes, 1999; Turechek and Miadden, 1999; Gent *et al.*, 2006 and 2007) and controlling the effects of aggregation in analysis of data from designed field experiments (Madden *et al.*, 2002), and also has been used in development of crop loss models (Gent *et al.*, 2008).

Many different methods have been used to characterize spatial aggregation for disease incidence data, including:

- Fitting of the beta-binomial distribution (Hughes and Madden, 1993; Madden and Hughes, 1995 and Gottwald *et al.*, 2007)
- Variance–mean relationships (Hughes and Madden, 1992; Madden and Hughes, 1995)
- Spatial autocorrelation (Campbell and Madden, 1990; Gottwald, 1995; Madden and Hughes, 1995 and Gottwal *et al*, 2007)
- Geostatistical methods (Chellemi *et al.*, 1988; Stein *et al.*, 1994; Gottwald *et al.*, 1996), and
- Distance class methods (Gray *et al.*, 1986; Nelson *et al.*, 1992; Gottwald, 1995; Nelson, 1995; Ferrandino, 1996; 1998).
- To quantify spatial paternity in botanical epidemiology the data's are usually collected through two ways.
- First, disease incidence is appraised within a sampling unit consist of a group of plants at a particular spatial point (often called a quadrat); a number of such quadrats would be randomly taken over space or time.
- Second, disease is evaluated on all individual plants or on a subset of plants; in this case, spatial location is recorded as well as disease incidence.

- Sometimes, a hybrid of these two methods can also be used, *i.e.* disease is assessed within a quadrat but a number of such quadrats is selected on a set of predefined spatial locations. Usually, the information extracted from the data is positively related to the amount of efforts put into the disease assessment (Maanen and Xu, 2003).

Basically for plant disease pathological and epidemiological studies spatial levels includes *N* sampling units (groups), *n*, individuals per sampling units, and *X* diseased individuals per sampling unit. Sampling units could be groups of plants (*e.g.,* quadrates), plants or plant units (*e.g.*, branches), individual plants (*e.g.*, *n* plants per sampling unit [quadrate]), leaves (*e.g. n* leaves per sampling unit [plant]).

Xu and Ridout have conducted two different studies on effects of initial epidemic conditions (initial number of infected plants and their spatial pattern), sporulation rate, and spore dispersal gradient on the spatiotemporal dynamics of plant disease epidemics. They demonstrate the importance of initial epidemics conditions and spore dispersal gradient in influencing intraclass correlation (based on 5 × 5 quadrat) and spatial autocorrelation (1998) and effects of quadrate size and shape, initial epidemic conditions and spore dispersal gradient on spatial statistics of plant disease epidemics (2000).

Study or quantitative evaluation of spatial dynamics related with plant diseases or with epidemic may be related or liked with certain issues:

- Is the disease is aggregated?
- How does the degree of aggregation fluctuate with time
- Is there any spatial dependence in development of disease?
- Can such spatial dependence be quantified?
- What is the rate of disease spread, disease gradient or spore dispersal gradient?
- What is the impact of disease aggregation on disease management and
- What factors have caused the observed spatial pattern

Disease incidence and its variance, intra-cluster correlation beta-binomial distribution, index of dispersion, power-low relationships and spatial autocorrelation are the major approaches generally utilize for such attributes of spatial analysis of plant epidemics.

Pathogen movement can result in heterogeneous distribution of the disease within the affected crop. Spatial patterns may be classified into uniform, random and clustered or clumped. In a uniform pattern there is a regular arrangement of infected and healthy plants. While random means that all distinguishable arrangements of infected and healthy plants are equally possible, or that in every point in the crop there is the same probability of a plant being infected. In a clustered pattern, every plant in the field does not have an equal probability of being infected so that a diseased plant increases the probability of nearby plants being infected (Trumper and Gorla, 1997).

Knowledge related with spatial pattern of pathogen and plant disease can be utilized in different ways. The detection of disease pattern in the field can be used to develop and efficient sampling plan. Basically it is well known fact that a larger aggregation requires larger sample sizes.

Spatial paternities can enhance our understanding of the dynamics of disease epidemics and determination of the mechanics of disease spread. For example, random pattern of infected plants would suggest that the pathogen does not spread within the field or in other words, no secondary transmission occur.

Characterization of spatial patterns and spatial dynamics of plant disease can provide insights into important epidemiological processes, especially those related to sources of inoculum, mechanisms of propagule dissemination, and reproductive strategies of the pathogen population (Wu and Subbarao, 2004).

For example spatial epidemiological aspects of brown rot development in fruit orchards have received considerably less attention than the aforementioned temporal dynamics. Two-dimensional spatial patterns, such as aggregation of symptomatic hosts at the orchard scale, have been investigated by van Leeuwen *et al.* (2000) and Xu *et al.* (2001) in pome fruits. In general, these studies showed that infected hosts were spatially clustered within orchard rows, with variable intensity of disease among individual hosts.

Although spatial patterns were more pronounced in pear (*Pyrus communis*) than in apple (*Malus domestica*) orchards, wounding of the fruit by birds, insect damage or growth cracks was considered an influential source of the spatial pattern in both species.

Everhart *et al.* (2011) proposed new approach to quantify spatial patterns of plant disease in complex fruit host canopies using point pattern analysis. They reported that symptoms of brown rot in *Prunus cerasus* caused by *Monilinia laxa* infection were generally aggregated within the canopy volume, but there was no consistent pattern for one symptom type to be more or less aggregated than the other. Analysis of spatial association among symptom types indicated that previous year's twig cankers may play an important role in influencing the spatial pattern of current year's symptoms. This observation provides quantitative support for the epidemiological role of twig cankers as sources of primary inoculum within the host.

In general a population will have of three basic spatial patterns: (1) regular (or even, uniform, negative contagious) pattern in which individuals within this population are uniformly spaced; (2) random (or chance) pattern in which all individuals have an equal chance of living anywhere within an area; and (3) clumped (or aggregated, patchy, contagious) pattern in which individuals have a higher probability of being found in some area than in others (Taylor and Taylor, 1978; Molles, 1999 and Condit *et al.* 2000).

These three basic spatial patterns of population can be commonly described by three probability distribution: (1) the regular pattern is quantified by the binomial distribution of which the variance (σ^2) is less than the mean (μ), *i.e.*, $\sigma^2/\mu < 1$; (2) the random one is described by Poisson with the variance equal to the mean, *i.e.* $\sigma^2/\mu = 1$; and (3) the clumped one is modeled by the negative binomial distribution that has the mean $\mu = kp$ and the variance $\sigma^2 = kp\,(1+p)$, and thus $\sigma^2/\mu = 1+p>1$ (Yin *et al.*, 2005).

Ludwig and Reynolds, (1988) have described them as

1. Poisson distribution ($\sigma^2 = \mu$) for random pattern
2. Negative binomial distribution ($\sigma^2 > \mu$) for clumped pattern, and
3. Positive binomial ($\sigma^2 < \mu$) for uniform pattern

There are many statistical tools available for quantification of spatial distribution. Madden *et al.* (1982) evaluated ordinary runs, original doublets, and corrected doublets for detecting the spatial pattern of plant virus diseases through simulation of random and clustered patterns. According to them ordinary runs test is the most satisfactory in terms of misclassification chance.

For ordinary runs test we can use various symbols, consider the following pattern of 10 symbol representing a crop row with disease (+) and healthy plants (-) + - - - + + - - ++. Here, there are five runs, reading left to right plants, the expected number of runs (E((U)) can be calculated through

$$E\,(U) = 1 + \frac{2m\,(N-m)}{N}$$

where m is the number of diseased plants, N is the total number of plants in the sample and U is the number of runs and the standard deviation of U can be quantify through

$$s(U) = \sqrt{2m\,(N-m)\frac{2m\,(N-m)-N}{N^2\,(N-1)}}$$

The observed number of runs will be less the E (U) if there is clustering of diseased plants and this can be tested a standard normal deviated calculated as

$$Z = \frac{U - E(U)}{s(U)}$$

A value of Z<-1.64 indicates clustering (P<0.05).

The null hypothesis (H_0) of runs test states that a number of observed runs is not significantly lower than the expected number of runs obtained from a randomly distributed population, and the alternative hypothesis (H_1) states that the observed number of runs is significantly lower than the expected number of runs obtained from a randomly distributed population.

Trumper and Gorla (1997) in their simulation study have concluded that when a random pattern was simulated, the probability of rejecting the null hypothesis was almost unaffected by the sampling unit size and slightly decreased with disease incidence. However when clustered patterns generated, the probability of error clearly decreased both with disease incidence and size of the sampling unit.

Gent *et al.* (2006) had performed the ordinary run anlaysis to characterize large scale patterns of diseased cones among plants in a transect and for this test they considered plant as diseased if at least one diseased cone was observed in the sampling unit.

4.1.1. Spatial Heterogeneity

Information's of spatial heterogeneity of disease incidence are very much useful for sampling methodology, treatments comparison, yield loss predictions and assessment of spatiotemporal disease dynamics. Hughes *et al.* (1997) suggested that spatial heterogeneity is also useful for precise determination of incidence at lower level in a spatial hierarchy (*e.g.* leaf) and prediction of incidence at a higher level (*e.g.,* plant) is dependent on the degree of heterogeneity of incidence at the lower scale.

In particular, as the degree aggregation increased at the lower scale, the incidence of disease at the higher scales decreases. Further, the relationships between incidences at two levels can be exploited to improve sampling methods and also to understand the temporal dynamics of disease at multiple hierarchical levels (McRobets *et al.*, 1996).

Characterization of spatial heterogeneity depends on the methods used for collection data on diseased and healthy individuals. If cluster sampling is used, where N sampling units are obtained and the disease status of every individual within the sampling units (groups) is determined then it represent that the data using discrete statistical distribution for random or over dispersed data. Madden and Hughes (1999) has explored the relationships between disease incidences at two levels in a spatial hierarchy.

In spatial hierarchy studies P_{Low} (expected probability that an individual at the lower level in a hierarchy is diseased) and this can be estimated directly from information on diseased individuals.

Similarly P_{High} (probability that a sampling units has one or more diseased individuals) and this can be directly estimated as the proportion of sampling units with one or more diseased individuals. In case of binomial distribution is appropriate, the above relationship between incidence at the two levels can be explain as

$$P_{high} = 1 - (1 - P_{Low})^n$$

$$P_{high} = 1 \prod_{i=0}^{n-1} \frac{1 - P_{Low} + i\theta}{1 + i\theta}$$

Where Π is the product function, *I* is an index of integers from 0 to n-1, and θ is the aggregation parameters of the distribution. The moment estimate of θ can be quantify by using variance S^2 of *X* (each value is an integer in the range of 0 to n).

$$\theta = \frac{S^2 - n\hat{p}_{Low}\,(1 - \hat{p}_{Low})}{n^2\,\hat{p}_{Low}\,(1 - \hat{p}_{Low}) - S^2}$$

White and Gilligan (1998) had quantified the involvement of spatial heterogeneity due interaction between the dynamics and movement of three species systems involving host, parasites and hyperparasites. They have constructed a model based on following assumptions (a) growth and movement for each species are not coupled, (b) movement is essentially random and (c) movement is not affected either by the presence of a second species or by the local density of the species. They utilized these assumptions with three nonlinear reaction diffusion equations which describe the rate of change in local densities of each species such that, the rate of changes in local density is equal to the growth and decay of the population plus the random movement and this related to host (H), parasite (P) and hyperparasite (Q)

$$\frac{\partial H}{\partial t} = A\,(H,P,Q) + D_H \frac{\partial^2 H}{\partial x^2}$$

$$\frac{\partial P}{\partial t} = B\,(H,P,Q) + D_P \frac{\partial^2 P}{\partial x^2}$$

$$\frac{\partial Q}{\partial t} = C\,(H,P,Q) + D_Q \frac{\partial^2 Q}{\partial x^2}$$

Where A, B and C represent the intra and interspecific dynamics for each of the host, parasite and hyperparasite, and D_H, D_P, and D_Q are diffusion coefficients. Interspecific components include birth, death or net growth rates. Their simulation study revealed that when there is strong density dependence limiting host growth, the pattern is stable over time, whereas with less stable underlying host growth, the pattern varies with time.

Related to spatial pattern in epidemic development the rate at which new infections are produced, can be quantify by following formula.

Rate = (Spatial decay due to location of inoculums away from host) (Time decay due to change in host susceptibility and inoculums infectivity) X (Delay in onset of infection).

4.1.2. Dispersal Gradient

Disease transmission depends on two primary processes: dispersal of infective proagules and successful infection of host by those propagules (Roche *et al.*, 1995).

In case of sessile hosts, the resultant process will be aggregated.

Dispersal process largely depends on distance from the source of inoculums and consequently it develops the disease foci (primary or secondary).

Separation of distance-based effects of dispersal from factors affecting successful infection is an important aspect in integration of spatial patterns into ecological models of disease.

Knowledge about the disperse ways of infectious units in space can be an essential component of spatially explicit epidemiological models. Success of long distance dispersed disease can be evaluated through evaluating the data of contact distribution (*i.e.,* the probability that a disease propagule released at one location will be deposited at another location).

Further, quantitative relationship between spore deposition and disease incidence is also an important attribute and generally a nonlinear relationship between number of spores and proportion of plants that become infects is likely because a minimum number of spores may be required for successful infection and, at some higher number of spores, increased spore deposition may not lead to an increase in frequency of infection.

Spore dispersal gradients and disease incidence gradients need not correspond to one another, as the conditions which are favorable for spore dispersal may not be favorable for infection.

Roche *et al.* (1995) with *Silene alba- Ustilago violacea* host pathogen system have concluded that the limited dispersal of spores is likely to be important in the aggregation of diseased plants in nature and such aggregation of inoculum in turn can increase or decrease disease spread depending on how much inoculums is needed to cause infection.

Willocquet *et al.* (2008) quantified the spore dispersal and disease gradients in powdery mildew of strawberry (*Fragaria x ananassa*) caused by *Podosphaera aphanis* under two production system (glasshouse and a plastic tunnel). They found disease gradient in the tunnel became shallower as epidemics developed and were associated with decreasing slope estimates overtime.

For gradient characterization (to test the effects of distance from source, orientation and their interaction of the number of lesion counted on trap plant) mixed model analysis of variance conducted by Schabenberger and Pierce (2002), Garrett *et al.* (2004) and Willocquet *et al.* (2008). For gradient analysis Willocquet *et al.* (2008) work can be use to quantify the number of powdery mildew lesions per trap plant with following equation

$$N_1 = 1n \left(\frac{N_{les} + 0.1}{N_{1t} \; x \; N_{SOU}} \right)$$

Where N_{les} is the number of lesions counted per trap plant, N_{1t} is the number of leaves per trap plant, and N_{SOU} is the number of spores counted per source plant after the exposure period.

Rieux *et al.* (2014) provided long distance wind dispersal of spores in a fungal plant pathogen and estimated anisotropic dispersal kernels (*i.e.,* distribution of dispersal distances). The magnitude and frequency of Long Distance Dispersal (LDD) events, may affects the shape of tail of the dispersal kernel.

Three complementary approaches are generally used to study the wind dispersed propagules such as pollen, seeds or spores. First approach relies on mechanistic models that include a fine modeling of physical process contributing to the release, transport and deposition of propagules (Tufto *et al.*, 1997; Aylor, 2003, Soubeyrand *et al.*, 2008 and Burie *et al.*, 2011). The second approach is a indirect approach related with genetic differentiation among populations or individuals. The third approach is based on direct measurements of dispersal distances by tracking propagule movement. In principle, dispersal kernels can experimentally be estimated from the real-time tracking of propaguls (Lagrangian methods) or from the amount and/or diversity of propagules observed at different distance from a source (Eulerian approach).

Basically Lagrangian methods have mostly been applied to animals or animal-dispersed propagules, using mark/recapture or tracking designs. They can hardly be applied to wind-dispersed pathogen propagules because of small size. On the other side, Eulerian approach is more suited in determining how many propagules originating from an isolated or identifiable source of inoculums arrive at specific distant location

Disease progress is generally measured as rate of changes over time and variation in levels of disease form a source is usually expressed as a gradient in space. In the time function, Vand der Plank (1963) used $r = (1/t)$ [logit (x_i)], where r is the average apparent infection rate, t is time in the interval between x_1 and x_2 (disease proportions), and logit $x = log_e (x/[1\text{-}x)]$.

For quantification of focal epidemics a Isopathic Velocity concept was provided where isopath is a contour in space of constant disease level and as focal epidemic progress, isopaths expands outward from center of the focus (Sackett and Mundt, 2005). The rate of movement of an isopath in particular direction it its velocity, expressed in units of distance per time (*e.g.*, metres/day). Basically epidemics with constant velocity nature are believed to expand as traveling wavers, with disease gradients (plots of disease levels *versus* distance from the source) that maintain a constant shape after the buildup period is past.

On the other hand, dispersive wavers are exhibited by epidemics with increasing velocity; disease gradients become shallower as the epidemics progresses (Ferraandion, 1993 and 1996).

The extend of disease from an primary source of infection depends on the dispersal of a propagule from an infected source to an uninfected host at some distance. The probability per unit area of target left that a spore released from a mother lesion will cause a daughter lesion at distance *r* is termed is contract distribution, CD(*r*).

Mollison (1977 and Ferraandion 1993) have imposed a condition in which an epidemic can take form of a traveling wave only if the tail of the contact distribution is an exponential order such that

$$\lim_{r\to\infty}[CD(r)\exp(kr)] \rightarrow o\ for\ some\ k > 0$$

Minogue and Fry (1983 and 1986) assumed that as propagules moved in both directions away from the source, a constant fraction, *p,* of these propagules was deposited per plant. The deposit on the nth plant, CD_n is $(CD_{n\,=\,p(1-p)}{}^{n}/(2\text{-}p)$. If the conditions expressed in above equation hold true, then the contact distribution can be characterized by the exponential length scale, $1/k_{max}$ in which k_{max} is the largest value of *k*.

4.1.3. Spatiotemporal Dynamics of Airborne Inoculum

For airborne inoculums, analysis of spore dispersal mechanism, spore and disease gradient, and long range spore dispersal are major approaches, while at filed scale, spatial analysis have primarily concentrated on soil borne inoculums (McCartney *et al.*, 1998).

Carisse *et al.* (2007) have identified the knowledge gaps related with quantitative aspects of spatial and temporal relationship that develop between airborne inoculums and disease intensity during the course of aerially spread epidemics. They have also identified the causes of such space which are related with difficulty in measuring airborne fungal spores and the lack of knowledge on the relationship between airborne inoculums and disease development

For example airborne inoculum plays a crucial role in the development of Botrytis leaf blight epidemic (cause by *Botrytis squamosa*) on onion. This pathogen overwinters as sclerotia and in spring, these sclerotia are serving as primary sources of inoculums germinate and produce conidia that are dispersed through wind.

Under favourable weather conditions, these conidia can infect onion leaves causing lesions and maceration of leaf tissue, resulting in leaf dieback and blighting. The secondary spread of the disease is caused by airborne conidia produced on necrotic tissues.

Sutton *et al.* (1978) have reported a significant correlation between airborne conidium concentration severity of this disease, while, Vincelli and Lorbeer (1988) reported that large increases in Botrytis leaf blight severity were associated with days with more than 10 conidia per m^3 of air.

Carisse *et al.* (2005) have assessed the usefulness of airborne inoculums measurements as an aid for improving Botrytis leaf blight management. Further, Carisse *et al.* (2007) have studied the spatiotemporal pattern of Botrytis leaf blight in spontaneous, unmanaged epidemics and under commercially managed filed conditions; they characterized the spatiotemporal structure of this epidemic with monitoring weekly lesion density (LD) and aerial conidium concentration (ACC).

From academic point of view its essential to describe their sampling methodology, which is very important for such types of airbiological works, they measured LD and ACC weekly in each quadrat (experimental plot 80 x 60 m and further split it into 48 quadrats of 10 x 10 m; commercial site 125 x125 m). In each quadrat LD and ACC were measured on six occasion and ACC was measured at the center of each quadrat using rotating-arm impaction spore samples placed at 10 to 15 cm above the canopy. Height of samplers was adjusted weekly according to onion growth and spraying equipment. The sampler ran from 10:00 a.m. to 12:00 a.m on each sampling day, authors have chosen this sampling period because of *B. squamosa* conidia concentration usually reach a peak between 9:00 a.m. and 12: a.m. After exposure, the number of conidia per rod was counted under a microscope and counts were transformed to conidium concentration of conidia per m^3. On each sampling day, the number of lesions per leaf was counted on all leaves of five randomly chosen plants per quadrat, and disease data were expressed as LD (average number of lesions per plant).

Taylor's power law was utilized to describe the relationship between mean (*M*) and variance (*V*) over the sampling periods for LD and ACC, separately. With using equation $1n\ (V) = 1n\ (A) + b1n\ (M)$, *V* is the sample variance, *M,* is the sample mean, and $1n\ (A)$ and *b* are parameters. In this method a random distribution is indicated by $1n\ (A) = 0$ and $b = 1$, however, when value of $b > 1$ indicates heterogeneity among quadrats. In (*A*) and *b* quantified through ordinary least square regression.

4.1.4. Basic Reproduction Number

The basic reproduction number (R_0) defined as the total number of infections arising from one newly infected individual introduced into a healthy (disease-free) host population.

It is also denote as generation to generation multiplication factor of the pathogen population at infinitesimally low pathogen density (means at a density when propagules produced by pathogen will not come in contact with host individuals previously infected by the pathogen).

For quantification of R_0 knowledge about pathogen life cycle and its interactions with the host plant are essential. R_0- having an example importance in epidemiological studies that includes:

- It defines the threshold, R_0 =1, between the epidemic (increase in pathogen density), R_0>1 and no epidemic, R_0<1. When each mother infection caused on average, more than one 'daughter' infection, it exceed 1, so the pathogen or disease population density will increase.
- It is helpful for comparsion of epidemiology, events in life cycles of a pathogen and the interactions with the host into one metric signifying the pathogen reproduction capacity.
- This also can be used to determine the ultimate or steady state value of the disease or pathogen density in host population (Jeger *et al.*, 2002 and Madden *et al.*, 2007).
- It is a simple tool to evaluate different management methods with one coherent experiment.

Vanderplank (1963) introduced the corrected basic infection rate, R_c which is the number of new infections caused per time-unit b one infective infection and *I* is the length of the infectious period which has the length of time. In botanical epidemiology the uses of R_0 have been reviewed by van den Bosch *et al.* (2008). They suggested that in complex dynamics due to presence of more than one host cultivar, large spatial scales with multiple fields and due to propagator host nursery system, the quantification of R_0 is very complicated. However, in plant pathology there basic procedure is being utilized for R_0, which are:

- Develop a nonlinear model for pathogen dynamics (generally having a system of nonlinear differential equations.
- Quantify the steady-state and,
- Derive a criteria from steady state that shows the combination of model parameter in which the pathogen population density is larger than zero.

However, van den Bosch *et al.* (2008) provided with an example that these methods do n167

ot necessary reflects the R_0. That is actually the expression of the basic reproduction number can only be determined retrospectively finding a biological interpretation for the components of the expression that convinces the researcher that it is basic reproductive rate.

4.2. WHAT IS HYPOTHESIS

The initial step in pattern detection in community ecology often involves testing the hypothesis that stated that distribution of the number of individual per sampling unit is random. If this hypothesis is rejected, then the distribution may be in the direction of clumped (usually) or uniform (rarely).

If the direction is toward a clumped dispersion, agreement with the negative binomial may be tested and certain indices of dispersion, which are based on the ration of the variance to mean, may be used to measure the degree of clumping. Various steps for detecting spatial pattern are presented in flow diagram (Fig. 1).

A benchmark hypothesis to be contrasted is the Complete Spatial Randomness (CSR) which asserts that the points are randomly and independently distributed in the region of interest. Rejection of CSR is a minimal prerequisite for any serious attempt to model an observed pattern as CSR operates as a dividing hypothesis between regular and aggregated patterns. Complete spatial randomness acts as a baseline hypothesis against which to assess whether observed patterns are "regular", "clustered" or "random" (Bailey and Gatrell, 1995).

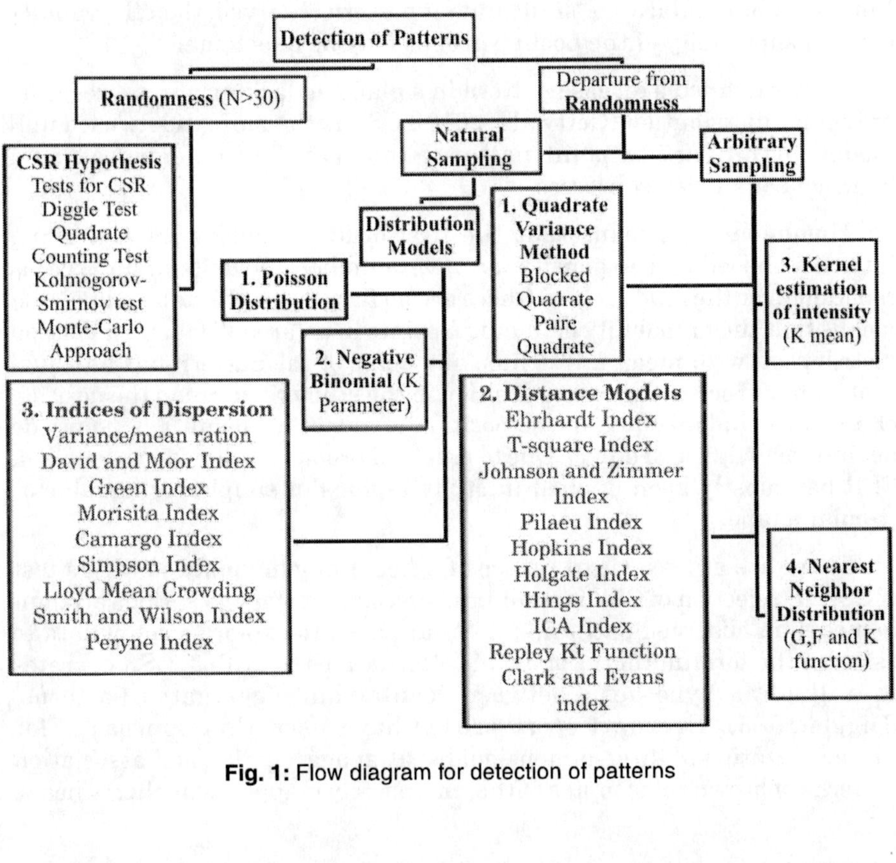

Fig. 1: Flow diagram for detection of patterns

Complete spatial randomness (CSR) is the simplest theoretical model for the quantitative description of a spatial pattern. For CSR, the events in region R constitute a partial realization of a homogeneous Poisson process (Diggle, 1983).

There are two important aspects of this model. Firstly, homogeneous implies that the intensity, λ is constant over the study region, so each of the n events is equally likely to occur at any point within R. Secondly, Poisson implies that there are no spatial interactions among the events, so the n events are located independently of each other.

A pattern for which CSR is rejected seldom merits any further formal statistical analysis. If a systematic pattern does exist, one may be interested in the particular nature of the departure from the CSR model. If applicable, one may also fit other

Spatial models to represent the clustering or regular pattern that exists. In this analysis of spatial patterns of homogeneous pooling groups, most patterns may be expected to show spatial clustering for a small distance and it might be interesting to determine over what spatial scale this occurs. But if CSR or regularity is the dominant pattern observed, then the validity of the homogeneity of the pooling groups may be questioned.

CSR is defined a disease pattern in a planting lacking any aggregation or regular distribution (Gottwald, 2010). CSR is the most widely used null model for univariate point patterns that can be implemented as a homogeneous Poisson process.

Homogeneous means that the first-order intensity λ (Measuring intensity) based on the density (or mean number of events in an area) is constant over the study region (there are no first-order effects), and Poisson means that the probability of finding k points in an area W follows a Poisson distribution with mean λW. Thus, any point of the pattern has an equal probability of occurring at any position in the study region, and the position of a point is independent of the position of any other point (*i.e.*, points do not interact with each other). Due to practical problems with edge correction, CSR has mostly been applied in study regions of simple rectangular or circular shape.

There are at least three reasons for beginning an analysis with a test of CSR: 1. rejection of CSR is a minimal prerequisite for any serious attempt to model an observed pattern; 2. tests are used to explore a set of data to assist in the formulation of plausible alternatives to CSR; 3. CSR operates as a dividing hypothesis between regular and aggregated patterns, (Diggle,1983). Everhart *et al.* (2011) have used this approache for characterization of three-dimensional spatial aggregation and association patterns of brown rot symptos within intensively mapped sour cherry hosts.

4.2.1. Heterogeneous Poisson Process

If a pattern is not homogeneous, the null model of CSR is not suitable for exploration of second-order characteristics (Measuring spatial dependence –based on distances of points from one another), and a null model accounting for first-order effects has to be used to reveal "true" second-order effects.

The heterogeneous Poisson process is the simplest alternative to CSR if the pattern shows first-order effects. The constant intensity of the homogeneous Poisson process is replaced by a function $\lambda(x, y)$ that varies with location (x, y), but the occurrence of any point remains independent of that of any other.

The intensity function $\lambda(x, y)$ determines the process completely, and numerical implementation of this null model is a matter of finding an appropriate estimate of the intensity function. The numerical approach suggests a simple method to implement the heterogeneous Poisson process using the moving-window estimate $\lambda^{R\wedge}$ of the intensity function $\lambda(x, y)$: a provisional point is placed at a random cell (x, y) in the study area, but this point is only retained with a probability given through $\lambda\hat{}R\ (x, y)$. This procedure is repeated until n points are distributed. The moving window estimator $\lambda\hat{}R\ (x, y)$ involves a decision on an appropriate radius R of the moving window. Because the bandwidth R is the scale of smoothing, possible departure from this null model may only occur for scales $r < R$, and for small moving windows it will closely mimic the original pattern, whereas a large moving window approximates CSR (Pélissier and Goreaud, 2001).

In some cases, the nature of the data and the strength of trends in the observed pattern may make such judgment relatively straightforward. In other cases, this may be difficult and open to debate and interpretation.

Typical biological situations for application of a heterogeneous Poisson process are presence of exogenous factors (*e.g.*, soil, topography, rocks, etc.) or obstacles that cause irregularly shaped study regions. In fact, a simple variant of the heterogeneous Poisson process can be used as alternative to avoid edge correction for homogeneous point patterns in irregularly shaped study regions: the intensity $\lambda(x, y)$ is zero outside the study region and constant inside.

4.2.2. Random Labelling

Random labeling is a somewhat different approach to correct for underlying environmental heterogeneity that can be used where a "control" pattern is available to act as surrogate for the varying environmental factor. The assumption of univariate random labeling is that the pattern of controls was created by the same stochastic process as the primary pattern ("cases").

Therefore, the n_1 cases represent a random subsample of the joined pattern of the n_2 control points and n_1 case points. The test is devised by computing the univariate K function for the observed cases, then randomly re-sampling sets of $n1$ points from the $(n1 + n_2)$ points of the cases and controls to generate the confidence limits. Note that this null model makes sense only if there are many more controls than cases. Univariate random labeling is closely related to bivariate random labeling (see below) and has been applied to investigate competitive thinning (*e.g.*, Kenkel 1988, Moeur 1993; Batista and Maguire 1998) under the null hypothesis that the survivors are no different from a random draw of the initial cohort.

4.2.3. Poisson Cluster Process

The Poisson cluster process explicitly incorporates a clustering mechanism. Parent events form a CSR process and each parent produces a random number of offspring according to a probability distribution f. Offspring are spatially distributed around their parent according to some bivariate probability density g. The final pattern consists of the offspring only.

To avoid edge effects, the parents must be simulated over a region larger than the study region but the offspring falling outside the study region are lost (Bailey and Gatrell, 1995).

Some authors advocate the use of more complex models such as Neyman–Scott and Thomas processes or Matern cluster processes (Diggle 1983; Cressie 1991; Batista and Maguire 1999; Dixon 2002; Wiegand and Moloney 2004). The Neyman Type A and Thomas distribution have been widely used in plant pathology (Stamatellos and Panourgia, 2005).

Both distributions have two parameters which provide measures of the cluster density and average number of individuals per cluster. There are two assumptions associated with these distributions. The first is that the clusters are randomly distributed and second that the number of plants per cluster follows a Poisson distribution. The distinction between these two distributions is that Neyman's distribution models only associate the offspring (new generation) with an individual cluster, while the Thomas distribution models deals with both the parent (mother host) and offspring. The probability function for the Neyman Type A distribution is given by Reich and Davis (1998)

$$P(X = 0) = \exp\{-m_1[1 - \exp(-m_2)]\} \text{ and}$$

$$P\ (X = k + 1) = \frac{m_1 m_2\, e^{-m_2}}{k + 1} \sum_{t=0}^{k} \frac{m_2^t}{t!}\, P(X = k - t), k = 0,1 \ldots..$$

Where m_1 is the number of cluster per unit area, m_2 is the number of individuals per cluster. The Thomas distribution is given by:

$$P\ (X = k) = \sum_{t=0}^{k} \frac{m^t}{t!} \frac{(t\lambda)^{k-t}\, e^{-t\lambda}}{(k - t)!,}\quad k = 1,2$$

Where m is the number of clusters per unit area, λ is the number of offerspring and (1+λ) is the number of individuals.

4.3. TESTING FOR COMPLETE SPATIAL RANDOMNESS

The hypothesis of complete spatial randomness for a spatial point pattern asserts that (i) the number of events in any region $A(A \subseteq R^m,\ m \in N)$ with volume $|A|$ follows a Poisson distribution with mean $\lambda|A|$; (ii) given n events in a region A, the coordinates of locations of these events $(\mathbf{x}_1, \ldots, \mathbf{x}_n)$ are an observed random sample from the uniform distribution on A. In (i), the constant λ is the *intensity,* or mean number of events per unit region.

Most analysis begins with a test of CSR, and there are several good reasons for this. Firstly, a pattern for which CSR is rejected scarcely merits any further formal statistical analysis. Secondly, tests are used as a means of exploring a set of data, rather than because rejection of CSR is of intrinsic interest. Thirdly, CSR acts as a dividing hypothesis to distinguish between patterns which are broadly classifiable as “regular” or “aggregated”.

There are few general methods for testing CSR in two dimensional region against the other assumption:

4.3.1. Quadrant Methods

In these methods, small plots (quadrants) are located in the study region:

(i) Traditionally, these plots are square, but they may be circular, triangular, hexagonal, or any other shape;

(ii) All quadrants must be of equal area;

(iii) Quadrants are usually placed randomly within the study region, but also may be located systematically and

(iv) Quadrants must not overlap.

4.3.2. Distance Methods

This method involves measuring distance either:

(i) Between events and neighbouring events;

(ii) Between sample points and neighbouring events.

Often the distribution theory of test statistic is intractable. Therefore Monte Carlo methods must be applied. Historically, the first CSR tests were Chi-square tests applied to quadrate counts, *i.e.* the number of events in disjoint cells. Then appeared many methods based on various distance measurements between events or between sampled points and the nearest event (Cressie, 1993). Other tests includes Diggle test, Quadrat counting test Kolmogorov-Smirnov test and Monte Carlo tests.

4.3.3. Diggle's Tests of Spatial Randomness

The tests relay on the comparison of an observed distribution of nearest-neighbour distance to the theoretical distribution of expected under the null hypothesis of complete spatial randomness. They involve two cumulative distribution function (c.d.f.s) of nearest-neighbour based on (Gignoux *et al.*, 1999).

(1) The distance w between each point of the observed spatial pattern and its nearest neighbour (Diggle's *G* function)

(2) The distance x between each point of a predefined sampling grid and the nearest point of the observed spatial pattern (Diggle's *F* function)

The theoretical cumulative distribution function, are

$$G_o(w) = 1 - e^{-\pi\lambda w^2}$$

$$F_o(x) = 1 - e^{-\pi\lambda x^2}$$

Where λ is the intensity of the homogeneous Poisson process and the index to H_0. These expression are compared to empirical c.d.f.s calculated from a sample of limited spatial extent. Points located near the border of the mapped plot might have their nearest neighbour outside the plot.

This edge effect causes a systematic error in the estimation of the empirical c.d.f.s. A classical edge-effect correction (EEC) derived from minus sampling, proposed by Diggle (1979) and called the border method by Ripley (1988), relies on the elimination of two groups of points, specifically.

(a) All points that lie closer to the border than to their nearest neighbour inside the plots are discarded,

(b) When estimating the empirical c.d.f.s at distance d, all point lying within the strip of width d from the border of the plot are discarded increasing d leads to discarding a large number of points.

In this method, the mean proportion of discarded point increases when sample size decreases. With the correction, the empirical c.d.f.s are (Diggle, 1983)

$$G_{(w)} = \# (w_i \leq w_i\, b_i > w)/\# \{b_i > w\}$$

$$F_{(x)} = \# (x_i \leq x_i b_i > x)/\# \{b_i\, x\}$$

Where w_i (respectively, x_i) is the distance of point i to its nearest neighbour and b_i the distance of point i to the nearest border of the sample plot. These estimators do not have the same properties (Ripley, 1988). F (x) is unbiased (since its denominator is a fixed number, E [# $\{x_i \leq x_i, b_i > x\}$]/ # $\{b_i > x\}$ = E[F (x)] = F (x)), while G (w) is only ration-unbiased (since it denominator is a random number, e [#$\{w_i \leq b_i > w\}$]/e[#$\{b_i\, w\}$]= g (w).

4.3.4. Quadrat Counting Test

A classical test for the null hypothesis of CSR is the X^2 test based on quadrat counts. The region W is divided into sub-regions ('quadrats') $B_1, \ldots, B_m$ of equal area. The main critique of the quadrat test approach is the lack of information.

This is a goodness of- fit test in which the alternative hypothesis H_1 is simply the negation of H_0, that is, the alternative is that "the process is not a homogeneous Poisson process and there are too many types of departure from H_0.

The usual justification for the classical X^2 goodness-of-fit test is to assume that the counts are independent, and derive a test of the null hypothesis that all counts have the same expected value. The Pearson X^2 test statistic is generally use for getting the significance of randomness.

$$\chi 2 = \sum_j \frac{(n_j - \frac{n}{m})^2}{n/m}$$

$$n = \sum_j n_j$$

The power of the quadrat test depends on the size of quadrats, and falls to zero for quadrats which are either very large or very small. The power also depends on the alternative hypothesis, in particular on the 'spatial scale' of any departures from the assumptions of constant intensity and independence of points. The choice of quadrat size carries an implicit assumption about the spatial scale. The test is two-sided: X^2 too large represent *aggregation* or heterogeneity while X^2 too small represent regularity.

4.3.5. Kolmogorov-Smirnov Test (K-S Test)

In statistics, the Kolmogorov–Smirnov test (K–S test) is a nonparametric test for the equality of continuous, one-dimensional probability distributions that can be used to compare a sample with a reference probability distribution (one-sample K–S test), or to compare two samples (two-sample K–S test).

The Kolmogorov–Smirnov statistic quantifies a distance between the empirical distribution function of the sample and the cumulative distribution functitons of the reference distribution, or between the empirical distribution functions of two samples. The null distribution of this statistic is calculated under the null hypothesis that the samples are drawn from the same distribution (in the two-sample case) or that the sample is drawn from the reference distribution (in the one-sample case).

In each case, the distributions considered under the null hypothesis are continuous distributions but are otherwise unrestricted.

The two-sample KS test is one of the most useful and general nonparametric methods for comparing two samples, as it is sensitive to differences in both location and shape of the empirical cumulative distribution functions of the two samples.

The Kolmogorov–Smirnov test can be modified to serve as a goodness of fit test. In the special case of testing for normality of the distribution, samples are standardized and compared with a standard normal distribution (Table 1).

This is equivalent to setting the mean and variance of the reference distribution equal to the sample estimates, and it is known that using these to define the specific reference distribution changes the null distribution of the test statistic: see below. Various studies have found that, even in this corrected form, the test is less powerful for testing normality than the Shapiro–Wilk test or Anderson–Darling test

The Kolmogorov-Smirnov test is defined by:

H_0: The data follow a specified distribution

H_a: The data do not follow the specified distribution

The Kolmogorov-Smirnov test statistic is defined as

$$D = \max(F(Y_i) - \frac{z-1}{N}, \frac{z}{N} - F(Y_i))$$

Where F is the theoretical cumulative distribution of the distribution being tested which must be a continuous distribution (*i.e.*, no discrete distribution such as the binomial or Possion), and it must be fully specified (*i.e.*, the location, scale and shape parameters cannot be estimated from the data).

K-S test in various fields of plant ecology have been utilize by various workers like Lescure *et al.* (1992); Koenig and Knops (2000); Lin *et al.* (2004); Camarero *et al.* (2005); Freita-Nunes and Rocha (2007); Carneiro *et al.* (2009); Li *et al.* (2010); Everhart *et al.* (2011); Gao *et al.* (2012); Levick *et al.* (2012) and Trifkovic (2013). This test used by Elizabeth *et al.* (2011) for powdery scab effect on potato *Solanum tuberosum* ssp. *Andigena* growth and yield while, Everhart *et al.* (2012) for spatio-temporal patterns of pre-harvest brown rot epidemics within individual peach host canopies

4.3.6. Monte Carlo Approach Concepts

4.3.6.1. *Random test based on Monte Carlo simulation*

A Monte Carlo procedure can be used to evaluate the departure of a pattern from CSR. The Monte Carlo procedure is performed to get a 95 per cent CI for $L(d)$ under CSR. If the observed $\hat{L}(d)$ at d is below the lower boundary of the 95 per cent CI, the spatial point pattern is considered regular rather than CSR; if the observed at d is above the upper boundary of the 95 per cent CI, the spatial point pattern is considered clustered instead of CSR.

4.3.6.2. *Need of alternative of Monte Carlo*

Marcon *et al.* (2013) have pointed out that with Monte Carlo procedure large point patterns are difficult to deal because computation time is roughly proportional to the square of the number of point multiplied by the number of simulation. They further clarified that this test is valid for one distance but using it repeatedly for all distances where the K function is calculated dramatically increases the risk to reject the null hypothesis by error. In continuation Marcon *et al.* (2013) introduced a new statistical test against complete spatial randomness.

Table 1: P Value for Kolmogorov-Smirnov Test

	p = 1%	p = 5%	p = 25%	p = 50%	p = 75%	p = 95%	p = 99%
n = 1	0.01000	0.05000	0.2500	0.5000	0.7500	0.9500	0.9900
n = 2	0.01400	0.06749	0.2929	0.5176	0.7071	1.0980	1.2728
n = 3	0.01699	0.7919	0.3112	0.5147	0.7539	1.1017	1.3589
n = 4	0.01943	0.08789	0.3202	0.5110	0.7642	1.1304	1.3777
n = 5	0.02152	0.09471	0.3249	0.5245	0.7674	1.1392	1.4024
n = 6	0.02336	0.1002	0.3272	0.5319	0.7703	1.1463	1.4144
n = 7	0.02501	0.1048	0.3280	0.5364	0.7755	1.1537	1.4246
n = 8	0.02650	0.1086	0.3280	0.5392	0.7797	1.1586	1.4327
n = 9	0.02786	0.1119	0.3274	0.5411	0.7825	1.1624	1.4388
n = 10	0.02912	0.1147	0.3297	0.5426	0.7845	1.1658	1.4440
n = 11	0.03028	0.1172	0.3330	0.5439	0.7863	1.1688	1.4484
n = 12	0.03137	0.1193	0.3357	0.5453	0.7880	1.1714	1.4521
n = 15	0.03424	0.1244	0.3412	0.5500	0.7926	1.1773	1.4606
n = 20	0.03807	0.1298	0.3461	0.5547	0.7975	1.1839	1.4698
n = 30	0.04354	0.3151	0.3509	0.5605	0.8036	1.1916	1.4801
n > 30	$Y_p - 1/\left(6\sqrt{n}\right) + o(1/n)$, where $y_p^2 = \frac{1}{2}in\left(1/(1-p)\right)$						
Y_p =	0.07089	0.1601	0.3793	0.5887	0.8326	1.2239	1.5174

Their new test returns the P value to reject the null hypothesis of independence between point locations. They have claimed that their test is more rigorous and faster than classical Monte Carlo simulations.

Earlier Hinrichesen *et al.* (2009) have used three tests statics using Monte Carlo Hypothesis, these tests are permutation based inference method that is commonly used in spatial statics. These include Cuzick-Edward' k-NN test, Moran's I test and Tango's MEET test:

1. Cuzick-Edward k-NN test

   This test includes following notations

   C_i = the observed number of cased in block group i

   n_i = the expected case in block group i

   H = the total number of block groups

   D_{ij} = the Euclidean distance between block group i and j

   S_i (k) = the area in the Smallest circle around i with at least k expected case

$$r = \frac{1}{H}\sum_{i=1}^{H} r_i, r_i = \frac{c_i}{n_i}$$

$$T_k = \sum_i c_i \sum_j c_j I\,(j \in S_i(k))$$

Where I is the indicator function such that I (true) = 1 and i (false) =0. Cuzick-Edwards' *k*-NN test was originally created to evaluate clusters in case-control data, but can easily be modified to handle aggregate data as well. Hinrichesen *et al.* (2009) defined *k*/N as a proportion of the population, where *k* is the number of cases, and used the following parameter values: *k*/N = 0.1%, 1%, 10%, 25% and 50%. Higher values of the test statistic indicate more clustering.

2. Moran I can be defined as

$$I = \frac{\sum_{i=1}^{H}\sum_{j=i+1}^{H}(r_i - \bar{r})(r_j - \bar{r})a_{ij}}{\sum_{i=1}^{H}(r_i - \bar{r})^2}$$

Where a_{ij} = 1 if j[^S_i (k); 0 otherwise. Moran's I is a correlation test between nearest neighbours, originally designed to evaluate continuous data. Modification to Moran's I have been used to assess spatial clustering and in particular Local Moran's I is used as a local indicator of spatial correlation (McLaughlin, 2007). Higher value of the test statistic indicate more clustering

3. Tango's MEET

Tango's (1995) have defined as

$$MEET = \min P(EE\ \lambda) > eet\ (\lambda) | H_0, \lambda)$$

Where

$$eet\ (\lambda) = \sum_i \sum_j \frac{e^{-4d_{ij}^2}}{\lambda^2}(c_i - n_i)(c_j - n_j)$$

and

$$EET(\lambda) = \sum_i \sum_j \frac{e^{-4d_{ij}^2}}{\lambda^2}(c_i - n_i)(c_j - n_j)$$

and C_i is the random number of cases in block group o as generated by the null hypyotesis. Tango's Meet was designed to extend a general spatial clustering test to one that does not reguire specification of the scale parameter value λ. smaller values of the test statistic indicate clustering.

4.3.7. Test Based on J Function

The interaction between points in a spatial point process can be measured by its empty space function *F*, its nearest-neighbour distance distribution function *G* (details of these functions will be discussed in following part of book), and by combinations such as the J-function J = (1 " *G*)/(1 " *F*).

The J function is identically equal to 1 for a Poisson process, and values of J(r) less than or greater than 1 are suggestive of clustering or regularity, respectively.

4.4. METHODS FOR ANALYZING SPATIAL PATTERN

In this section some of the methods related with spatial data analysis are comparatively described.

Spatial patterns are considered as a manifestation of biological processes, reflecting the interactions among different determining factors. Spatial patterns provide information regarding the mechanisms that bring about these patterns, which can then be applied to improve sampling techniques, and to develop strategies for the management of our environments.

Methods to detect and assess spatial pattern and to quantify spatial distribution and association among different elements in an ecological system, have received increasing attention in ecological research since the beginning of modern ecology.

Methods for analyzing spatial patterns have been developed in a wide variety of disciplines, such as plant communities, statistics, forestry, disease assessment and geostatistics, over the past decades (Clark, 1979, Campbell and Noe 1985, Madden and Hughes 1995, Real and McElhany 1996, Liu 2001).

Commonly used approach to determine spatial patterns is to compare the observed frequency distribution with theoretical frequency distributions such as Poisson, binomial, negative binomial, Neyman type A, and beta-binomial distributions (BBD) (Hughes and Madden 1995, Madden and Hughes 1994).

Based on the best fit of observed frequencies to these distributions, a spatial pattern is considered aggregated, random or uniform (Campbell and Noe, 1985, Campbell and Madden, 1990, and Madden and Hughes, 1995).

It is generally accepted that for count data, a good fit to the Poisson distribution suggests random distribution (Fisher, 1925), while a good fit to the negative binomial distribution implies heterogeneity (Madden and Hughes, 1995).

Similarly, for a binary variable a good fit to the binomial distribution indicates homogeneity while a good fit to the beta-binomial distribution suggests heterogeneity (Madden and Hughes, 1995).

The frequency distribution approach has also been extended to calculate indices of aggregation or dispersion. The indices that have been commonly used include, Fisher's variance-to-mean ratio (Fisher, 1925), David and Moore's index of clumping (David and Moore, 1954), the slope of the log (variance) to log (mean) line in the Taylor's empirical power law (Taylor 1961), Morisita's index, I_δ (Morisita 1962), Lloyd's mean crowding and indices of patchiness (LIP) (Lloyd, 1967), the parameter k of the negative binomial distribution (Elliott, 1977), and the parameter θ of beta- binomial distribution (Irwin, 1954; Griffiths, 1973, Hughes and Madden, 1993, Madden and Hughes, 1994 and 1995).

Because the methods based on frequency distributions ignore the information about the locations of sample sites and their spatial relationship, they reflect only the relationship among individuals within sample units.

If the presence of one individual enhances the presence of other individuals within the same sample unit, then the frequency distribution would be determined as aggregated; otherwise it follows a uniform or random distribution. The aggregation determined using frequency distribution based methods has often been referred to as heterogeneity (Madden and Hughes, 1995). Other methods that take into consideration the location of sample sites have also been developed for characterization of spatial patterns. These methods include: spatial analysis by distance indices (SADIE), methods based on quadrat variance, spatial auto-correlation, distance based joint-counts/network method and geostatistics.

SADIE (Perry and Hewitt 1991, Alston 1994, Perry 1995) determines an index of aggregation by comparing the observed spatial arrangement with other arrangements derived from it, such as those where individuals are arranged as crowded as possible, those as random as possible, and those as regular as possible. Aggregation detected by SADIE method may be nonrandom in two different ways, nonrandom frequency distribution of counts regardless of the locations of sample units, and the nonrandom spatial arrangement of the sample units.

Perry and Hewitt (1991) defined 'move to crowding' as the minimal total number of moves required to move all individuals, one by one and step by step, so that all the individuals finish in the same sample unit. 'Move to randomness' was also defined similarly, and a test based on the ratio of 'move to crowding' and 'move to randomness' was proposed as an index for detecting aggregation in spatial patterns. Lately, 'distance to regularity' (D) was suggested as a replacement to 'move to randomness' (Alston 1994).

The approach was further extended to map two-dimensional data, and a new method was developed to estimate the initial focus of a cluster based on the movements (Perry 1995).

SADIE program, which determines the shortest possible distance to move spatially referenced data from one location to another in order to achieve regularity (regularity is defined as the data arrangement in which each quadrat has the same number of individuals).

These distances are summed to provide an overall distance to regularity (observed distance to regularity). The ratio between the observed and expected distance to regularity is the index of aggregation (I_a). When the ration of I_a is <1, =1 or >1, the pattern of distribution is considered regular, random or aggregated, respectively. Chi-Square test was used to determined whether I_a was significantly different from 1.

Spatial autocorrelations and partial auto-correlations were calculated for unmanaged epidemics for each sampling date and for both variables. Autocorrelograms were generated for different proximity patterns (rook, bishop and queen).

Their results revealed the random pattern of disease (for LD) early in the season, followed by an aggregated pattern in the second part of the epidemic. The index of aggregation for ACC was recorded significantly greater than 1. Positive trends in partial autocorrelation were observed, they concluded that LD and ACC were not aggregated in early stage of epidemics, when both disease intensity and airborne conidia concentration were low.

The use of an airborne inoculum threshold as a tool for management of Botrytis leaf blight implies that monitoring can be achieved by only one or few samplers per field or farm. From a management standpoint, it is not necessary to know the exact amount of disease or airborne inoculum present, knowing whether disease exceeds the action threshold is sufficient for decision making. Thal and Campbell (1986) studied spatial pattern of disease severity data for alfalfa leaf spot caused primarily by *Leptosphaerulina brisiana.*

Blocked quadrat variances (BQV) (Greig-Smith, 1952) method determines cluster size by identifying the peak of mean "local variance" (mean squared difference between adjacent blocks) as quadrat sizes increase by combining two contiguous quadrats to form a new quadrat.

The two major drawbacks with the BQV method are that the block sizes must be some power of 2, and that the starting position of blocking can affect the results. The two-term local quadrat variance (TTLQV) method (Hill, 1973) offers more flexibility to the block size, and reduces the effects of the starting position. However, quadrats are still blocked along the belt transect and edge effects are important. Therefore, the variance is affected not only by distance (spacing) but also by the block size.

This disadvantage was overcome in the paired-quadrat variance (PQV) and random paired quadrat variance (RPQV) methods, in which the variance was calculated for all possible paired quadrats along the belt at a given spacing for the former or for randomly selected paired quadrats for the latter (Goodall, 1974).

The quadrat variance methods provide informative results of aggregation and the size of clumps and are independent of the assumption of distributions. However, the calculations become complicated for two-dimensional data sets.

Spatial autocorrelation, distance based joint-counts methods and geostatistics are all similar. Spatial autocorrelation can be defined as a spatial property that the presence of some quality at one sample site (quadrat, or location) makes its presence at proximal sample sites more or less likely (Cliff and Ord, 1973).

Proximity can be determined by the connection between the two samples sites, such as distance, lag distance, vector or other criterion. Unlike methods based on negative binomial and Poisson distributions, which can only be used for discrete count data, the spatial autocorrelation analysis can be conducted on any type of data, continuous or discrete.

Data used in spatial autocorrelation can be point based, area based, or even without information on the exact shape of areas. The two most common coefficients in spatial autocorrelation analysis are Moran's *I*, which is based on sum over the cross product of deviations from mean, and Geary's *c,* which is based on sum of the squared difference (Moran, 1950, Geary, 1954).

Autocorrletion will further discussed in subsequent chapter. Joint-counts method, which has also been referred as distance-based methods and neighbour networks, is an analysis for discrete data based on point-pairs (Campbell and Madden, 1990, Real and McElhany, 1996, Liu, 2001). There are a number of ways for point-pairing, such as Gabriel connectedness (Gabriel and Sokal, 1969), the nearest neighbor (Clark and Evans, 1954, Pielou, 1959), and one- or two-dimensional distance class (Gray *et al.*, 1986, Nelson *et al.*, 1992, Ferrandino, 1996).

After establishing connectedness, the number (or average distance) of each type of joint is counted and this observed value is then compared with the expected one under the assumption of randomness.

Geostatistical techniques, originally developed for use in mining (Clark, 1979), can be used for both continuous and discrete variables, and require less stringent assumptions of stationarity compared with spatial autocorrelation techniques. In geostatistics, spatial dependence can be analyzed directionally or omnidirectionally, and represented as semivariogram or correlogram. Semivariogram calculates semivariance against each lag distance (a vector h), in which semivariance is defined as half of mean squared difference between all point pairs (p, p+h) separated by h. In a correlogram, the linear (Pearson's) correlation coefficients between point pairs separated by a vector h are plotted against vector h, or lag distance at different directions.

4.4.1. Poisson Probability Distribution

For a randomly dispersed population of organisms, the Poisson model ($\sigma^2 = \mu$) gives probabilities for the number of individuals per sampling units, provided the following conditions hold:

1. Each natural sampling units has an equal probability of hosting an individual.
2. The occurrence of an individual in a sample unit does not influence its occupancy by another, and
3. Each sampling unit is equally available, and
4. The number of individual per sampling unit is low relative to the maximum possible that could occur in the sampling unit.

To compute these probabilities, estimation of the mean number of individuals per sampling unit are required and these are assess in following steps

4.4.1.1. *State the hypothesis*

Hypothesis stated that the number of individuals per sampling units are from a poisson distribution (variance equal to mean)

4.4.1.2. *The frequency distribution F*

The frequency data, which consist of the number of individuals per sampling unit, are summarized as a frequency distribution, that is, the number of sampling unit with x = 0, 1, 2, ... r individuals.

A rule of thumb is that the minimum number of sampling unit (N) should be greater that 30 (Ludwig and Reynold, 1998), If N is less than this than we should use other spatial indices developed and reported by several researches.

4.4.1.3. *The poisson probabilities P(x)*

The probability of finding x individuals in a sampling unit, that is P(x), where x = 0, 1, 2 .... r individuals, is given by the Poisson series.

$$P\,(X = x) = \frac{\lambda^{x} e^{-\lambda}}{x!}; x = 0,1,2$$

Where e is the base of the natural logarithms (2.718) and x! is the factorial of x (for example, for x =3, x! = (3) (2) (1) = 6. λ is the mean number of /sampling unit. The above model was fitted by using the estimation of λ *i.e.* by obtaining the probabilities thorough recurrence relation λ=*x*

$$P\,(X = r + 1) = \frac{\lambda}{(r + 1)} * P\,(X = r); r = 1,2 \ldots\ldots\ldots where\ P\,(X = 0) = \varepsilon^{X}$$

4.4.1.4. *Goodness-of fit test statistic* χ^2

The chi-square goodness-of fit test is used to ascertain how well the observed frequency (F_x – Step b) compare to expected frequencies (E_x–Step d). The chi-square test statistic (χ^2) is computed as

$$\chi^2 = \sum_{x=0}^{q} [(F_x - E_x)^2 / E_x]$$

This test statistic is compared to a table of chi-square probabilities with q-2 degrees of freedom. If the chi-square test statistic is greater than the tabular chi-square at some selected level of probability (*e.g.*, 5%), we conclude that it is improbable that the frequency distribution follows a Poisson series and, hence, reject the null hypothesis.

4.4.1.5. *An index of departure from poisson*

The simple index of departure from Poisson would be based on the degree of overlap or concordance of the observed distribution and the Poisson distribution of the same μ (Hurlbert, 1990). It is defined as:

$$D_p = 1 - \sum_{k=1}^{w} \min(\pi_{kp}, \pi_{ka})$$

Where: π_{kp} =qkp/Q, π_{kp} =qka/Q; w = largest number of individuals observed in any quadrate; k = number of individuals per quadrate and q_{ka} = number of quadrates containing k individuals, in the observed or actual distribution,

$$q_{kp} = Q\frac{\mu^k e^{-\mu}}{k!}$$

= number of quadrates expected to contain k individuals, in a Poisson distribution with mean = μ

$$Q = \sum_{k=0}^{W} q_{ka}$$

= total number of quadrates, in the sampling universe or study site.

μ = X/Q = mean number of individuals per quadrate

$$X = \sum_{i=1}^{Q} X_{!} = \sum_{k=1}^{W} kq_{ks}$$

Total number of individuals present in the Q quadrates, the sampling universe, and $x_!$ = number of individuals in the ith quadrate. This index, D_p, varies from 0 to 1, assuming the former value when the observed distribution conforms exactly to the Poisson, and the latter value (actually always slightly less than 1) when a single quadrate contains all X individuals present in the population. Geometrically it represents 1.0 minus the area of interaction of the π_{kp} and π_{ka} distribution with respect to k (Hurlbert, 1990).

4.4.2. Binomial Distribution and Negative Binomial Distribution

Beta-binomial distribution provides a good description of populations of diseased individuals in many patho-systems (Hughes and Madden, 1993; Hughes *et al.*, 1995; Hughes and Maddan, 1998; Tanne *et al.*, 1996; and Xiao *et al.*, 1997 and Jones *et al.*, 2011).

Beta-binomial and binomial distribution are useful tools to represents functional relationships between disease incidences at two levels in a spatial hierarchy.

By using these functions, incidence of disease at the higher level can be predicted, without curve fitting, based solely on the incidence and degree of aggregation at the lower level.

If incidence is binomially distributed (random), it can take the reciprocal approach and predict incidence at the lower level based on incidence at the higher level.

This is particularly useful in the context of group testing and sampling (Madden and Hughes, 1999) where the disease status of each groups (sampling unit) is determined and the results from the sampling units are used to calculate the proportion of diseased individuals at the lower level.

This group testing approach is known to be very efficient for estimating incidence at the lower level under many conditions. In general data belongs to beta-binomial distribution (over dispersed or aggregated), this approach is not so useful, because inherently this function cannot be rearranged to predict incidence at the lower level based on the higher level incidence.

Xu and Ridout (2000) advocated that beta-binomial distribution is more appropriate than the negative binomial distribution for examining spatial pattern of disease incidence for the presence of aggregation.

The negative binomial distribution is mathematically related to the Poisson and logarithmic series (Yin *et al.*, 2005).

The negative binomial model is probably the most commonly used probability distribution for clumped. When two of the conditions associated with the use of Poisson model are not satisfied that is (1) each natural sampling unit has an equal probability of hosting an individual and (2) the occurrence of an individual in a sampling unit does not influence its occupancy by another. This usually leads to a high variance-to mean ratio of the number of individuals per sampling unit, this suggests that a clumped pattern may exist.

The negative binomial has two parameters (1) μ, the mean number of individuals per sampling unit and (2) k, a parameter related to the degree of clumping. Disease incidence is appraised of the probability (π) of a plant or other plant entity being diseased.

This probability related with three elements of epidemics *i.e.,* host, environment and pathogen. In an unrestricted random sample, in which of n plants is identified for disease, the estimate of this probability is simply the total number diseased (X) divided by $n(\pi = X/n)$ and the probability of a plant not being diseased is $1\text{-}\pi$.

If the probability of a plant being diseased is not constant, then the binomial is not appropriate for representing the distribution of diseased entities per sampling unit.

On the other hand poisson distribution is the appropriate model for a random pattern and it can obtain from the binomial when n goes to infinity and $n\pi$ (mean μ) is constant. In poisson distribution the mean and variance are equal. When the beta-binomial fits disease incidence data, researcher can use the estimate of θ as index of aggregation. Madden and Hughes (1995) have provided the graphical relationships (Fig. 2) between various variable of discrete distributions.

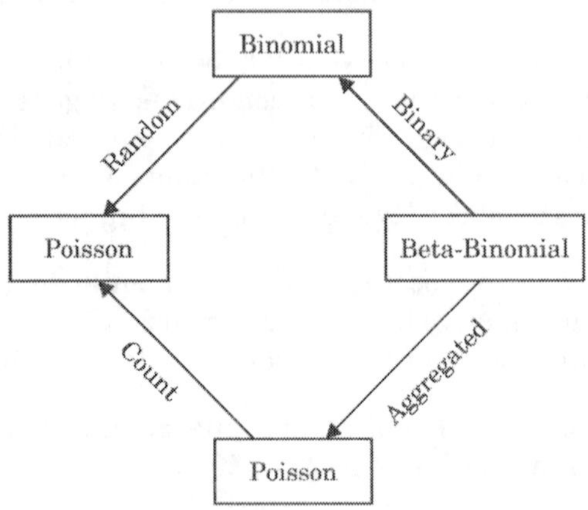

Fig. 2: Relationships among the four discrete distributions

The binomial distribution has a single parameter π which represent the probability for each plant or plant part to be diseased, independently of other plants or plant parts. The beta-binomial distribution has two parameters, p, the expected probability of disease, and θ, a measure of variation in disease incidence among quadrats.

A good fit to the binomial distribution would suggests a random spatial pattern of disease incidence, while a good fit to the beta-binomial would suggest an aggregated spatial pattern of disease incidence (the greater the θ, the more aggregated the disease (Gosme *et al.*, 2007).

Neyman's C(α) test can be use to determine if the beta-binomial distribution provides a better fit to the data than the binomial distribution. For comparison of the aggregation between quadrats of different sizes.

Gosme *et al.* (2007) have used index of dispersion D for quantification of size, shape and intensity of aggregation of take all disease during natural epidemics in second wheat crop caused by *Gaeumannomyces graminis* var. *tritici*

$$D = \frac{1 + n\ x\ \theta}{1 + \theta}$$

Where n is the number of units in a quadrat and θ the dispersion parameters of the beta-binomial distribution.

4.4.2.1. *State the hypothesis*

The hypothesis to be tested is that the number of individuals per sampling unit follows a negative binomial distribution, and , hence, a non-random or clumped pattern exists.

Failing to reject this hypothesis, the ecologist may have a good empirical model to describe a set of observed frequency data although this does not explain what underlay causal factor might be responsible for the pattern. In other words, we should not attempt to infer causality solely based on our pattern detection approach

4.4.2.2. *The negative binomial probabilities P(x)*

The probability of finding x individual in a sampling unit, that is P(x), where x = 0,1,2......r individual is given by

$$P(X) = \left[\frac{\mu}{\mu + k}\right]^x \left[\frac{(k + x - 1)!}{x!\,(k - 1)!}\right]\left[1 + \frac{\mu}{k}\right]^{-k}$$

The parameter μ is estimated from the sample mean (x ᒻ). Parameter k is a measure of degree of clumping and tends towards zero at maximum clumping. An estimation for K is obtained using the following equation

$$log_{10}\left[\frac{N}{N_0}\right] = \acute{K} log_{10}\left[1 + \left(\frac{\bar{X}}{\acute{K}}\right)\right]$$

Where N is the total number of sampling unit in the sample and N_0 is the number of sampling unit with 0 individual. First, an initial estimation of K is substituted into the right-hand side (RHS) of this equation and the resultant value is compared to the value of the left-hand side (LHS). If the RHS is lower than LHS, a higher value of K is then tried, and again, the two sides are compared.

This process is continued in an iterative fashion (appropriately selecting higher or lower value of K) until a value of K is found such that the RHS converges to the same value as the LHS. A good initial estimation of K for the first iteration is obtained from:

$$\acute{K} = \frac{x^2}{s^2 - x}$$

Where s^2 is the sample estimate of variance. When the mean is small (less than 4) is an efficient way to estimate of K.

On the other hand, if the mean is large (greater than 4), this iterative method is efficient only if there is extensive clumping in the population. Thus, if both the population mean (x) and the value of $\acute{K}$ are greater than 4 the last equation is actually preferred over 2nd equation for estimation of $\acute{K}$ (Southwood, 1978).

Once the two statistics, x and $\acute{K}$ are obtained, the probabilities of finding x individuals in a sampling unit, that is, P(x), where x = 0,1,2,....r individuals, are computed as

$$P(0) = [\bar{x}/(\bar{x}+\acute{K})]^{o}\ [(\acute{K}+0\text{-}1)!/(0!(\acute{K}\text{-}1)!)][1+(\bar{x}/\acute{K})]^{-\acute{K}}$$

$$= [1+(\bar{x}/\acute{K})^{-\acute{K}}$$

$$P(1) = [\bar{x}/(\bar{x}+\acute{K})]^{1}\ [(\acute{K}+1\text{-}1)!/(1!(\acute{K}\text{-}1)!)][1+(\bar{x}/\acute{K})]^{-\acute{K}}$$

$$= [\bar{x}+(\bar{x}+\acute{K})]\ (\acute{K}/1)P(0)$$

$$P(2) = [\bar{x}/(\bar{x}+\acute{K})]^{2}\ [(\acute{K}+2\text{-}1)!/(2!(\acute{K}\text{-}1)!)][1+(\bar{x}/\acute{K})]^{-\acute{K}}$$

$$= [\bar{x}+(\bar{x}+\acute{K})]\ (\acute{K}+1)/2]P(1)$$

$$P(r) = [\bar{x}/(\bar{x}+\acute{K})]^{r}\ [(\acute{K}+r\text{-}1)!/(r!(\acute{K}\text{-}1)!)][1+(\bar{x}/\acute{K})]^{-\acute{K}}$$

$$= [\bar{x}+(\bar{x}+\acute{K})]\ (\acute{K}+r\text{-}1)/r]P(r\text{-}1)$$

4.4.2.3. *The expected ngative binomial frequencies, E_x*

As calculated for the Poisson distribution, the expected number of sampling units containing x individuals is obtained by multiplying each of the negative binomoial probabilities by N, the total number of sampling units in the sample.

The number of frequency classes, q is also determined as calculated for the Poisson model.

4.4.2.4. *The k exponent of negative binomial distribution (k)*

The parameter k of the negative binomial distribution is one measure of aggregation that can be used for species having clumped or aggregated spatial pattern. This index should only be used when the data fit the negative binomial distribution.

When k values are low and positive (k < 2) they indicate a highly aggregate arrangement; k values ranging from 2 to 8 indicate moderate aggregation; and values higher than 8 (k > 8) indicate a random arrangement (Elliott, 1979; Southwood, 1978). k values were calculated by the moments methods (Anscombe, 1949), given by:

$$k \quad \frac{\bar{X}^2}{s^2 \quad \bar{X}} \qquad \ldots 1$$

and later by the maximum likelihood method, since it is more accurate (Elliott, 1979). According to Bliss and Fisher (1953), this method consists in finding the value that balances both members of the equation:

$$N \ln(1 \quad \frac{\bar{X}}{k}) \quad \overset{nc}{\underset{i\ 1}{}} (\frac{A(X_i)}{k \quad X_i})$$

Where N = number of sampling units; 1n = Napierian logarithm; $\bar{x}$ = sampling mean; = k value estimate; xi = number of species found in sampling untits; A (xi) = sum of frequencies of values higher than xi; nc = number of classes in the frequency distribution.

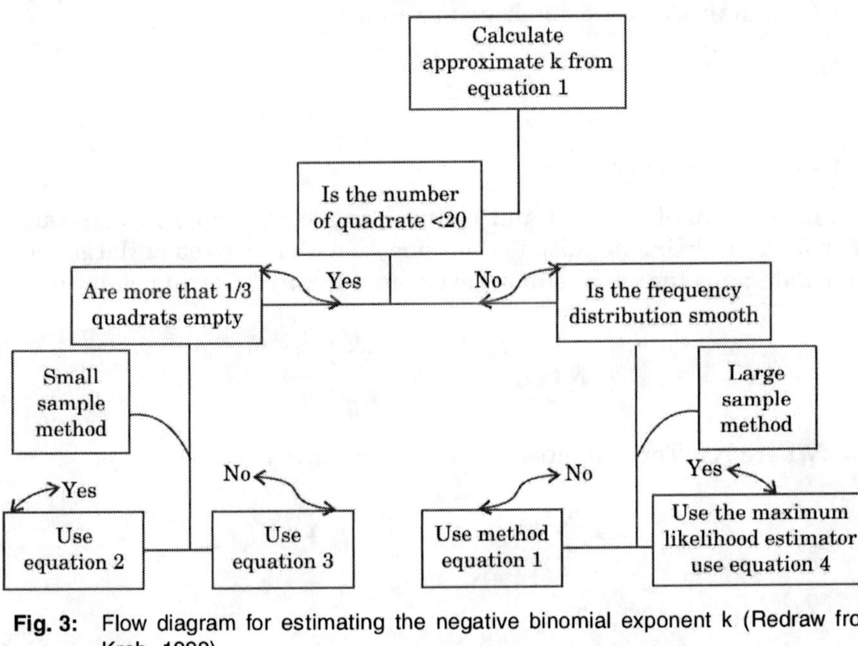

Fig. 3: Flow diagram for estimating the negative binomial exponent k (Redraw from Kreb, 1999)

The estimation method is interactive, by trial and error, until equality is achieved between both members of the equation, with a previously-established margin of error. The calculation can be started from the k value obtained by the method of moments, thus allowing convergence to be achieved more quickly (Costa *et al.*, 2010). Kreb (1999) have differentiated the k exponent as per small or large sample size (Fig. 3, redraw from Kreb, 1999).

4.4.2.4.1. *Small sample size*

Number of quadrats < 20 and counts are too few to arrange in a frequency distribution.

(a) More than 1/3 of the quadrats are empty calculate an estimate of k by following equation

$$log_e\left(\frac{N}{n_0}\right) = \acute{k} log_e\left(1 + \frac{\bar{x}}{\acute{K}}\right) \quad ...2$$

Where N = total number of quadrate counted

N_0 = Number of quadrats containing zero individuals

$\bar{x}$ = Observed mean

K = Estimate of the negative binomial exponent

(b) Less than 1/3 of the quadrats are empty

$$\acute{K} = \bar{x}^2 - \left(\frac{s^2/n}{s^2 - \bar{x}}\right) \quad ...3$$

4.4.2.4.2. *Large sample size*

Number of quadrats >20 and counts can be arranged in a frequency distribution. Frequency distribution smooth with no extremely large counts: calculate a maximum likelihood estimate for k by following equation.

$$N\, log_e\left(1 + \frac{\bar{x}}{\acute{K}}\right) = \sum_{i=0}^{\alpha}\left(\frac{A_x}{\acute{K} + x}\right) \quad ...4$$

Where N = Total number of quadrats counted

$$A_x = \sum_{j=x+1}^{\alpha}(f_j) = f_{x+1} + f_{x+2} + f_{x+3} + f_{x+4} + \cdots ...$$

J= a counter (0,1,2,3...)

X=a counter (0,1,2,3....)

F_x = observed number of quadrate containing x individuals

4.4.2.4.3. *Tests for goodness of Fit*

There are few approached to testing the null hypothesis that a observed counts do not depart from those predicted by a negative binomial distribution.

4.4.2.4.4. *Chi-squared Goodness of Fit*

This test should be used whenever the observed data can be arranged in a frequency distribution (n>2).

$$\chi^2 = \sum\left(\frac{(observed\ frequency - expected\ frequency)^2}{Expected\ frequency}\right)$$

If the observed 5^2 value is less than the critical value, we can accept the null hypothesis that the negative binomial distribution is an adequate description of the clumped pattern in our data.

4.4.2.4.5. *U-Statistic goodness of fit test*

This test compares the observed and expected variance of the negative binomial distribtuiobn and has more statistical power than the chi-square goodness of fit test (Kreb, 1999). U = observed variance-expected variance.

$$= S^2 = -\left(\bar{x} + \frac{\bar{x}^2}{k}\right)$$

The expected value of U is zero. To test if t he observed value of U is different from zero, calculate the standard error of U as follows:

$$a = \bar{x}/1$$

$$b = 1+a$$

$$c = \frac{ba^4}{[b\ (log_{eb)-a}]^2}\left[b^{(1+\frac{\bar{x}}{k})} - (\bar{x}+b)\right]$$

then

$$S.E.(\hat{U}) = \sqrt{\frac{1}{n}\left\{2\,\bar{x}(\bar{x}+a)(b)\left[\frac{(b^2)(\log_e b) - a\,(1+2a)}{(b)\log_e b) - a}\right] + c\right\}}$$

If the observed value of U exceeds 2 standard errors of U then the null hypothesis is rejected that the negative binomial is an adequate fit to the observed (at $\alpha = 0.05$).

4.4.2.5. *An alternative parameterization of the negative binomial distribution*

Poisson process is a commonly used starting point for modelling stochastic variation of ecological count data around a theoretical expectation. However, data typically show more variation than implied by the Poisson distribution. Such overdispersion is often accounted for by using models with different assumptions about how the variance changes with the expectation.

The choice of these assumptions can naturally have apparent consequences for statistical inference.

The negative binomial distribution is defined by two parameters, the mean and a measure of dispersion k. Bliss and Fisher (1953) indicate the dispersion parameter is best estimated using equation [1] below.

This equation, which produces the maximum likelihood estimate, is more accurate than equations that estimate k using the first (mean) and second (variance) moments or estimates based on the proportion of zero samples (Bliss and Fisher, 1953). Methods of estimating k are outlined with some examples in Bliss and Fisher (1953); Southwood (1978) and Krebs (1999).

$$N \log e[1+(\bar{x}/k)]=\_ [Ax/(k+x)] \qquad \ldots 1$$

In equation [1] x ˪is the mean and N is the number of samples with *G. morhua*. The procedure for estimating k in equation [1] is to substitute parameter estimates until the difference between the two sides of the equation is zero, or very close to zero. This is done iteratively and in the case of equation [1] produces the maximum likelihood estimate. Lloyd's (1967) index of patchiness P measures intensity of aggregation and is defined by two parameters, the mean ($\bar{x}$) and variance (s^2).

$$P=1+[(s^2-\bar{x})\ \bar{x}^2] \qquad \ldots 2$$

This index measures how many times as crowded an individual is on average than it would have to be if the same population had a random (non aggregated) distribution (Lloyd, 1967). The variance of the negative binomial is related to the mean and k as

$$s^2=\bar{x}+(\bar{x}^2/k) \qquad \ldots 3$$

When the formula for variance [3] is substituted into Lloyd's index of patchiness [2], patchiness simplifies to

$$P=1+(1/k)$$

Where k is estimated using [1]. k has long been known as a measure of dispersion. Its use as an index of dispersion in ecology (David *et al.*, 2003).

4.4.2.6. *Linden and Mantyniemi, approach*

Linden and Mantyniemi, (2011) have proposed a parameterization of the negative binomial distribution, where two overdispersion parameters are introduced to allow for various quadratic mean–variance relationships, including the ones assumed in the most commonly used approaches.

The negative binomial distribution has been parameterized in a number of different ways in the statistical and applied literature. Perhaps the most common way to parameterize is to see the negative binomial distribution arising as a distribution of the number of failures (X) before the rth success in independent trials, with success probability p in each trial (consequently, $r \geq 0$ and $0 \leq p \leq 1$) . In such a case the probability mass function can be expressed as

$$P\left(X=\frac{x}{r},p\right)=\frac{\Gamma(x+r)}{x!\,\Gamma(r)}p^{r}(1-p)^{x}$$

And the random variable X has the expectation (theoretical mean) $\mu = r(1-p)/p$ and variance $\sigma^2 = r(1-p)/p^2$. Corresponding to this parameterization, parameters r and p can be written in term o mean and the variance as

$$r=\frac{\mu^2}{\sigma^2-\mu}$$

$$p=\frac{\mu}{\sigma^2}$$

Linden and Mantyiemi (2011) have utilized above two equations as a starting point for a more flexible parameterization of the negative binomial distribution that can be used to represent various kind of overdispersed Poisson process. By writing the variance as a quadratic function of the mean

$$\sigma^2=\omega\mu+\theta\mu^2$$

Diverse mean variance relationships are obtained using the parameters ω and the θ, hereafter called overdispersion parameters. This process is defined as long as $\sigma^2 > \mu$, or phrased in terms of the overdispersion parameters, when $\omega+\theta\mu>1$. With any valid combination of expectation E(X) and overdispersion parameters, the parameters r and p of the negative binomial distribution can be derived as

$$r = \frac{\mu}{\omega - 1 + \theta\mu}$$

$$p = \frac{1}{\omega + \theta\mu}$$

And then be substituted in to equation 1 in order to obtain the relevant likelihood. The negative binomial distribution can also be seen as explicit overdispersed Poisson process, where the Poisson intensity is drawn from a gamma distribution (Gelman, *et al.* 2004).

4.4.2.7. *Frequency distribution models for spatial patterns of plant abundance*

Frequency distributions of plant density, cover, biomass per unit area, and height, as measures for expressing biological abundance and biological dominance of plant species, have been used to describe species composition and spatial patterns of species in the studies of plant communities (Chen *et al.*, 2006; 2008a, 2008b, and Damgaard, 2009).

Frequency distribution models includes:

(1) the number of individuals per unit area, expressed using the negative binomial distribution (Fisher, 1941) and the Poisson distribution (*e.g.*, Greg-Smith, 1983);

(2) the number of plant occurrences in binary counts (presence/absence) per unit area, expressed as the beta binomial distribution (Skellam, 1948) and the binomial distribution (*e.g.*, Greig-Smith, 1983);

(3) biomass per unit area, expressed as the gamma distribution (May, 1973; Shiyomi *et al.*, 1984); (4) plant cover per unit area, expressed as the beta distribution (Chen *et al.*, 2006).

Models (1) and (2) are probability distributions with a discrete variate for count data, and models (3) and (4) are density functions with a continuous variate. These models are based on some simplified assumptions about biological processes. Spatial heterogeneity, *e.g.*, patchy/uniform, coarse/dense, and aggregated/random, is an important property of organisms in communities.

Various measures of heterogeneity have been developed in population ecology (*e.g.*, David and Moore, 1954; Morisita, 1959; Lloyd, 1967; Iwao and Kuno, 1971), based on the number of individuals in a population.

Shiyomi and Yoshimura (2000) derived, to examine populations, a basic formula that can be modified for use with any of the above plant measurements.

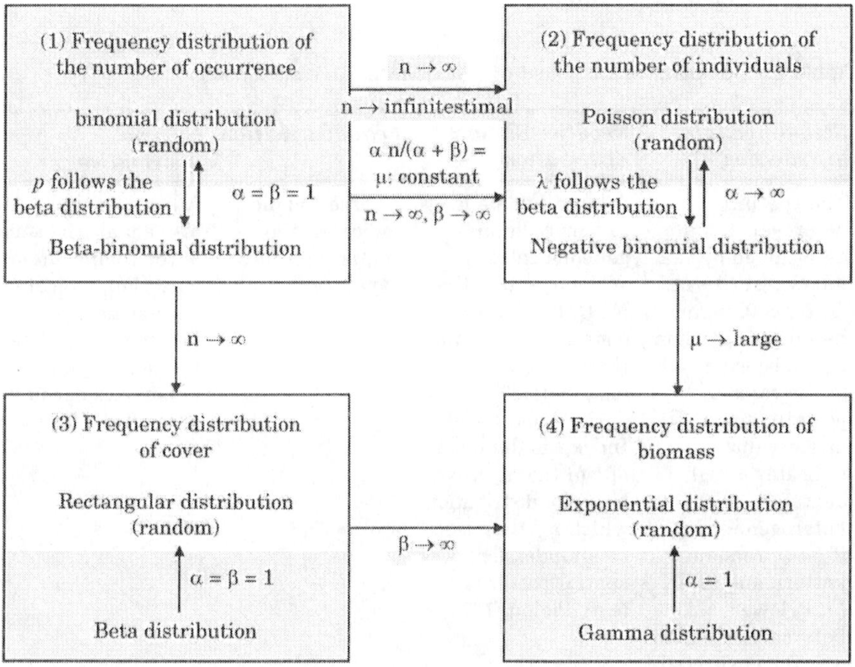

Four models for expressing frequency distributions of the number of binary occurrences, the number of individuals, cover, and biomass, and the relationships between them. Arrow: the asymptotic derivation of distribution; p: probability that species A occupies a cell, which is one of n cells contained in a quadrat; μ: mean in the beta-binomial and negative binomial distributions; α and β: parameters of the distributions; λ the mean of the Poisson distribution (Adopted from Chen *et al.*, 2008a)

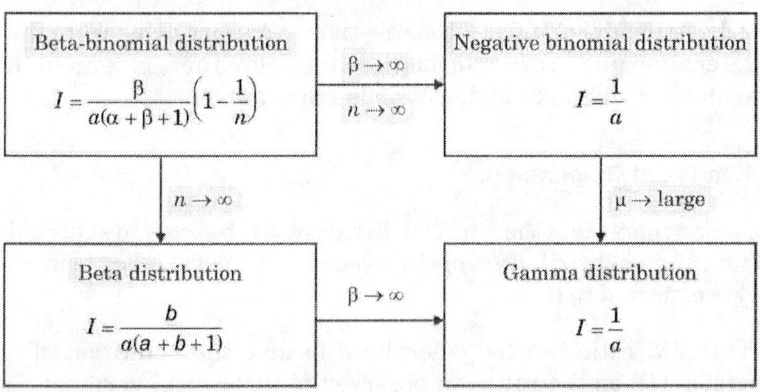

Four spatial heterogeneity indices for the number of binary occurrences, the number of individuals, cover, and biomass, and the relationships between them. Interpretation of above these spatial heterogeneity indices are summarized in Table 2.

Table 2: Comparative statement of various binomial distributions.

Beta-Binomial distribution	*Negative-Binomial distribution*	*Beta-distribution*	*Gamma distribution*
The spatial heterogeneity is determined by the following criteria: (1) if $I > 0$, higher heterogeneity than would be expected in the random pattern, and a larger value indicates a high degree in heterogeneity; (2) if $I = 0$, random pattern and (3) if $I < 0$, lower heterogeneity than would be expected in the random pattern, and a large negative value indicates a low degree of heterogeneity.	A large I value leads to a more hetero-geneous pattern. When $\alpha \rightarrow \infty$ in the NBD, the Poisson distribution (denoting the random distribution) is derived and $I \rightarrow 0$. Index I is the inverse of k of the negative binomial distribution, which is often used as an index for spatial patterns (*e.g.*, Pielou, 1977).	A large I value indicates high-spatial hetero-geneity	A large I value, which also equals the square of the coefficient of variation, indicates high-spatial heterogeneity. The frequency distribution of biomass per unit area often follows the GD

4.5. SPATIAL INDICES FOR NATURAL SAMPLING

After detecting the pattern of a species it is necessary to measure the degree of aggregation and in order to measure the degree of aggregation there are large number of indices of dispersions are available.

4.5.1. Index of Dispersal (I_D)

This index represents the variance to variance (observed to expected ration) and for random distribution of observed ration the variance to mean ration (VM) is expected to be 1.

This VM ratio can be generalised to give the definition of index of dispersion (D) as the ration of observed to theoretical variance. D =1 for random data, *D*>1 for over-dispersed data and *D*< 1 for over dispersed data.

For binary data (0,1) such as disease incidence this *D* can be calculated by following formula

$$D = \frac{S^2}{np\,x\,(1-p)}$$

Where *np x (1-p)* is the expected variance of incidence data assuming that the binomial distribution fits the data, and S^2 is the observed variance and further, chi-square test is utilize to determine whether to reject or accept the null hypothesis (*i.e.* disease is randomly distributed).

This index was used by many workers like Bassanezi *et al.,* 2003 and 2005a; Bassanezi and Laranjeira 2007; Batista *et al.*, 2008a; 2008b and Spolti *et al.*, 2012).

Dispersion of a population can be classified through a calculation of variance to mean ratio; namely: s^2/m. An I_D value equals to 1.0 indicates a random distribution; zero (or <1) indicates uniform while, more than one indicates the clumped distribution. Test of significance of I_D can be tested with chi-square test (χ^2).

According to Ludwig and Reynolds (1988), for small sample size (N<30), χ^2 is a good approximation with N-1 degrees of freedoms. If the value for χ^2 falls between the χ^2 tabular values at the 0.975 and 0.025 probability levels (P>0.05), agreement with a random distribution is accepted (*i.e.*, $s^2 = \bar{x}$). On the other hand, values for χ^2 less than the 0.975 probability level suggest a regular pattern (*i.e.*, $S^2 < \bar{x}$), whereas χ^2 values greater than the 0.025 probability level suggest a clumped pattern (*i.e.,* $S^2 > \bar{x}$).

4.5.2. Intra-cluster Correlation (ρ)

This measures the tendency of the species within a sampling unit (quadrat) to have a similar disease status.

Positive value of ρ indicate aggregation of disease and this can be calculated directly from the quadrat data as

$$\rho = 1 - \frac{1}{p\,x\,(1-p)NT\,(T-1)} \sum_{i=1}^{N} Y_i\,(T - Y_i) \qquad \ldots 1$$

Where Y_i is the number of infected plants in the *i th* quadrat, N is the total number of quadrats and T is the number of plants in a quadrat and is same for all quadrats. This index can also be quantify through *D* by using $\rho = (D-1)/(n-1)$ and in addition, if the beta-binomial distribution fits the data, this index is related to θ as $\rho = \theta/(1+\theta)$.

4.5.3. Index of Clumping (I_C)

David and Moore (1954) proposed a modification of I_D, *i.e.*, index of clumping (I_C). $I_C = S^2/X-1$. The negative value of this index predicates uniform distribution where as a value of 0 indicates random distribution and a value equal to n-1 maximum clumping.

David and Moore (1954) proposed a test for the significance of difference between samples, *e.g.*, for the same species in two different habitats. It depends on the same number of quadrates being used in each case and takes the form of calculating

$$\omega = -\frac{1}{2}\log\frac{v_1\,\lambda_2}{\lambda_1\,v_2}$$

Where λ_1, λ_2 are the observed means and v_1, and v_2 are observed variances of the two sets of data. If ω lies outside the range of $\pm 2.5/\sqrt{N-1}$), N is the number of quadrates in each set, a significant difference in index of clumping is indicated.

4.5.4. Green Index (GI)

As compared to I_D and I_C indices, Green index (GI; Green, 1966) better measures dispersion since it is independent to the number of individuals (Ludwig and Reynolds, 1988).

$$GI = \frac{\left(\frac{S^2}{x}\right) - 1}{n-1} = \frac{IC}{n-1}$$

Where S^2 is variance, $\bar{x}$ is mean and n is number of quadrate. GI varies between 0 (for random) and 1 (for maximum clumping). This index is a usual index for measuring the degree of aggregation. The upper value for a test of randomness at the significance level of α is determined by following formula.

$$GI\ (1-\alpha) = \frac{\frac{(X^2_{(1-\alpha)})}{(n-1)} - 1}{nm-1}$$

4.5.5. Lloyd's Mean Crowding (M^*)

Michereff *et al.* (2011) utilized the Lloyd's Index of Patchiness (LIP) for spatial pattern analysis and they utilize following equation.

$$LIP = \frac{\left[\left(\bar{x} + \left(\frac{S^2}{\bar{x}}\right) - 1\right)\right]}{\bar{x}}$$

Where x bar is the mean severity of the disease from selected plants using chosen number of leaves per plant and S^2 is the sample variance. Values of LIP lower than, equal to or greater than 1.0 indicate regular, random or patchy spatial pattern and the significance value can be checked with chi-square test. Ideal sample size (n) can be estimate for each crop area based on the coefficient of variation of the mean (CV_X) and random pattern of diseased plants with equation

$$n = \frac{S^2}{(x^2 . CV_x^2)}$$

Mean crowding (x^*) was proposed by Lloyd to indicate the possible effect of mutual interference or competition among individuals (Mordi-Vajargah *et al.* 2011). Theoretically mean crowding is the mean number of other individuals per individual in the same quadrate:

As an index, mean crowding is highly dependent upon both the degree of clumping and population density.

To remove the effect of changes in density, Lloyd introduced the index of patchiness, expressed as the ratio of mean crowding to the mean. As with the variance-to-mean ration, the index of patchiness is dependent upon quadrate size. Lloyd's index > 1 indicates aggregated dispersion, and Lloyd's index < 1 indicates regular dispersion. Adapted by Nordmeyer (2009), Mordi-Vajargah *et al.* (2011) and Sule *et al.* (2012).

$$M^* = X + \left(\frac{S^2}{\mathrm{X}}\right) - 1$$

Where X = mean density and S^2 is the variance

4.5.6. Morisita's Index of Dispersion

Morisita's index of dispersion (I_δ) has been extensively used to evaluate the degree of dispersion/aggregation of spatial point patterns (Morisita, 1959; Tsuji and Kasuya, 2001).

This index is based on random or regular quadrate counts, and is closely related to the simplest and oldest measures of spatial pattern, the variance: mean ratio n (Dale *et al.*2002) and to other dispersion indices. Because Morisita's index can be calculated for different quadrate sizes, the scale of analysis is not inherent, and it can be used to investigate pattern over a range of densities and scales (Kristensen *et al.*, 2006).

The value of this index equal to 1 indicates random, more than one indicates aggregated, and less than one uniform distribution. Suppose n quadrates are sampled and xi represents the number of individuals in the ith quadrate.

Define δ as the probability that individuals of a randomly drawn pair will come from the same quadrate. If the xi came from a random or poisson distribution, the expected value of δ is $1/n$.

Then the index $I\delta$ can be defined as the ratio of δ to its expected value assuming a random distribution. Values for $I\delta$ may range from 0 to n and N is the total number of individuals sampled in n quadrate. $I\delta$ is calculated as follows and used to classify population.

Schuh *et al.* (1986) have used this approach for analysis of spatial patterns in sorghum downy mildew infected with *Peronosclerospora sorghi*, similarly Thal and Campbell (1986) have used this for analysis of disease severity data for alfalfa leaf spot caused by *Leptosphaerulina briosiana.* They reported following formula for computation.

$$I_\delta = [(\Sigma(x^2) - \Sigma x/((\Sigma x)^2 - \Sigma x)]n$$

Where, x is the disease incidence and n is the number of sampling units. Through I_5 they also quantified the index of clumping IC = (I_5 *for quadrate of area x*)/ (I_5 *for quadrate of area of 2x*). Murolo *et al.* (2014) have used this tool for bois noir grapevine disease caused by stolbur phytoplasma

$$I\delta = n\frac{\Sigma xi(xi - 1)}{N(N - 1)}$$

To determine if the sampled population significantly differs from random, the following large sample test of significance can be used: $Z = (I\delta\text{-}1)/(2/nm2)(1/2)$, where m = mean population density per leaf in each sampling date and n = the number of sample units.

By comparing the value of Z with tabulated values for a random distribution and reject the hypothesis that the sampled population is dispersed $|Z| > z(\alpha/2)$ randomly if (Pedigo and Buntin, 1994). Poole, (1974) and Freitas-Nunes and Rocha (2006) have used the following F-test for significance of I_δ

$$F = \frac{I_\delta \left(N - 1_{+n} - N\right.}{n - 1}$$

Where I_δ is the calculated vale of the index of dispersion of Morisita, n is the number of sites, N is the total number of individuals found in all of the n site

4.5.6.1. *Standardized Morisita index*

Krebs (1999) suggest using an improved Morisita's index, which converts it to a scale ranging from -1 to +1 and is independent of sample size.

A minimum of 50 quadrats is recommended, but 200 or more quadrats with a highly aggregated population.

Krebs (2000) has a program for the Morisita index and a standardized Morisita's index, listed under spatial pattern-indices of dispersion.

According to Kreb, 1999 calculate the Morisita index of dispersion as above and two new indices, the uniform index and the clumped index can be calculated as fallows:

4.5.6.2. *Uniform index (M_U)*

$$\text{Uniform Index } (M_U) = \frac{\chi^2_{0.975} - n + \sum x_i}{(\sum x_i) - 1}$$

Where = value of chi-squared from the table with (n-1) degree of freedom that has 97.5% of the area to the right, X_1 = number of organism in quadrate I N = number of quadrats.

4.5.6.3. *Clumped index (M_C)*

$$\text{Clumped Index } (M_C) = \frac{\chi^2_{0.025} - n + \sum x_i}{(\sum x_i) - 1}$$

Where = value of chi-squared from the table with (n-1) degree of freedom that has 2.5% of the area of the right. Then calculate the Standardized Morisita's Index (MIS) by one of the following equation;

- When $I_\delta \geq M_C > 1.0$ then

$$\text{MIS} = 0.5 + 0.5 \left[\frac{I_\delta - M_C}{n - M_C}\right]$$

- When $M_c > I_\delta \geq 1.0$ then:

$$MIS = 0.5\left[\frac{I_\delta - 1}{M_C - 1}\right]$$

- When $1.0 > I_\delta > M_U$ then:

$$MIS = 0.5\left[\frac{I_\delta - 1}{M_U - 1}\right]$$

- When $1.0. > M_U > I_\delta$ then;

$$MIS = -0.5 + 0.5\left[\frac{I_\delta - M_U}{M_U}\right]$$

MIS > 0 = clumped or aggregated distribution; MIS = 0 = random distribution; MIS < 0 = uniform distribution or; +1 = maximum aggregation; 0 = random distribution and -1 = perfectly uniform distribution

4.5.7. Cole's Index of Dispersion (I)

Cole (1946) developed the formula to study the dispersion of the population in nature.

Cole's index value is greater than the value of maximum regularity and randomness, and indicates the aggregative nature of dispersion. I<0, it represented uniform distribution, I =0 indicate random one, that I>0 revealed aggregative one. Adapted by Danielson and Gaines (1987).

$$I = \frac{\Sigma(x)^2}{(\Sigma x)^2}$$

4.5.8. Kuno Index (Ca)

This index was provided by Kuno and Iwao (1968), which calculated by the formula Ca = $(S^2/X\text{-}1)/X$; where X represents the mean and S indicates variance. When Ca<0, it indicated uniform distribution, Ca=1 represented randomness and Ca>1 was aggregation.

4.5.9. Taylor's Power Law

The Taylor's power law (Taylor, 1961, 1984) is based on the empirical relationship between the mean (m) and the variance S^2 that are related by the exponential function

$$S^2 = am^b$$

The parameters a and b are population parameters. Parameters b is considered to be an intrinsic measure of population aggregation varying continuously from zero for regular distribution (S^2 =a with a < 1) through 1 for random distribution (S^2 = m with a =1) to "for strongly contagious distribution (Jimenez *et al.*, 2001).

The parameter a is a scaling factor depending on the size of the sample unit (Elliot, 1971) and the exponent b is used as an aggregation index. Parameter b is assumed to be independent of the mean so that it is a value measuring aggregation without being confused by changes in population mean density.

The parameters a and b are determined by linear regression of sample means and variance after the mean and variance estimates were log-transformed.

$$\log S^2 = \log a + b \log m$$

For disease incidence data, the binary form of Taylor's power law provides a simple model with only two parameters to detect and quantify aggregation at the scale of individual sampling units.

Taylor's power law relates the observed variance (v_{obs}) to the population mean (m) for count data with no upper limit through a power function $V_{obs} = am^b$ where a and be are parameters and particularly for count data, when $V_{obs} = m$ (*i.e.*, a =b=1) the data can be described by the poison distribution, suggesting a random pattern.

Parameter b is the rate of change of log variance with log mean and describe as index of aggregation. However, for binary data, the binomial is the appropriate distribution for a random spatial pattern in which variance of $V_{bin} = np\,(1\text{-}p)$, where, n indicates the number of observations (individuals) within a sampling units and p is the probability of plant being diseased.

Gent *et al.* (2008) have suggested that the estimate of is equivalent to disease incidence, the mean proportion of plant diseased and accordingly the binary power law can be expressed as

$$V_{obs} = A_x V_{bin}^b = A_x \, [n\,\hat{p}\,(1-p)]^b$$

Turechke and Maddan (1999) have provided its log transformation which expressed as

$$1n\,(V_{obs}) = 1n(A_x) + bin\,(v_{bin})$$

In above equation, conditions with A_x =1 and b=1 indicates in a random pattern of diseased individuals, however, when A_x >1 and b =1, disease incidence has an aggregated pattern that is not dependent on *p*, while *b*>1 indicate that aggregation is systematically related to *p,* which is relevant to most plant diseases at multiple scales (Turechek *et al.*, 2001 and Xu and Madden, 2002).

Further, Madden and Hughes (1995) have suggested that binary power law provides a appropriate method of quantifying and characterizing aggregation of disease form multiple data set when they are collected at a single or over several time periods.

Gent *et al.* (2008) have quantified the incidence of hop powdery mildew on leaves, caused by *Podosphaera macularis*, variability in heterogeneity of disease and the estimated binary power low parameters were also quantified.

They calculate the spatial analysis of data sets at level of individual rows (row level) and multiple rows within a field level. Disease incidence assessed using a cluster sampling design (Hughes *et al.*, 1996). In cluster sampling, the sampling units not the individual (plant), but a group (cluster) of individuals, with disease incidence data and each individual in a cluster is classified as healthy or diseased. For such technique, the lower limit is 0 (diseased individuals) per cluster, as well as there is an upper limit also. In their study the degree of aggregation of disease incidence was quantified by using the parameter of θ of the beta binomial distribution, which provides a measure of variation in disease incidence per sampling units (Madden and Hughes, 1999). In the similar study the index of dispersion (*D*) was quantified by dividing the observed variance of diseased leaves (V_{obs}) by the theoretical variance for a binomial distribution (V_{bin}).

$$V_{obs} = \sum \frac{(X_i - \hat{p}n_i)^2}{(N-1)}$$

$$V_{bin} = n\hat{p}\,(1-\hat{p})$$

When θ=0 or D=1 the pattern of diseased plants is random, with aggregation indicated when D>1 or θ>0 and the degree of aggregation directly proportional to the magnitude of the statistic. Binary form of power law relationship relate the observed variance (V_0) of disease incidence between quadrats to the variance expected (V_R) if a binomial distribution is assumed for the number of diseased plants per quadrat

$$V_0 = AV_R^{\beta} = A\left[\frac{p\,(1-p)}{n}\right]^{\beta}$$

Where p is the disease incidence and n is the total number of plants per quadrats. When V_0 is greater than V_R it indicates aggregation. This variance/variance ratio is called index of dispersion (D) and provides a measure of the degree of aggregation of a diseased plants.

The value of D is determined jointly by A, n I and β. According to Xu and Ridout (2000) if D is plotted against p, parameter A affects only the amplitude of the curve, whereas β determines its shape. The greater the value of A, the greater the value of D.

The greater the value of β, the lower the value of D, since $p(1-p)/n<1$. If $A<1$, β must be less than 1 to satisfy the condition *i.e.* $V_0/V_R = A[p\ (1-p)/n]^{\beta-1} >1$. When $\beta =1$, D is constant and equal to A and when both A and $\beta =1$, there is no aggregation or over-dispersion ($D=1$). When $\beta >1$, D increased with p, reaches a maximum, $A[1/(4n)]^{\beta-1}$. K is an another parameter which indicates the intraclass correlation and it measures the tendency of the plants within a sampling quadrat to have a similar disease status. Positive value of k indicate aggregation of disease and it related with the binary forms of power law and index of dispersion.

$$K = (D-1).(n-1)$$

Gosme *et al.* (2007) have used power law analysis for quantification of size, shape and intensity of aggregation of take all disease during natural epidemics in second wheat crop caused by *Gaeumannomyces graminis* var. *tritici*. They used binary form of Taylor's power law to relate heterogeneity of disease incidence by taking the natural logarithm of observed and theoretical variance, power function transformed into a linear function expressed as

$$1n\ (V_{obs}) = a + b\ 1n\left[\frac{p(1-p)}{n_h}\right]$$

Where V_{obs} is the observed variance of the root disease incidence per plant over the 300 plants in a plot and a and b are the intercept and slop of a straight line, respectively and p is the ratio of number of diseased roots to the total number of roots in the plot and n_h is the harmonic mean of the number of roots per plant. If $a=0$ and $b=1$, diseased root have a random spatial pattern that could be described by the binomial distribution. If $b=1$ and $a>0$, diseased roots would have an aggregated distribution, and, if $a>0$ and b#1, the degree of aggregation would change systematically with incidence (Turechek and Mahaffe, 2004).

4. 5.10. Iwao's Method

The Iwao's patchiness regression method quantifies the relationship between the mean crowding index (m^*) and the mean (m) by the following formula

$$M^* = \alpha + \beta m$$

Where m^* was determined as $[m(S^2/m\_1)]$ (Lloyd, 1967). The intercept (_) is the index of the basic component of a population or basic contagion (where α <, = and > 0 represent regularity, randomness, and aggregation of populations in spatial patterns, respectively), and the slope (β) is the density contagiousness coefficient interpreted in the same manner as *b* of Taylors regression.

4.5.11. Camargo's Index (E')

Developed by Camargo, 1993. Adopted by Payne *et al.*, (2005) and Chalcraft *et al.* (2009). This index is relatively unaffected by sites with very few organism, and is unaffected by site richness (Krebs, 1999).

$$E' = 1 - \sum_{i=1}^{S} \sum_{j=i+1}^{S} \left(\frac{P_i - P_j}{S} \right)$$

Where P_i = proportion of individuals of a species at site i; P_j = proportion at site j; and s = total number of sites in sample. Index value ranges from 0 (patchy) to 1 (even).

4.5.12. Simpson's Measures of Evenness, $E_{1/D}$ (Variant of Simpson 1949)

This evenness index is a variant of the reciprocal Simpson index, D (Simpson 1949). Index values ranges from near 0 (1/s) (Patchy or skewed) to 1 (even), and the index is relatively unaffected by sites with very few individuals. Adopted by Payne *et al.* (2005).

$$E_{1/D} = \frac{1/D}{S}$$

Where D= Σp_1^2 and P_i = proportion of total individuals of a species at site i, and s = number of sites in the samples. McGarigal and Mark proposed a modified Simpson Diversity Index (MSIDI) for distribution analysis. MSIDI = 0 when the landscape contains only 1 patch (*i.e.*, no diversity). MSIDI increases as the number of different patch types (*i.e.*, patch richness, PR) increases and the proportional distribution of area among patch types becomes more equitable.

$$\text{MSIDI} = 1n \sum_{i=1}^{m} P_1^2$$

4.5.13. Smith and Wilson Index (E_{var})

$$E_{var} = 1 - \frac{2}{\pi} arctan \frac{\left\{\sum_{i=1}^{s}[1n(n_1) - \sum_{i=1}^{s} \frac{1n(n_j)}{s}]\right\}^2}{S}$$

Where n_i = number of individuals of a species at site in i in the (i = 1,2,.3,4......s), s = total number of sites in sample. This evenness index ranges from 0 (patchy or skewed) to 1 (even).

The index is mathematically symmetrical (equally sensitive to sites with many organism and those with few organism), and estimates evenness at occupied site only. Adapted by Pyane *et al.* (2005).

4.5.14. Dispersion Index (D_L = Payne, 1997)

This index is the variant of J' of Pielou (1975), and based on H' of Shannon (Shannon and Weaver, 1949).

$D_L = H'/1n\ S_s$ = where H can be calculated as

$$H' = \sum_{i=1}^{s} P_i\, 1n(p_i)$$

Where, *pi* proportion of individuals of a species at site *i*, *s* 5 number of sites occupied by that species, *Ss* total number of sites surveyed. Values range from 0 (patchy) to 1 (even).

This index is asymmetrical (somewhat sensitive to sites with fewest hypothetical organisms), and mathematically incorporates sites surveyed but with zero individuals, unlike the original *J*9 (Pielou 1975).

4.5.15. Cassied R.M. Index

Cassie´s (1962) index of patchiness, *P*, evaluate dispersion patterns and calculated by $P = (s - \bar{x})/\bar{x}^2$, where x is the mean larval count per sample divided by the average volume filtered (ind/m3) and *s*, the variance.

Cassie´s index represents a measure of the intensity of pattern (*e.g.* degree of aggregation) that is independent of the mean. Therefore, it is possible to compare spatial patterns when means differ among samples. p>0 indicates patchy distribution (assuming the Poisson distribution). This was adopted by Matheven *et al.* (2003), Williams *et al.* (2009) and Williams and Sais, (2010).

4.5.16. Doublet and Runs Analysis

Doublet and run analysis has been used to detect whether diseased plants in a homogeneous field have a random or nonrandom occurrence within plant rows. In doublet analysis the observed number of adjacent diseased plants is compared with the number expected if the disease were randomly distributed in the field.

If the observed number is greater than the expected number, contagion within the field is also suspected.

Johnson *et al.* (1988) have used this approach to examine direction of spread in the direction of each axis of the rectangular hop yards, in the direction of the prevailing wind, and in the direction perpendicular to the prevailing wind. Through the double analysis they reported aggregation in most location because the number of occurrences of two adjacent diseased plants was greater than the expected number.

4.6. ARBITRARY SAMPLE ANALYSIS

4.6.1. Spatial Point Analysis

A spatial point pattern can be defined as a set of locations, irregularly distributed within a region of interest, which have been generated by random mechanisms (Diggle, 1983).

Commonly, spatial point pattern analysis is used to measure how individuals are located with respect to each other over the horizontal space within a region of interest. There are three fundamental spatial point patterns: complete spatial randomness (CSR), regularity and clustering.

This classification may be oversimplified, but can be useful at an early stage of spatial analysis (Diggle, 1983). It is also helpful to formulate a model for the underlying spatial point process. Researchers have developed many methods of spatial point pattern analysis.

These methods can be classified into different categories based on various criteria. For example, Cressie (1993) classifies the methods into five broad categories: quadrate methods, distance methods, kernel estimators of intensity, nearest neighbor distribution functions and K -function analyses.

Each method has its advantages and problems, and works well for particular situations or datasets. Detailed descriptions of these methods and the associated test statistics can be found in the texts such as Ripley (1981), Cliff and Ord (1981), Diggle (1983) and Cressie (1993).

Spatial point patterns are based on the coordinates of events such as the locations of outbreak of a disease. It is also possible that they include attribute information such as the time of outbreak occurrence.

Data on point patterns can be based on a complete map of all point event or a sample point pattern. The basic interest of a spatial point pattern analysis will be to detect whether it is distributed at random or represents a clustered or regular pattern.

This spatial technique are aimed at explaining an observed point pattern and typically involve comparison with the model of CSR. A point pattern generated by a random spatial process should follow a homogeneous poisson process. This implies that every event has an equal probability of occurring at any position in the study area and occurrence is independent of the location of any other event, hence the absence of first and second order effect.

A spatial point pattern can be quantified in term of the intensity of the process by using its first order properties, measured as the mean number of events per unit area. Second order properties or spatial dependency are analyzed on the basis of the relationship between pairs of points or area. First order properties describes the way in which the expected value (mean or average) of the process varies across space, while second order properties describe the covariance (or correlation) between values of the process at different region in space. More formally, first order properties are described in term of intensity, λ (s), of the process, which is the mean number of events per unit area at the point s

First order effect for point patterns can be examined with using two techniques (a) quadrat counts and (b) kernel estimation.

The first one involves dividing the area into sub-regions of equal size-quadrats and produce a summary statistic on the basis of the number of counts per quadrat. The counts are then divided by the size of the area. This technique provide an indication of the variation of the intensity of the underlying process in space, however one major disadvantage of the technique is associated with its property through which it aggregated the information into area type data that can result in loss of information.

Kernel estimation is a technique which uses the original point location to produce a smooth bivariate histogram of intensity. On the other hand, second order properties can be investigated using the distance between the points, particularly nearest neighbour distance.

Information about the whether infection occurs at random is important to an understanding of the epidemiology of the infection by soil borne pathogen. The need to determine if clusters or random exist in a given data set has prompted the development of several techniques that come from a field of study called as point pattern analysis (Kamu *et al.*, 2015).

There are many point pattern analysis or statistical techniques that can be utilized to identify the spatial pattern of plant dieseas and the simplest theoretical model for a point pattern analysis is that of Complete Spatial Randomness (CSR). Basically CSR relay on the fact that the events are distributed independently according to a uniform probability distribution over the region. It describes that pattern of point would occur by chance in a completely undifferentiated environment (Mathur, 2015). CSR can be generated by means of two assumptions

1. All places are equally likely to be the recipient of event or case, and
2. All cases are located independently of one another

There are a range of methods available to test for CSR. Some are based on quadrat counts such as the index of dispersion test, while other use nearest neighbor distances such as the Clark-Evans test or the K function test.

In analyzing spatial point patterns, sample points near the boundaries of the study region may have their neighbors located outside the region. Thus, the distance to these neighbors and any statistics involving these neighbors may be unobservable. This causes the so-called edge effects.

If such edge effects are not taken into account, considerable bias can be introduced into the computation of spatial statistics for the sample points as well as the region (Moeur, 1993; Haase, 1995).

There are four main ways to deal with edge effects (Cressie, 1993) – (1) Buff zone method (Sterner *et al.,* 1986): A buffer (guard) zone is established inside the perimeter of the study region. The sample points within the buffer (guard) zone are not used for the spatial pattern analysis, but are used as the nearest neighbours of the selected points in the study area. (2) Border method (Diggle, 1979; Ripley, 1988; Gignoux *et al.* 1999): This method corrects edge effects by eliminating all sample points that are closer to the border than to their nearest neighbours inside the study region. (3) Toroidal edge correction method (Ripley, 1981):

It treats a rectangular study region as a torus, where sample points near two opposite edges are considered close. This is equivalent to making eight identical copies of the study region and placing these copies around the original one and (4) Weighting method (Ripley, 1976; Getis and Franklin, 1987; Andersen, 1992). The weight, say *w* (*s i* , *s j*), is equal to the proportion of the circumference of a circle centred at *s i* (the location of sample point *i*) passing through *s j* (the location of sample point *j*), and that is inside the study region.

This correction is based on the assumption that the region surrounding the study area has a similar point density and spatial pattern to the study area. Details about these methods are being discussed in chapter 3.

A fundamental property of every point process is its intensity the expected number of events per unit area at the point s. The simplest spatial process is complete spatial randomness (CSR), termed the homogeneous Poisson process, with intensity k. It has two important properties (Stoyan and Penttinen, 2000).

1. The number of events in any region A (with area $|A|$) follows the Poisson distribution with a mean of $\lambda|A|$.
2. Given n events in A their positions follow an independent sample from the uniform distribution on A.

The first property states that under CSR the intensity of events will not vary across the region, and the second that events do not interact in the region (Diggle 2003).

Commonly, observed events are then tested for departure from this CSR distribution. At a given scale, event sets may exhibit departure from CSR as either (i) clustering (aggregation in the bivariate case), or (ii) regularity (segregation), with CSR acting as the 'dividing line' between the two (Diggle, 2003).

Spatial point patterns can be characterized in terms of their first-order and second-order properties, each the focus of different analyses. First-order properties are related to the mean number of events per unit area, while second-order properties are related to the (co)variance of the number of events per unit area.

Thus, in the same way that the mean (μ) and variance (□α) are the first and second moments of a regular probability distribution, the density (λ) and covariance structure (k) are the first and second moments for a two-dimensional distribution.

Finally, point processes can be described in terms of their stationarity and/or isotropy. A stationary process is one where all the statistical features of the process are the same at any location. Processes which are invariant under translation (moving very event the same distance in the same direction) are termed stationary or homogeneous; this means that the underlying characteristics of the pattern, such as the mean (first-order stationarity) and/or variance (second- order stationarity) of a variable are constant over the area under study. Those which are invariant to rotation are termed isotropic—that is, the characteristics of the pattern are the same in any direction (Guttorp, 1991; Dale, 1999; Diggle, 2003).

A spatial point pattern may be characterized in terms of so-called first-order and second-order effects (Bailey and Gatrell, 1995). The first-order properties relate to global or area-wide variation in the mean number of occurrences of the process in space. They can be more formally described in terms of the intensity, λ, which is the mean number of events per unit area at the point *s* (Bailey and Gatrell, 1995). The intensity is defined as:

$$\lambda_{(s)} = \lim_{ds \to 0} \left\{ \frac{E(Y(ds))}{ds} \right\}$$

Where ds is an area of a small region around the point s, E is the expectation operator, and Y (ds) is the number of events in this small region (Shu and Burn, 2003). The second order properties, or spatial dependence, of a spatial point process involve the relationship between numbers of events in pairs of areas in R (Bailey and Gatrell, 1995). This can be formally described by the second-order intensity, $\lambda(s_{-i}, s_j)$ of the process. Its mathematical form is defined as

$$\gamma_{(s_i s_j)} = \lim_{d_{si}\, ds_j \to 0} \left\{ \frac{E(Y(ds_i)Y(ds_j))}{ds_i\, ds_j} \right\}$$

Various methods have been developed to examine the first-order and second order effects. Generally kernel estimation uses for the first order properties and the Ripley's K function uses as to estimate the second-order properties. Various test statistics relalted with first and second order test are presented in Table 3 (Adapted from Perrey *et al.*, 2006).

Table 3: Description of methods of point patterns analyses (Perry *et al.*, 2006)

Test statistics	*Description*	*References*
First order test		
• First order nearest-neighbour[u]	Distance to closest event and deviation from theoretical expectation under CSR	Clark and Evans (1954)
• Nearest-neighbour contingency table[b]	Contingency table test based on first-order nearest neighbour distances	Pielou (1961); Dixon (1994)
• Diggle's F(x)[u] and	Cumulative frequency of distances to	Diggle (2003)

Table 3: *Contd.*

Table 3: *Contd.*

Test statistics	*Description*	*References*
G (w) (refined nearest-neighbour	closest event and deviation from theoretical expectation under CSR	
• N-order nearest neighbour	Distance to 1 to n closest events and deviations from expectation under CSR	Davis *et al.* (2000); Thompson (1956)
• All events to all events	Distance from each event to every other event, binned into distance classes and analysed as absolute frequencies	Galiano (1982); Dale (1999
Second-order tests		
• Ripley's K (t) and l (t) transformation	Number of events within a circle of radius sequentially larger t from each focal event, and deviation from expectation at each tunder CSR. Haase (2001) has modified this method so it can test for anisotropic associations.	Ripley (1977); Lotwick and Silverman (1982); Haase (2001)
• Neighbourhood density function (NDF (t))	Similar to Ripley's K except non-cumulative, *i.e.*, distance classes are annuli, not circles. The NDF is also known as the pair correlation function (Stoyan and Stoyan, 1994) and the O-ring statistic (Wiegand and Moloney, 2004).	Ward *et al.* (1996); Condit *et al.* (2000); Stoyan *et al.* (1994)
In-homogeneous tests		
• In-homogeneous forms of Ripley's K-function and NDF	Forms of Ripley's K and the NDF (or PCF)which do not assume first order homogeneity in the study region. Intensity is estimated at each event and this estimate is used to calculate the inhomogeneous form of the statistics	Baddeley *et al.* (2000)
• Local tests Getis and Franklin's L (d)	Localised version of Ripley's K in which L(d) is calculated for each event individually, providing information concerning local trends in pattern (*e.g.*, areas of aggregation *vs.* areas of regularity in the same plot)	Getis and Franklin (1987)
• Spatial analysis by distance indices (SADIE) SADIEM[u]	Calculates the 'distance' from the event set to regularity by moving events until a regular Voronoi tessellation (approximating a hexagonal lattice) is achieved; allows calculation of index of aggregation (Ip), test of departure from CSR, and diagnostic-exploratory data analysis	Perry (1995)

u—Univariate event sets only, b—bivariate event sets only

Cressie (1993) has reported detailed formula and theoretical distributions of the 13 statistics that includes (1) Pielou's (1959) index of non-randomness, (2) Clark and Evans's (1954) index, (3) Hopkins's (1954) aggregation index, (4) Skellam's (1952) test statistic, (5) Pollard's (1971) test statistic, (6) Holgate's (1965) test statistic I, (7) Holgate's (1965) test statistic II, (8) Byth and Ripley's (1980) test statistic, (9) Diggle's (1977) test statistic, (10) Diggle, Besag and Gleaves's test statistic (Diggle *et al.*, 1976), (11) Besag and Gleaves's (1973) test statistic I, (12) Besag and Gleaves's (1973) test statistic II and (13) Cormack's (1979) test statistic.

4.6.2. Quadrant Count Method

The Quadrant Count Method can be described simply as partitioning the data set into n equal sized sub regions; these sub regions called as quadrants.

Each quadrant counting the number of events that occur and it is the distribution of quadrant counts that will serve as our indicator of pattern.

The choice of the quadrant size can greatly affect our analysis, where large quadrants produce a coarse description of the pattern.

If the quadrant size is too small then many quadrants may contain only one event or they might not contain any events at all. Rule of thumb is that for the area of a square is twice the expected frequency of points in a random distribution (*i.e.*, $2\frac{Area}{n}$), where n is the number of points in the sample size. After partitioning the data set into quadrants, the frequency distribution of the number of points per quadrant has been constructed. The Mean and Variance of the sample are then computed to calculate the Variance-to-Mean Ratio (VTMR). The following is the way we will interpret the VTMR of a sample:

If VTMR>1, the pattern is clustered. This implies that the data set has one or more groups of points in clusters and large areas of maps without points.

The region might look like Fig. 4.

If VTMR<1, the pattern is regularly dispersed implying the events are distributed more or less regularly over the region. A regularly dispersed area might look like Fig. 5.

If VTMR=1, the pattern is random. This implies the data set has no dominant trend towards clustering or dispersion. A random pattern may look like Fig. 6.

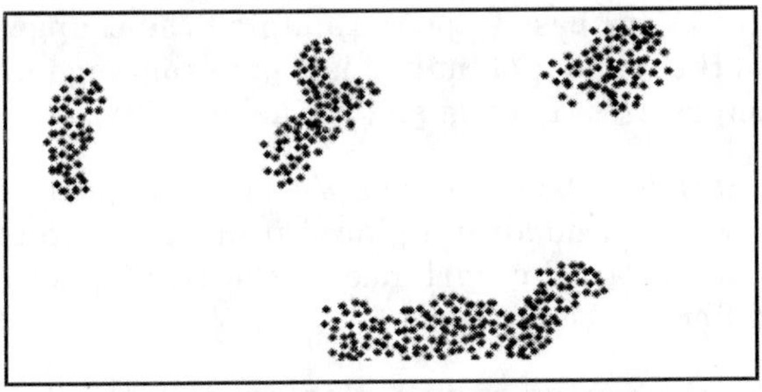

Fig. 4: Clustered pattern

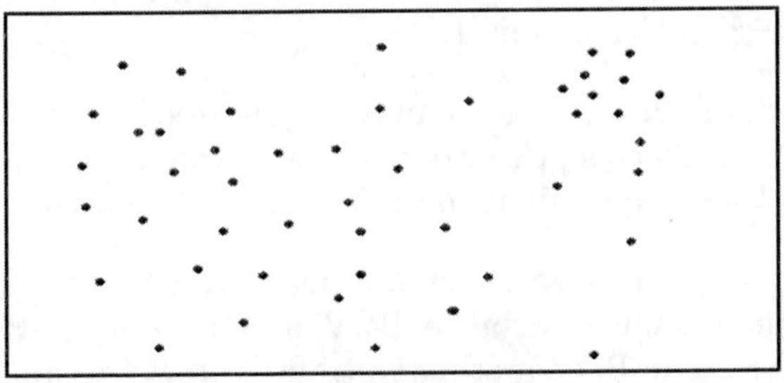

Fig. 5: Regularly Pattern

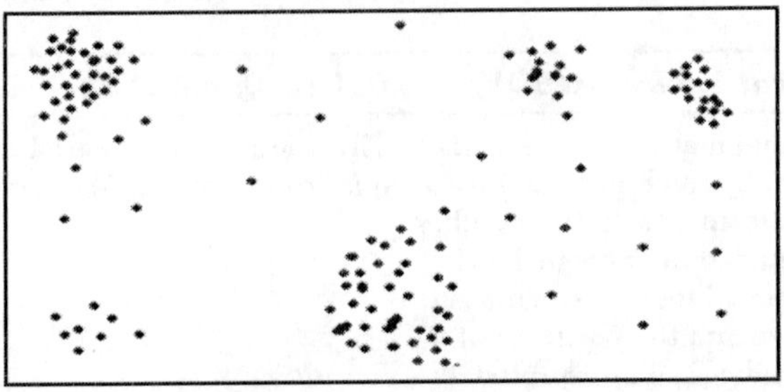

Fig. 6: Random Pattern

4.6.3. Quadrat Variance Methods

According to Ludwig and Reynolds (1988) Quadrat variance method concerned with spatial pattern of individual of species that are found continuously across a community (*e.g.* hosts in a forest).

In this method some type of arbitrary sampling unit chosen to obtain a sample and consequently results may be obtain a sample and consequently results may be influenced by size and shape of sampling unit. To address this problem, methodologies have been developed that allows us to examine the effect that varying the size of the sampling unit has on the detection of some underlaying pattern. Collectively these are called quadrate – variance methods.

These methods are based upon examining the changes in the mean and variance of the number of individuals per sampling unit over a range of different sample sizes (Ludwig and Reynolds, 1998).

Data are gained from belt transects of contiguous quadrats (*i.e.,* a joined or continuous series of quadrats) placed in a linear manner across the population of interest. The variance of the number of individuals is calculated at different "block sizes".

The block sizes are obtained by combining progressively the *N* quadrats (therefore increasing the theoretical sample unit size) in a prescribed manner (Ludwig and Reynolds, 1998).

In Populations displaying a patchy or clumped distribution (such as many holothurian species), the variance peak (maximum variance) can be interpreted as being equivalent to the radius of the patch.

Two types of quadrat variance methods are generally utilize. One is based on Blocked-Quadrat Variance (BQV) and the other on Paired-Quadrat Variances (PQV). The BQV methods utilize changes in quadrate size *via* the blocking or combining of adjacent or abutting quadrats to identify the pattern intensity. The PQV methods utilize changes in quadrat spacing.

Blocked Quadrat Variance (BQV)	***Paired Quadrat Variance (PQV)***
BQV is the original method for analysis of spatial pattern by combining quadrats into blocks. In this approach, the number of quadrat is included in only one block. The variance among blocks is calculated for each block size and the positions of peaks in the variance are interpreted as reflecting scale of patterns in the data.	This method uses spaced individual quadrats from a grid or transect.
In this method the variance is determined for the powers of 2 as well as limitations on the necessary rectangular shape of certain block sizes (*e.g.*, a block size of 8 must be 2 × 4) (Dale, 1999).	It calculates the average squared difference between quadrats at each of a range of distance, b.
Two avoid this limitation; Hill (1973a) developed another blocking scheme, the two-term local quadrat variance (TTLQV). Dale, 1999) have further defined the properties of Three-term local quadrat variance (3TLQV); Four-term local quadrat variance (4TLQV); Nine-term local quadrat variance (9TLQV) and Two-term local quadrat covariance (TTLQC).	This version uses all the possible pairs of quadrates. Peak in the plot of the variance are indicative of scales of pattern in the data.

4.6.3.1. *Blocked quadrat variance (BQV)*

Through block quadrate variance method we can compute the variance of the number of individuals at different block size. According to Ludwig and Reynold (1998) the block size are obtained by combining progressively the N Quadrate in following manner. A belt of transect of eight continues quadrate represent the block size one.

1	2	3	4	5	6	7	8

At block size two adjacent quadrate are joined and forming four rectangular quadrates that are twice as block size one.

1,2	3,4	5,6	7,8

1,2,3,4	5,6,7,8

The next block size is 4, formed by joining adjacent quadrats of blocks size 2. Since adjacent pairs are joined, the BQV block sizes increase in power of 2 (*i.e.,* $2^1 = 2$, $2^2 = 4$, $2^3 = 8$, $2^4 = 16$) and so on. For block size one the variance can be calculated as

$$\text{Var (X)1} = \left(\frac{2}{N}\right)\left\{\left[\frac{1}{2}(x_1 - x_2)^2\right] + \left[\frac{1}{2}(x_3 - x_4)^2\right] + \cdots \ldots + \left[\frac{1}{2}(x_{N-1} - x_N)^2\right]\right\}$$

The data contained in adjacent pairs of quadrate are combined to give oblong two-quadrat block, which are twice as large (but half as numerous) as the original one –quadrate blocks. The variance at block size 2 can be calculated as

$$\text{Var (X)2} = \left(\frac{4}{N}\right)\left\{\left[\frac{1}{4}(x_1 + x_2 - x_3 - x_4)^2\right] + \left[\frac{1}{4}(x_5 + x_6 - x_7 - x_8)^2\right] + \cdots \ldots + \left[\frac{1}{4}(x_{N-3} + x_{N-2} - x_{N-1} - x_N)^2\right]\right\}$$

Variance at higher block sizes (*i.e.* 8,16,32,64 and so on) are computed using computations analogous to first equation. The largest block size can be determined by the size (or length) of the belt.

The variance then plotted as function of block size and peaks in the plot are interpreted as indicative of scales of patterns. This method can only examine the block size that are power of two (Dale, 1999) and clearly relies on the transect length being chosen appropriately as a number that itself a power of 2.

This property becomes a greater liability as the length of transect or the largest block size examined. This method has a number of other limitations, for instance, the results depend to some extend on the starting position of the blocking and the variance peak can occur at a block size smaller than then the actual scale of pattern (Errington, 1973; Dale 1999).

Another drawback is that the size of the blocks is confounded with the distance between their centers (Goodall, 1974 Dale 1999). These concerns gave rise to a number of new methods that avoided some of the problems of BQV.

The Two-Term Local Quadrat Variance analysis (TTLQV) is a modification of the basic blocked quadrat variance (BQV) methods for spatial pattern analysis that are limited to powers of 2 (Ludwig and Reynolds, 1988).

The TTLQV uses variance data in a similar manner to the BQV method although it has a more refined "blocking scheme" in its calculation to overcome the BQV's limitation.

4.6.3.2. *Two-term local quadrat variance (TTLQV)*

Described by Turechek and McRoberts (2013) particularly for considerations of scale in the analysis of spatial pattern of plant disease epidemics.

This a method of spatial pattern analysis that used blocks of adjacent quadrats from a transect of contiguous quadrats.

It calculates the average squared difference between pairs of adjacent blocks for a range of block size. Peaks in the plot of the variance are indicative of scales of pattern in the data. TTLQV has been shown to detect both the scale and the intensity of spatial pattern (Usher, 1975; Ludwig and Goodall, 1978).

Unlike earlier methods, TTLQV is not restricted to detecting pattern on a scale of $2n$ blocks (Hill, 1973; Ludwig and Reynolds, 1988). However, TTLQV has some limitations. Pielou (1977) and Galiano *et al.* (*e.g.* 1987) showed that TTLQV analysis actually reports the average of the patch and gap sizes.

They showed that transects with homogeneous patch size and heterogeneous gap size will result in variance peaks at differing block sizes. To allow for the detection of patch size *per se*, Dale and MacIsaac (1989) developed a method based on combinatory that, subsequent to quadrat-variance techniques, differentiates patch and gap sizes.

In addition, Errington (1973) suggests that TTLQV will often result in block sizes that are smaller than the actual scale of pattern when examining artificial data in which the scale of pattern is known.

Currently, TTLQV analysis has only been guided by two rules (1) do not interpret a peak at the first block size, and (2) only interpret block sizes that are less than one tenth (or one half) of the total transect length. Jonathan *et al.* (1998) suggest that TTLQV is subject to additional guidelines. In particular, block sizes 1 ($\alpha = 0.5$), 2 ($a = 0.25$), and 3 ($\alpha = 0.11$) are not significant below their respective α -levels. In other words, block size 2 can only be considered if the examiner has *a priori* accepted an α -level of 0.25 or higher.

Again, block size 3 can only be considered if the examiner has *a priori* accepted an α -level of 0.11 or higher. It is not until block size 4 ($\alpha = 0.05$) that the probability of accepting randomness as pattern begins to conform to the standard α -level of 0.05. As one may expect, however, there are exceptions to this 'rule'. In particular, block sizes 2 and 3 could be interpreted if analysis of several transects of similar data resulted in identical information.

For example, the probability that a local maximum in variance occurs at block size 2 on each of three independent transects reduces the possibility of a Type II error. If the random probability of a variance peak occurring at block size 2 is 0.25, then the probability of that happening on three independent transects is 0.25, 0.25, 0.25 = 0.016 (1.6%) – all transects must result in exactly the same variance peak for this multiplicity of probabilities to hold true.

This caveat supports the supposition that multiple transects result in more reliable data interpretation. However, multiplicity of transects does not aid in the interpretation of block size 1, as actual patch may still be smaller than the quadrat size for such transects.

If one holds to the strict 10% interpretation rule of Ludwig and Reynolds (1988), only the four block sizes in a transect of 40 quadrats should be interpreted

Variance in TTLQV at block size 1 can be calculated as

$$\text{Var (X)1} = \left[\frac{1}{(N-1)}\right]\left\{\left[\frac{1}{2}(x_1 - x_2)^2\right] + \left[\frac{1}{2}(x_2 - x_3)^2\right] + \cdots \ldots + \left[\frac{1}{2}(x_{N-1} - x_N)^2\right]\right\}$$

and at block size 2 by

$$\text{Var (X)2} = \left[\frac{1}{(N-3)}\right]\left\{\left[\frac{1}{4}(x_1 + x_2 - x_3 - x_4)^2\right] + \left[\frac{1}{4}(x_2 + x_3 - x_4 - x_5)^2\right] + \cdots \ldots + \left[\frac{1}{4}(x_{N-1} + x_{N-2} - x_{N-1} - x_N)^2\right]\right\}$$

The other forms of this method are comparatively described below.

3TLQV	*4TLQV*	*9TLQV*	*TTLQC*
This method uses blocks of adjacent quadrate from a transect of contiguous quadrats. It calculates the average squared difference between a block and the sum of the two adjacent blocks on either side of it for a range of block size. Peaks in the plot of variance are indicative of scales of pattern in the data.	Four-term local quadrat variance is the two-dimensional equivalent of TTLQV, for studying the spatial pattern of single species, based on square blocks of quadrats from a grid of quadrats. It calculates a variance for each of a range of block sizes and peaks in the variance are interpreted as indicating scales of patterns	Nine-term local quadrat variance is the two-dimensional equivalent of 3TLQV, for studying the spatial patterns of a single species, based on square blocks of quadrats from a grid of quadrats. It calculates a variance for each of a range of block sizes and peaks in the variance and interpreted as indication scales of pattern.	Two term local quadrat covariance is a method of joint pattern analysis for two species that uses blocks of quadrats from a transect. It is based on two-term local quadrat variance (TTLQV) and calculates covariance for a range of distances. Positive and negative peaks in the plot of the coavariance are indicative of scales at which the two species are positively or negatively associated.

Variance for 3 TLQV, 4TLQV, 9TLQV can be calculated as

$$3TLQVV_3(r) = \sum_{i=1}^{n+1-3r}\left(\sum_{j=i}^{i+r-1n} x_j - 2\sum_{j=i+r}^{i+2r-1} x_j + \sum_{j=i+2r}^{i+3r-1n} x_j\right)/8r(n+1-3r)$$

$$4TLQV_4(r) = \sum_{i=1}^{n+1-2r}\sum_{j=1}^{n+1-2r}\frac{D_{ij} + D_{i+rj} + D_{ij+r} + D_{i+rj+r}}{32r^3\,(n+1-2b)(m+1-2r)}$$

$$9TLQV_9(r) = \sum_{i=b+1}^{n+1-2r}\sum_{j=b+1}^{m+1-2r}\frac{[T_r\,(i,j)]^2}{72r^3\,(n+1-3r)(m+1-3r)}$$

The scales of heterogeneity in a landscape detected by pattern analysis are important characteristics of the landscape, as they affect the responses of organisms to their environment and other ecological processes.

The scale of pattern is defined as half the average distance between patch centres. The scale of pattern of each cover type (plant species, litter and bare soil) in the plots was analyzed using four-term local quadrat variance, 4TLQV, an extension of the two-term local quadrat variance (TTLQV) method used to analyze scale of pattern in one dimensional transects.

The local variance in 4TLQV is calculated based on the total cover in each of four mutually contiguous square blocks, each consisting of 1, 4, 9, 16... of the original sample quadrats. The original quadrats are combined into square blocks, which also form a square of four blocks for 4TLQV. The variance is essentially the squared difference between the total cover in one block and the average total cover in the adjacent three blocks. This calculation is then performed for a range of block sizes. Peaks in 4TLQV indicate the scale of pattern in the data. To confirm that these peaks did not occur by chance, randomization tests were performed to construct a null variance model and the associated 95% confidence intervals.

Montane *et al.* (2011) have proposed the alternative of this approach and according to them although a given 4TLQC analysis can find significant peaks at different block sizes, however, their new approach Relative Index of Shrub Encroachment Susceptibility (RISES) is based on the first significant peak (*i.e.* the one at the smallest block size). RISES can be calculated for each plot using the following equation:

$$RISES = \frac{\sum s_i \text{ X } (\text{B}_{max} - \text{B}_1)\text{X} c_i}{(B_{max} - 1)X\ c_{grass}}$$

For each cover type (i) in a particular plot, si takes the value +1 or -1 depending on whether cover type i was positively or negatively associated with particular species. When no association was found between particular and cover type i, s_i takes the value 0.

B_{max} is the maximum block size (expressed as number of quadrats) used in the 4TLQC calculations, whereas Bi is the smallest block size (expressed as number of quadrats) in which 4TLQC found a significant association. c_i is the cover (%) for the cover type i in the plot, while cgrass is the cover in % for the grassland surface in the plot, and cgrass~Pci.

RISES summarizes the results of the covariance analyses by creating a weighted sum of the scales at which the first significant (positive or negative) peaks are found in plots of covariance as a function of block size (scale).

The index takes values close to +1 when there are only strong positive associations with all cover types, and values close to –1 when there are only strong negative associations found. Values close to 0 can indicate either a mixture of positive and negative associations, or a neutral susceptibility to shrub encroachment.

4.6.3.3. *Paired quadrat variance (PQV)*

In the PQV method, pair of quadrats are selected at a specified spacing or distance along the belt to estimate variance. Since the quadrats are of fixed size, the spacing or distance between the quadrats is the only component contributing to the variance, rather than both spacing and size as observed in BQV.

Spacing	***Quadrats***
1	(1) (2) (3) (4) (5) (6) (7) (8)
2	(1,2) (2,3) (3,4) (4,5) (6,7) (7,8)
3	(1,3) (2,4) (3,5) (4,6) (5,7) (6,8)
4	(1,4) (2,5) (3,6) (4,7) (4,8)
5	(1,5) (2,60 (3,70 (4,80
6	(1,6) (2,7) 3,8)
7	(1,7) (2,8)

Variance at a spacing of 1 can be calculated as

$$\text{VAR (x)1} = \left[\frac{1}{(N-1)}\right] + \left\{\left[\frac{1}{2}(x_1 - x_2)^2\right] + \left[\frac{1}{2}(x_2 - x_3)^2\right] + \cdots \left[\frac{1}{2}(x_{N-1} - x_N)^2\right]\right\}$$

Variance at spacing of 2

$$\text{VAR (x)2} = \left[\frac{1}{(N-2)}\right] + \left\{\left[\frac{1}{2}(x_1 - x_3)^2\right] + \left[\frac{1}{2}(x_2 - x_4)^2\right] + \cdots \left[\frac{1}{2}(x_{N-2} - x_N)^2\right]\right\}$$

Computation analogous to first equation are used to calculate the variances at greater spacing, up to a selected spacing less than the maximum spacing of N-1 (10% of N is recommended, Ludwig and Reynold, 1988). After that plot the variance obtained by the PQV method against the quadrat spacing.

In the PQV method, variances are calculated for all possible paired quadrats at a given spacing. The quadrat pairs are selected at random from all the quadrat pairs possible at a given spacing.

Random selection is without replacement, so that, once the quadrats are selected for pairs they are not used again at this spacing or at any other spacing selected. The advantage of this procedure is that the variance estimates at each spacing are independent and statistical probabilities can be computed to test for significance differences in variances at these different spacing.

4.6.3.4. *Interpretation of variance plot*

A plot of the variances against the series of block seizes or spacing for wither the TTLQV or PQV methods. The graph can be interpreted as follows (Ludwig and Reynold, 1988).

1. *Random*: If the individual are dispersed randomly over the study area the variances will randomly fluctuate with the different block size or spacing (Fig. 7).
2. *Uniform*: If the individual are regularly dispersed, the variance will be low and will tend not to fluctuate at different block sizes or spacing (Fig. 7).
3. *Clumped*: If the individual are clumped, the variance will tend to peak (Fig. 6) at a block size or spacing equivalent to the radius of the clump mean area. If the clumps themselves are regularly patterned, the variance peak will decrease to a low value as the block size or spacing increases beyond the peak (Fig. 7). If the variance peak is high and sharp, this indicates a clumped pattern of high intensity and strong grain or in other words, a pattern with tight, distinct clumps and large open spaces between the clumps. However, if the peak is low and occurs over a number of block sizes or spacing, the pattern is low intensity and the clumps are not well defined.

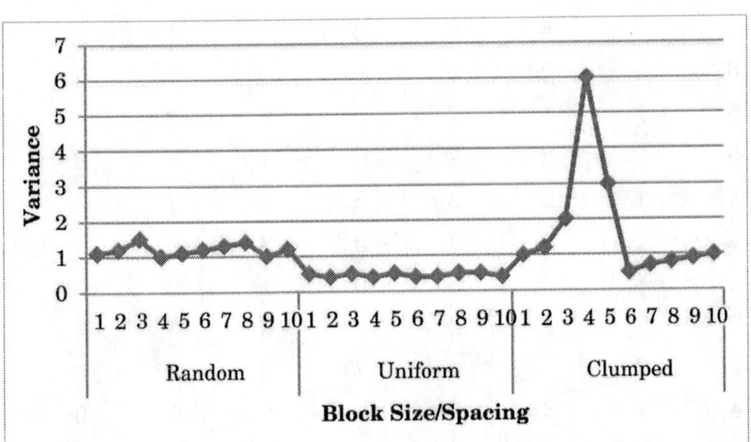

Fig. 7: Interpretation of variance plot

Development in the use of quadrat variance methods have been aimed at overcoming drawbacks in the BQV method. The most serious drawback in the BQV method was the limitation of variance estimates to block sites in successive powers of 2 (*e.g.*, 1, 2, 4, 8, 16, 32, and 64). Hills (1973a) TTLQV method efficiently overcomes this drawback and it is recommended as replacement for the original BQV method. The TTLQV method accurately detects both the scale and intensity of pattern (Ludwig and Reynold, 1988); it even detects clumping when two level or scales of pattern exist.

4.6.3.5. Working example for calculation of BQV (Through TTLQV) and PQV

For our example we assume that in an grassland study in Thar desert. A belt transect 10 m wide by 2000 m long was sampled, and the number of species were counted with each 10 m section of the belt, giving a total of 200 sampling unit.

0	1	5	4	7	0	2	5	4	6
1	2	4	7	0	2	5	8	0	3
6	4	2	1	7	0	2	5	8	0
1	1	2	5	8	0	2	3	0	1
4	2	5	6	2	4	7	8	9	2
1	5	8	1	1	4	7	2	5	3
2	0	0	1	4	7	4	2	5	5
7	3	6	2	4	7	8	2	3	1
1	4	7	5	2	3	5	4	2	4
7	8	1	2	3	9	8	5	1	7
3	6	1	4	7	8	9	2	1	4
0	0	2	8	7	4	1	5	9	3
2	4	1	2	5	3	1	7	3	7
0	2	1	4	7	3	3	6	1	0
2	1	4	8	7	0	1	2	5	3
9	7	2	1	8	2	5	3	1	4
7	2	2	5	4	7	8	3	6	9
1	4	7	8	9	9	2	1	0	4
7	0	0	1	2	3	1	4	8	2
0	0	1	4	5	2	1	4	3	2

As a rule, we can calculate the variance up to 20 only (*i.e.* 10% of N, the number of sampling unit; Fig. 8).

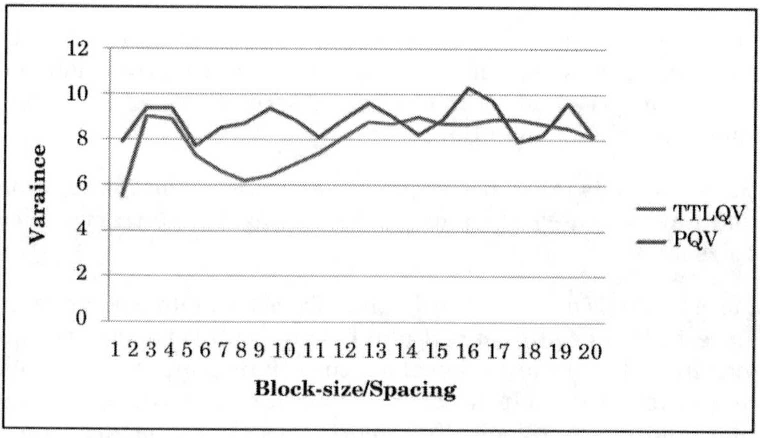

Fig. 8: Working example for interpretation of BQV

Although several researchers have attempted objective methods for evaluating the significance of peaks (*e.g.,* chi-square, Greig-Smith *et al.,* 1963; randomization tests, Mead, 1974; 95% confidence intervals, Greig-Smith, 1979; mean square, Carpenter and Chaney, 1983) there is no agreed upon method.

Also, while some researchers have found that the smallest block size corresponds to some readily recognized features of the system, it has commonly been accepted that a variance peak found at the first block size is the result of sampling in quadrats that are larger than the 'actual' scale of pattern.

In addition, Ludwig and Reynolds (1988) suggest that variances should not be analyzed for blocks > 10% of the number of quadrats when using a single transects of data. Although most researchers interpret block sizes up to one half of the transect length (*e.g.* Carpenter and Chaney, 1983; Ver Hoef *et al.,* 1989; Edwards, 1994), Ludwig and Reynolds (1988) suggest that such computations lack precision.

Variance peaks at intermediate block sizes (> 1 and < 10% or half the transect length) are thought to reveal meaningful interpretations of the data.

4.6.4. Distance Models

Distance sampling (or called plot less sampling) is widely used in forestry and ecology to study the spatial patterns of plants. Numerous mathematical models based on distance sampling have been developed since the 50's.

These models depend partly or wholly on distances from randomly selected points to the nearest plant or from a randomly selected plant to its nearest neighbour.

The majority of the models are based on the assumptions that (1) the population of interest is randomly distributed (Poisson distribution) within an infinitely large area and (2) an observed distribution is a realization (or part) of the theoretical population.

Distance methods make use of precise information on the locations of events and have the advantage of not depending on arbitrary choices of quadrat size or shape.

Distance methods measure the distance from a sampling point (or plant) to the nearest plant or n*th* nearest plant. The results of such a technique can provide important information about the relationships between plants. Distance methods can help determine whether plants are growing in discernible (and often ecologically important) patterns or are randomly dispersed. Many inter and intra-specific plant relationships are difficult to observe without using distance based sampling techniques.

With the nearest individual method, random points are located in a stand, the distance from each sampling point to the nearest plant is measured, the species is identified, and its basal area is measured. Only one measurement is made from each random point, and all distances for all species are summed and divided to yield one average distance (Barbour, 1987).

In the random pairs method, a line is taken from a point along a transect to the nearest plant. Perpendicular to the line and passing through the point is an exclusion line. The distance from this individual to the nearest neighbor that is on the same side of the exclusion line is measured. A difficulty in using these methods is that if the density is low, it may be impractical to search beyond a certain distance for the nearest individual. Distance sampling is considerably more efficient that quadrat sampling when individuals are sparse and widely scattered, since searching and counting individuals in a large quadrat is very time consuming.

Distance sampling refers to a class of methods where the basic sampling unit is the sample point and information is obtained from distances to the nearest events, *e.g.*, the first and second nearest events from the sampled points. The distribution of these distances under several spatial point processes generating the point pattern determines the performance of statistical tests and estimation procedures.

Most of the distance sampling methods are based on the random selection of m points with uniform distribution in a sampling frame S lying slightly within the region A in order to avoid edge effects. Some distance methods can be based on systematic samples of m points, rather than random samples.

4.6.4.1. *T-square index*

N sampling points (O) are selected at randomly with in population. Two distances are measured (Fig. 9): the distance (x) from the random point to nearest plant (P), and the distance (y) from P to the nearest neighbouring plant (Q) beyond the line at P perpendicular to O-P.

If the distribution of plants is random then the expected square of points –to individual distances (x) will be approximately one-half of the expected square of nearest neighobour distances y). From this index of spatial pattern (C) is derived as ration of squared x and y distance as:

$$C = \frac{\sum_{i=1}^{N} \left[\frac{x_i^2}{(x_1^2 + \frac{1}{2} y_1^2)} \right]}{N}$$

Variance of C of approximately 0.5 indicate random pattern, significantly less than 0.5 indicate uniform pattern, and significantly greater that 0.5 indicate clumped pattern. The significant departure of C from 0.5 is based upon calculating z as:

$$z = \frac{C - 0.5}{\sqrt{\frac{1}{(12N)}}}$$

At p = 0.05, z = 1.96, so calculated test values exceeding this critical value indicates a significant departure of C from 0.5. Other critical values of z can be obtained from a probability table for the standard normal distribution.

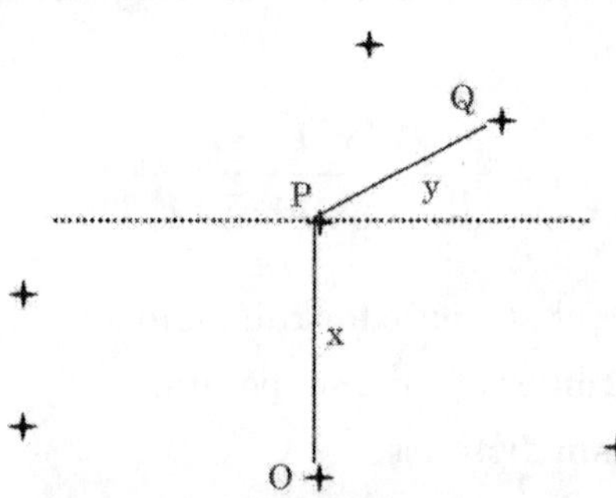

Fig. 9: Illustration of T-square method

Working example

Here we use the field data of distance sampling carried out in grazing land. Contrived set of 5 point to individual (x_1) and individual to neighbor (y_1) distance pair (in metre).

Points	1	2	3	4	5
X_1	4	3	7	8	6
Y_1	2	5	4	2	2

Compute the C using equation

$$C = \frac{\left[\frac{4^2}{4^2+\frac{2^2}{2}} + \frac{3^2}{3^2+\frac{5^2}{2}} + \frac{7^2}{7^2+\frac{4^2}{2}} + \frac{8^2}{8^2+\frac{2^2}{2}} + \frac{6^2}{6^2+\frac{2^2}{2}}\right]}{5}$$

$$C = \frac{\frac{16}{18} + \frac{9}{21.5} + \frac{49}{57} + \frac{64}{66} + \frac{36}{38}}{5}$$

$$C = \frac{0.88 + 0.415 + 0.859 + 0.969 + 0.997}{5} = 0.8128$$

Significance of C using equation

$$z = \frac{0.8128 - 0.5}{\sqrt{1\,(12x5)}} = \frac{0.3128}{7.74} = 0.040$$

This computed z value for C is less than the critical z value of 1.96 and we accept the underlying hypothesis that the distances were measured between points and individuals within a underlying random pattern. Hines has proposed a most powerful test statistic for T-square sampling procedure (Kreb,1999)

$$h_T = \frac{2n[2\Sigma(x_i^2) + \Sigma(z_i^2)]}{[(\sqrt{2}\,\Sigma x_i) + \Sigma z_i]^2}$$

Where h_T = Hines test statistic for randomness of T-square data

n = Sample size (number of random points)

x_i = Point to organism distances

z_i = T-square organism to neighbour distances

This test statistic is evaluated by referring to the critical values in Table 4. Low values of h_T indicate a uniform pattern, and high values indicate aggregation. In a random world h_T is 1.27, and smaller value indicates a uniform pattern while larger values indicate clumping.

Table 4: Critical values for the Hines test statistics h_T which test the null hypothesis that spatial pattern is random in a population sampled with the T-square sampling procedure.

	Regular alternative				***Aggregated alternative***			
n	**0.005**	**0.01**	**0.025**	**0.05**	**0.005**	**0.01**	**0.025**	**0.05**
5	1.0340	1.0488	1.0719	1.0932	1.4593	1.5211	1.6054	1.6727
6	1.0501	1.0644	1.0865	1.1069	1.4472	1.5025	1.5769	1.6354
7	1.0632	1.0769	1.0983	1.1178	1.4368	1.4872	1.5540	1.6060
8	1.0740	1.0873	1.1080	1.1268	1.4280	1.4743	1.4743	1.5821
9	1.0832	1.0962	1.1162	1.1344	1.4203	1.4633	1.4539	1.5623
10	1.0912	1.1038	1.1232	1.1409	1.4136	1.4539	1.4456	1.5456
11	1.0982	1.1105	1.1293	1.1465	1.4078	1.4456	1.4384	1.5313
12	1.1044	1.1164	1.1348	1.1515	1.4025	1.4384	1.4319	1.5189
13	1.1099	1.1216	1.1396	1.1559	1.3978	1.4319	1.4261	1.5080
14	1.1149	1.1264	1.1439	1.1598	1.3936	1.4261	1.4209	1.4983
15	1.1195	1.1307	1.1479	1.1634	1.3898	1.4209	1.4098	1.4897
17	1.1292	1.1399	1.1563	1.1710	1.3815	1.4098	1.4008	1.4715
20	1.1372	1.1475	1.1631	1.1772	1.3748	1.4008	1.3870	1.4571
25	1.1498	1.1593	1.1738	1.1868	1.3644	1.3870	1.3768	1.4354
30	1.1593	1.1682	1.1818	1.1940	1.3565	1.3768	1.3689	1.4197
35	1.1668	1.1753	1.1882	1.1996	1.3504	1.3689	1.3625	1.4077
40	1.1730	1.1811	1.1933	1.2042	1.3455	1.3625	1.3572	1.3981
45	1.1782	1.1859	1.1976	1.2080	1.3414	1.3572	1.3528	1.3903
50	1.1826	1.1900	1.2013	1.2112	1.3379	1.3528	1.3377	1.3837
75	1.1979	1.2043	1.2139	1.2223	1.3260	1.3377	1.3260	1.3619
100	1.2073	1.2130	1.2215	1.2290	1.3189	1.3289	1.3189	1.3492
150	1.2187	1.2235	1.2307	1.2369	1.3105	1.3184	1.3105	1.3344
200	1.2257	1.2299	1.2362	1.2417	1.3055	1.3122	1.3055	1.3258
300	1.2341	1.2376	1.2429	1.2474	1.2995	1.3049	1.2995	1.3158
400	1.2391	1.2422	1.2468	1.2509	1.2960	1.3006	1.2960	1.3099
500	1.2426	1.2454	1.2496	1.2532	1.2936	1.2977	1.2936	1.3059

[a] n = Sample size. Low values of h_T indicate a clumped pattern. *Source*: Hines and Hines, 1979

4.6.4.2. *Hines and Hines index (1979)*

Hines and Hines (1979) adapted the Eberhardt statistic by adding the T^2 information. The modified statistic is.

$$HI = 2m \frac{\Sigma_1 (2u_i + t_i)}{\Sigma_1 (\sqrt{2u_i} + t_i)^2}$$

4.6.4.3. *Johnson and Zimmer index (1985) of dispersion*

Johnson and Zimmer (1985) have proposed a new distance index of dispersion. This index is based on point-to individual distances.

Given a sample of N points with distances x_i from the ith point to the nearest individual, then the distance index of dispersion (I) is defined as the ration of the sum of squares of the squared distances to the square of the sum of squares of the distances. This method adopted by Gonzalez-Andujar *et al.* (2007); Mohebbi *et al.* (2012).

$$I = (N+1)\frac{\sum_{i=1}^{N}(x_1^2)^2}{\left[\sum_{i=1}^{N}(x_i^2)\right]^2}$$

Johnson and Zimmer have shown that I has an approximate excepted value of 2 for a random pattern, less than 2 for uniform pattern, and greater than 2 for clumped pattern. They further show that *I* converges to normaility at moderate sample sizes.

$$Z = \frac{\mathrm{I} - 2}{\sqrt{4(N-1)/(N+2)(N+3)}}$$

The z value is compared to a table of critical values for the standard normal distribution to obtain the level of significance of any departure from randomness. This index was adopted by Askari *et al.* (2013).

Working example

Here we use the field data of distance sampling carried out in grazing land. Contrived set of 5 point to individual.

Points	1	2	3	4	5
X_1	4	3	7	8	6
Y_1	2	5	4	2	2

- Step 1. The computation of *I* requires the sum of squares of the squared point-to individual distances:

$$\sum_{i=1}^{N}(x_1^2)^2 = (4^2)^2 + (3^2)^2(7^2)^2(8)^2(6^2)^2 = 8130$$

- Step 2

$$\left[\sum_{i=1}^{N}(x_i^2)\right]^2 = 16+9+49+64+36 = 174^2 = 30{,}276$$

$$I = \frac{(5+1)(8130)}{30{,}276} = \frac{6X8130}{30{,}276} = \frac{48780}{30{,}276} = 1.61$$

This index is near 2 the expected value for a random pattern

- Step $$Z = \frac{1.61 - 2}{\sqrt{4(5-1)/(5+2)(5+3)}}$$

$$Z = \frac{-0.389}{\sqrt{4(4)/(7)(8)}} = \frac{-0.389}{\sqrt{16/(56)}} = \frac{-0.389}{0.543} = 0.728$$

- The z value is compared to a table of critical values for the standard normal distribution to obtain the level of significance of any departure from randomness. The probability of obtaining a value of z =0 .728 by chance alone is much greater thatn P = 0.05. Thus, the hypothesis of a random pattern is accepted.

4.6.4.4. *Eberhardt's index* (I_E)

This index provides an analogous measure of distribution to that of the variance-mean ration. It is calculated as the mean of the squared distances divided by the square of mean distances

$$I_E = \frac{n\sum(d^2)}{(\sum d)^2}$$

Where each value d is the distance from each random point to the closest individual of each species of interest. Intuitively, if the plants are clumped, we would except some distances to be very small, and some to very large.

This would give a high variance (a variance is squared distance) and therefore a high value of I_E. However, if plants were evenly distributed, most of the distances should be fairly similar, the variance low, and the index small. Randomly spaces plants have an intermediate value.

The denominator of the $I\text{-}_E$ equation is a scaling factor, which corrects for mean plant density (Low density plants will have long distances on average and high density plants will have short distances). Adopted by Mohebbi *et al.* (2012).

Eberhardt found that, regardless of sample size, randomly distributed items produce $I_E = 1.27$. Thus

I_E <1.27	Indicates hyper-dispersion;
I_E =	Indicates randomness; and

$I_E > 1.27$ Indicates clumping.

4.6.4.5. *Hopkins index*

Utilized by Poushinsk and Basi (1984) used this index for study of distribution and sampling of soybean plant infected with *Pseudomonas syringae pv. glycinea.*

Hopkins *et al.* (1954) have suggested a coefficient of aggregation (CA) for detecting non-randomness based upon the squared distance (x) from a random point to the nearest plant, and the squared distance (y) from a randomly chosen individual to its neighbour .

$$CA = \frac{\Sigma x^2}{\Sigma y^2}$$

Value of CA= 1 indicate a random distribution, >1 a clumped distribution, and <1 a regular (uniform) distribution. CA is dependent upton random samples and the mean CA from at least 20 iterations has been found to product stable estimation (Briggs and Gibson, 1992).

The significance of the departure of CA values from 1 is assessed by calculating the parameters = CA/ (1+CA) which has a mean of 0.5 for random distribution. Armstron (2006) have proposed a significance test for this index

$$t = 2\left|\frac{CA_2 + 1)}{CA_{2+2}\ 0.5}\right|\sqrt{2n+1}$$

This method adopted by several (workers like Leps and Kindlmann, 1987; Counts and Martinez, 1998; Shaukat *et al.*, 2009; Shaukat *et al.*, 2012; Mohebbi *et al.*, 2012. and Khojasteh *et al.*, 2013). for plant distribution while Armstrong, (2006) use this index for measuring spatial arrangement pattern of pathological lesion in histological section of brain tissue.

4.6.4.6. *Hopkins-Skellam aggregation index*

Hopkins and Skellam (1954) rely on the fact that whether the spatial distribution is random, the distribution of the distances between points and a host is identical to the distribution of distances between host.

This is defined by the equation

$$A = \frac{\sum_{i=1}^{n} \omega_i^!}{\sum_{i=1}^{n} \omega_i^! + \sum_{i=1}^{n} \omega_i^!}$$

Here *ωi* is the quadratic distance from sample point to the nearest host and is the quadratic distance from host to the nearest neighbouring host.

Numahara *et al.* (2009) have used a slight deviate form of this index, which can be written as follows:

$$\text{HSI} = \frac{\sum_{i-1}^{N} r_i^2}{\sum_{i-1}^{N} r_2^2}$$

r_1 = Distances from point chosen at random to its largest individual

r_2= The distance from an individual chosen at random to the nearest individual

N = Sample size (Individual number)

HSI=< 1= regular

HSI = 1 = random

HSI = >1= Aggregated.

4.6.4.7. *Pervasiveness index*

Engeman *et al.* (2001) have proposed a modification of Hopkins index that can be describe as follows

$$\text{IP} = (1/n)\text{"}(W_1/W_2)$$

where *n* is the number of plots, *w*1 is the square of the distance from a tracking plot to the nearest plot with swine tracks, and *w*2 is the square of the distance from that nearest plot to its nearest plot with tracks. When the pattern is entirely random, IP = 1. If the plots with tracks show aggregation (localized concentrations), then IP > 1. If tracked plots show a systematic pattern, then IP < 1.

There are limitations to the T-square index, Hopkins CA and other statistical indices based upon nearest neighbor and random point to host/ pathgoen distances. Based upon nearest neighbor distances these indices do not consider the range of spatial scales in the data set.

The second general problem with these method is that the random points will tend to fall in the space between tightly packed clumps of individuals that may occur. This means that the random point-to-point distances will tend to most frequently sample plants on the edge of clumps as opposed to the plants within clumps.

The result is that edge individuals of clumps will be indistinguishable from single individuals and indications of aggregation in the mapped dataset may be underestimated. Finally, these techniques ignore scales and do not provide a measure of the scale of spatial pattern, that is, the size of clumps.

An advantages of the T-square index and Hopkins CA over some other indices is that knowledge of the density of individual in the study area does not have to be known or measured.

4.6.4.8. *Pielou-Mountford aggregation index*

This is defined by the equation

$$\alpha = \frac{1}{N}\pi(\frac{N}{P})\sum_{i=1}^{N}\omega_i'$$

Where *n* is the stem number per plot, *N* is the number of sample points, *P* is the plot area and ω'i is the quadratic distance from sample point to nearest host. $\alpha = 1$ mean value; $\alpha > 1$= aggregation and $\alpha < 1$ = regularity. This method adopted by Zahradnik *et al.* (2010); Vacek *et al.* (2010) and Bilek *et al.* (2011).

4.6.4.9. *Hamill and Wright test*

Pielou's S is a scales less measure and provides no information about how far away one plant type may be from another.

For example, a measure of tendency of spore to be located near or from maternal infected host provides important information on plant neighborhood. Hamill and Wright (1986) provides a test of the null hypothesis: are juveniles located at random with respect to adult?.

Their measure compares the observed cumulative distribution of distances from juveniles to the nearest adult.

This distribution is then compared to the null hypotheses distribution using a two sided-Kolmogorov-Smirnov test. This test compares the maximum difference between the observed and null distribution with a critical value based upon the numbers of distances used. For a descriptive purposes the observed and null distribution can be graphed.

In a x and y plot the cumulative distribution function of juvenile to nearest adult distances under the random hypothesis is

$$F(s) = \frac{1}{A}\int_{x=0}^{x_{max}}\int_{y=0}^{y_{max}} I(s)dy\,dx$$

Where A is the total area of the plot and I(s) is the indicator function:

$$I(s) = \begin{cases} 1\,Min \\ 0\,Otherwise \end{cases} \left\{\sqrt{(x-a_i)^2 + (y-b_i)^2}\ i = 1,2,K,n\right\} \le s$$

4.6.4.10. *Holgate's index*

The Holgate index is based on the ratio of the distance from a random point to the first and second nearest neighbours. The Holgate was designed to have the same distributions as the Hopkins' index.

$$A = \sum\left(\frac{d^2}{d_1^2}\right)/n - 0.5$$

Where d = distance measure, n = number of observation or plot For test of significance Armstrong proposed a t test for it.

T = %A%/"n/12

This index was adopted by Dazza (2010) and Mohebbia *et al.* (2012).

4.6.4.11. *Pielou's index of segregation (S)*

The *Segregation index* of Pielou (S) is often used, and is based on the relationship between the observed mixture and that expected under random conditions. Thus a distinction can be made between aggregation (intimate mixture, association between species), segregation (spatial separation) or neutral relation (random mixture) between two host species.

When calculating S for two species, A and B, the number of hosts of species A and species B which have a conspecific nearest neighbour has to be determined (a and d), as well as the number of hosts of species A or B which are neighbours to species B or A (b and c) (see Table 5).

Table 5: Pielou's Index of Segregation

Species	***Number of nearest neighbours of***		
	Species A	Species B	Sum
Species A	a	B	M
Species B	c	D	N
Species A + Species B	v	W	N

In stands with more than two species, the sum of a and b, or c and d, does not necessarily have to equal the number of hosts of species A and species B, respectively. This is because each host can be a neighbour to a host other than either of the two species being examined. Using the parameters in Table 5, the segregation index S can be calculated as follows:

$$S = 1 \frac{N\ (b+c)}{(v.n + w.m)}$$

The segregation index S by Pielou (1961) provides a spatially explicit measure for host species diversity, which considers the ratio of the observed probability that the reference host and its nearest neighbour belong to different species along with the same probability for completely randomly distributed or independent species attributes. S, which takes values between "1 and 1, indicates spatially segregated host species for S > 0, uncorrelated. If the two host species A and B are spatially separated, the number of observed mixed pairs will be smaller than would be expected if host were randomly mixed, and S approaches the value 1. The index S equals 0 if both species are randomly mixed within the stand. Negative values for S are possible, but according to Pielou (1977) this is unlikely in plant communities, as this would indicate association between the two species. The S-index most frequently shows values close to zero indicating a random mixture of hosts.

The only S-value for which a significant difference from zero can be assumed with sufficient certainty is that for the mixture between pedunculate oak and red oak. The high index value suggests spatial separation between these two species, which is logical. As S is calculated using only one nearest neighbour for each host, this effect is reflected in negative index values. In conclusion, negative values for S in Ravels forest clearly do not indicate association between species, but only prove an artificially induced alternation of species over small distances. (Adopted by Kint *et al.*, 2000; Sterba, 2008; and Motz *et al.*, 2010.).

4.6.4.12. *Individual-centered analysis (ICA)*

Podani and Czαrαn (1997) introduced a new technique, the so-called Individual-centered analysis (ICA), that also uses variable ready to analyse pattern but, unlike the *K*-function, focuses on the diversity of species combinations rather than on one or two-species patterns.

Individual-centered analysis may therefore be expected to detect non-random multi-species patterns in point maps of plant communities.

Recently, Individual-centered analysis (ICA) has been introduced as a new tool for analysis of multi-species co-occurrence patterns. Bassdorf *et al.* (2000) have tested this technique by applying it to spatial data from grazed and ungrazed shrub communities in the semi-arid Great Karoo, South Africa. They shows that the interpretability of ICA results is improved when the actual observed and expected frequencies of species combinations are added. The main strength of ICA is that it has the potential to detect association patterns that involve more than two species.

4.6.4.12.1. *Main steps of ICA (Podani et al., 1998)*

1. The sampling design involves the determination of exact position of all plants within a preferably rectangular study area.
2. The density of each species is determined, thus allowing calculation of the probability, that a given species occurs in a circular plot of a given radius, under the null hypothesis of complete spatial randomness (CSR). In other words, the random reference is that the pattern of each species results form a two-dimensional Poisson point process, thus assuming independence (lack of association) among species.
3. Based on species probabilities, the likelihood that a given species combination appears in a circular plot of a given radius is obtained by simple multiplication. The higher this likelihood the lower our 'surprise' that the given florula actual appears. On the other hand, if a florula with a low likelihood is found, something 'unexpected' happened as a result of the deviation from randomness.
4. The likelihood values pertaining to species combinations that actually occur find around plant individuals are averaged to obtain the weighted *mean compositional information*, MCI. The expectation of this under CSR is derived from Shannon's entropy function and the departure of MCI form from the expectation is calculated (ΔMCI). A graph showing ΔMCI as a function of plot radius is used to characterize multi-species pattern.
5. Species combination with low probabilities of being detected have high values of compositional information, where as combinations that should apprear frequently under CSR have low compositional information. From the values of compositional information of all plants the program calculates the mean compositional information (*MCI*) of whole pattern of certain radius can be calculated as

$$MCI_{(t)obs} = \sum_{k=i}^{N(i)} -1n \prod_{i=1}^{s} q_i\ (t) / N\ (t)$$

Where i the probability of presence of species *i,* whenever species *i* is included in a circle of radius *t* around individual k, and $q_i (t) = 1\text{-}p_i (t)$ if species *i* is not present in the respective circle. The product over all *s* species – the probability of the species combination around individual *k-* is transformed by the negative logarithm giving higher value to unlikely combination. Mean compositional information is the average of the values calculated for all $N'(t)$ individuals around which circles of radius *t* can be drawn. The expected value of *MCI* under CSR is given by the Shannon entropy function.

$$MCI_{(t)exp} = -\sum_{i=i}^{2^s}\left(\prod_{i=1}^{s} q_i (t) \ 1n \prod_{i=1}^{s} q_i (t)\right)$$

Where 2^s is the number of possible combination of s species. Comparison of expected and observed values provides a measure ΔMCI of the deviation of observed pattern from CSR.

$$\Delta MCI_{(i)} = MCI_{(t)obs} - MCI_{(t)\,exp}$$

Where $\Delta MCI > 0$ indicate a higher diversity of species combination that expected under CSR, and $\Delta MCI < 0$ if the diversity of species combination is lower than expected under CSR. In addition, the individual contribution of each species to overall ΔMCI is calculated.

Individual-centred analysis provides ΔMCI Values for the whole community and for each species in it. It does not tell, however which particular species are responsible for an observed deviation from CSR by occurring more or less frequently than expected around a particular other species at some scale.

4.6.4.13. *Clark and Evans index*

Clark and Evans (1954) were the first to suggest a method for analyzing pattern for spatial maps. They used the distance from an individual to its nearest neighbor as the relevant measure. These distances are measured for all individuals in the spatial map, and thus there are no problems with random sampling because it is a complete enumeration of this local population.

Poushinsk and Basi (1984) used this index for study of distribution and sampling of soybean plant infected with *Pseudomonas syringae pv. glycinea,* similarly, Palacio-Bielsa *et al.* (2012) used this index to quantify the spatial spread of *Erwinia amylovora* on pome fruit.

$$R = Index\ of\ Aggregation = \frac{\hat{r}_A}{\hat{r}_E}$$

$$\hat{r}_A = Mean\ distance\ to\ nearest\ neighbor = \frac{\sum r_1}{n}$$

Where r_1 = distance to nearest neighbour for individual I N = number of individuals in study area

$$\hat{r}_E = Expected\ distance\ to\ nearest\ neighbor = \frac{1}{2\sqrt{\rho}}$$

$$\rho = Density\ of\ Organism = \frac{Number\ of\ plants\ in\ study\ area}{Size\ of\ the\ study\ area}$$

Usually, the interpretation of R values is as follows: R>1 if the pattern has a tendency of regularity, R= 1 if it is completely random (Poisson), and R<1 if there is clustering in the patter. *R* approaches an upper limit around 2.15.

4.6.4.13.1. *Statistical test for Clark and Evans index*

$$z = \frac{\hat{r}_A - \hat{r}_E}{S_r}$$

Where z = Standard normal deviate;

S_r = Standard error of the expected distance to nearest neighbour $= 0.26136\sqrt{n\rho}$;

n = Number of individuals in study area; ρ = density of individuals in study area

4.6.4.13.2. *Modified Clark and Evans index*

Donnelly (1978) proposed modification for this index. In this method the average distance between host and their nearest neighbours is compared with the expected mean distance. The index is calculated according the following formula.

$$R = \frac{\hat{r}_A}{\hat{r}_E} = \frac{\frac{1}{N}\sum_{i=1}^{N} r_1}{0.5x\left(\frac{A}{N}\right)^{1/2} + 0.514x\frac{P}{N} + 0.041\,x\frac{P}{N^{1/2}}}$$

Where A = area, m^2; N = number of hosts; P= perimeter, (m) This index was adopted by Kint *et al.* (2000), Pommerning (2002); Motz *et al.* (2010), Sterba (2008) Saunders and Wagner (2008); Szmyt (2010); and Balanda (2012).

4.6.4.14 *Contagion index*

As a single host based alternative of the aggregation index, Godow *et al.* (1998) developed the contagion variable or neighbourhood pattern W_i to define the degree of regularity of spatial distribution of host positions in a forest.

Unlike the index of Clark and Evans, this variable is a single-host parameter, Assuming complete regularity of the positions of the n nearest neighbours around a reference host *i*, the expected standard angle α_0 between two neighbours would be equal to $360^0/n$. For example, $\alpha_0 = 90^0$ in a constellation involving four neighbours (Figure 10).

Each pair of neighbours shares two angles, α and β, with α+β = 360 and α≤β. Contagion is defined as the proportion of α angles which are smaller than the standards angle α_0

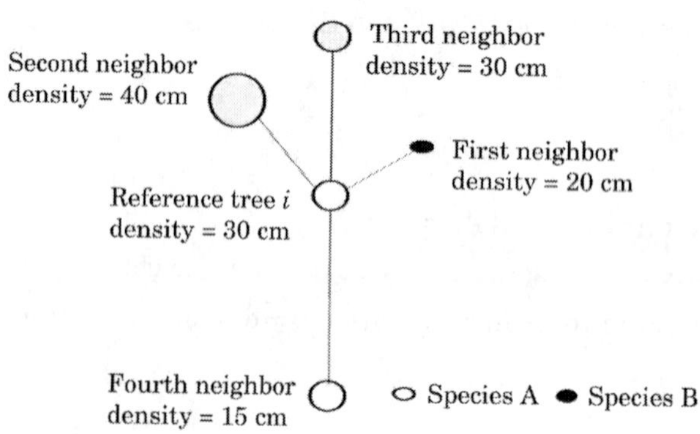

Fig. 10: Two of the angles between the 4 neighbors and the reference host are smaller than the standard angle (random arrangement) Redraw from Pommerening (2002)

$$W_i = \frac{1}{n}\sum_{i=1}^{n} W_i \quad \text{Where } W_i = \{1, \text{the } \alpha - \text{angle i is the smaller than } \alpha_0 \quad 0, \text{otherwise} \quad W_i \in \{0,1\}$$

In the structural of group of four W_i can be visually assessed in the field by comparing the actual angle α with and angle of 90^0. Thus a quick decision can be made in the forest on whether α is smaller than 90^0 or not.

If host positions are recorded W_i can be derived from the hosts coordinates. In a constellation involving four neighbours, W_i can assume five possible values 90, 0.25, 0.5, 0.75 and 1.0.

The average contagions (W) may be used to classify the point pattern into the categories 'regular', 'random' and 'clumped' (Gadow *et al.* 1998).

4.6.4.15. *Ripley's function*

A large number of methods for the analysis of point pattern data have been developed in a wide range of scientific fields. First-order statistics describe large-scale variation in the intensity of points in a study region, whereas second-order characteristics are summary statistics of all point-to-point distances in a mapped area and offer the potential for detecting both different types and scales of patterns.

Second-order analysis based on Ripley's *K*-function is increasingly used in ecology to characterize spatial patterns and to develop hypothesis on underlying processes.

Over the last several years, methods based on Ripley's *K*-function have undergone a rapid development (*e.g.*, Haase, 1995; Podani and Czaran, 1997; Coomes *et al.*, 1999; Goreaud and Pelissier, 1999; Dale and Powell, 2001; Haase, 2001; Pelissier and Goreaud, 2001; Dungan *et al.*, 2002; Freeman and Ford, 2002; Perry *et al.*, 2002) and are now being widely used, especially in plant ecology (*e.g.*, Haase *et al.*, 1996; Ward *et al.*, 1996; Haase *et al.*, 1997; Martens *et al.*, 1997; Jeltsch *et al.*, 1999; Larsen and Bliss, 1998; Pancer-Koteja *et al.*, 1998; Wiegand *et al.*, 1999; Barot *et al.*, 1999; Chen and Bradshaw, 1999; Coomes *et al.*, 1999; Mast and Veblen, 1999; Kuusinen and Penttinen, 1999; Bossdorf *et al.*, 2000; Camarero *et al.*, 2000; Condit *et al.*, 2000; Grau, 2000; He and Duncan, 2000; Kuuluvainen and Rouvinen, 2000; Lookingbill and Zavala, 2000; Wiegand *et al.*, 2000; Haase, 2001; Fehmi and Bartolome, 2001).

Ripley's K function is a frequently used method for studying the spatial pattern of mapped point data in ecology (Hasse, 1995; Podani and Czaran 1997 and Dale, 1999). Applications includes the analyses of distribution patterns of *Eucalyptus* species (West, 1987), *Pinus ponderosa* (Getis and Franklin, 1987). *Pinus banksiana* (Kenkel, 1988) Mixed deciduous-coniferous forest (Szwagryk and Czerwczak, 1993), sub-tropical hosts (Debski *et al.*, 2000) and for *Acacia* (Spooner *et al.*, 2004). The advantage of K-function analysis is that it uses all point-point distances, not just the nearest neighbour distances, to show spatial clustering at various scales of pattern, and the distance where clustering or over-dispersal becomes significance (Bailey and Gatrell, 1995; Haase, 1995, Moeur, 1997, Li and Zhang, 2007 and Szmyt, 2010).

Ripley's K-function is a cumulative distribution equation that calculates point densities within an entire circle of a given radius (Chen *et al.*, 2011). The characteristic of the K-function confounds the effect at large distance with the effect at small distances and complicates the overall interpretation of the data (Condit *et al.*, 2000 and Schurr *et al.*, 2004). Ripley's function can be estimated for both uni-variant as well as for bivariant case.

4.6.4.15.1. *Uni-variant case (i.e. within a class of points)*

To estimate the Ripley function, K(t), a circle of radius t is placed around each point, and the number of neighbouring points within the circle is countered. The estimation of equation is as follows.

$$\acute{K}(t) = \frac{A}{n^2}\sum_{i=1}^{n}\sum_{j=1}^{n}\frac{1}{W_{ij}}\,I_t\,(u_{ij})$$

A = area, m^2; N = number of plants; u_{ij} = distance between 'i'th host and 'j'-th host; W_{ij} = Weighting factor for reducing edge effect which can be set at 1 when the value of t describes a circle that remains within the study area. Oterwise, W is the reciprocal of the proportion of the circle's circumference lying within the study area (Gibson, 2002 and Salas *et al.*, 2006). I_t = is a counter variable whic takes the value of 1 if the distance between i and j is less than t, otherwise the value 0. Term I_t- (u_{ij}) indicates that the interpoint (point-to plant) distance are summed for each radius, t. This distance can be calculated as (Gibson, 2002).

$$u_{ij} = \sqrt{(x_i - x_j)^2 + (y_i - y_j)^2}$$

Besag (1977) proposed to use the L' (t) modification against K'(t) that stabilizes its variance and linearizes its curve:

$$L'(t) = \sqrt{\frac{K'(t)}{\pi} - 1}$$

In uni- variant case (*i.e.* within a class of points), L(t) = 0 for randomly distributed population the function of. For aggregated population the L(t)>0, and L(t)<0 if the population is arranged regularly.

Significance of departures from CSR (complete spatial randomness=null hypothesis) is evaluated using Monte Carlo test (Haase, 1995, Szmyt, 2010).

For the bi-variant (*i.e.* between two class of points) case L(t)>0 indicates that there are on a average more points of pattern 2 with distance t that point of pattern 1, as one would expect under independence. This outcome indicates attraction (competition) between the two pattern up to distance r. Similarly values of L(t)<0 indicate repulsion (facilitations) between the two patterns up to distance t and (L(t) = 0 indicates independent pattern (Wiegand and Moloney, 2004 and Chen *et al.*, 2011) .

Edge effect can be correct by using following expression proposed by Haase (1995)

$$W_{ij} = 1 - \frac{2\,Cos^{-1}\left(\frac{e_1}{u_{ij}}\right) + 2Cos^{-1}\left(\frac{e_2}{u_{ij}}\right)}{2\pi}$$

Where Wij is the correction factor of ith border host and its jth neighbour host, e_1 and e_2 the distances to the two closer borders from the ith hosts, and cos^{-1} refers to the inverse cosine function.

Besag (1977) and Ward and Ferrandino (1999) have also provided the correction factors. Besag (1977), in his discussion of Ripley's paper, underlined that their correction gave an excessive weight to the farthest neighbours.

Ward and Ferrandino (1999) introduced a global correction, arguing that local correction methods depend on the number and position of points close to the borders, thus introducing more variability in *K*'s estimator.

They proposed to evaluate the expectation of the number of points concerned with edge-effect correction for a given, compute the correction for them and finally calculate the global underestimation of the number of point pairs which only depends on the domain's geometry. They denoted *KA* (*A* for analytical) their estimator of *K* defined by:

$$\acute{K}_A(r) = \frac{1}{\lambda(N-1)C(r)} \sum_{i=1}^{N} \sum_{j=1, i\neq j}^{N} c\,(i,j,r)$$

C (r) is global correction factor that can be denote by

$$C(r) = 1 - \frac{4}{3\pi}\left(\frac{r}{L} + \frac{r}{W}\right) + \left(\frac{11}{3\pi} - 1\right)\left(\frac{r^2}{WL}\right)$$

4.6.4.15.1.1. *Bivariate case (i.e., between two classes of points)*

For example following two groups of hosts can be selected for bivarate version of Ripley's K(t) function.

1. Live hosts < 15 m tall (group 1, the shortest height class *versus* live hosts 35 m and taller (group 2, the tallest height class),
2. All live hosts *versus* standing dead hosts, and
3. Live hosts (group 1) *versus* standing dead hosts.

The bivariate version of Ripley's k(t) function can be used to detect the degree of repulsion, attraction, or indifference between the two groups (Andersen, 1992; Haase, 2001, Salas *et al.*, 2006.). The estimators of $K_{12}(t)$ and $K_{21}(t)$ of the pooled distribution for each of the two groups can be calculated as follows:

4.6.4.15.1.2. *Statistics for $K_{12}(t)$*

The interaction between group 1 and 2 calculated as

$$\acute{K}_{12}(t) = \frac{A}{n_1 n_2} \sum_{i=1}^{n_1} \sum_{j=1}^{n_2} \frac{1}{W_{ij}} I_t(u_{ij})$$

Where n_1 and $_{n2}$ are the number of hosts in groups 1 and 2 respectively, and the u_{ij} is the distance (m) between the ith subject hosts of group 1 and j^{th} hosts of group 2.

4.6.4.15.1.3. *Statistics for $K_{21}(t)$*

The interaction between groups 2 and 1 can be calculated by using

$$\acute{K}_{21}(t) = \frac{A}{n_1 n_2} \sum_{i=1}^{n_1} \sum_{j=1}^{n_2} \frac{1}{W_{ij}} I_t(u_{ji})$$

and finally, the bivariate index can be calculated through

$$\acute{K}_{12}(t) = \frac{n_2 \acute{K}_{12}(t) + n_1 \acute{K}_{21}(t)}{n_1 + n_2}$$

By a similar argument as before, we can also calculated

$$L_{12}(t) = \sqrt{\frac{\acute{K}_{12}(t)}{\pi}} - 1$$

Spooner *et al.* (2004) have point out that despite the wide adaptability of this function in studying the mapped spatial data, the application of this method to point patterns on road networks is inappropriate, as the K-function assume an infinite homogeneous environment in calculating the Euclidean distances.

For univariate network K-function network links are defied as L_T, and the total network distance as $|L_T|$ and the formula can be written as

$$\breve{k}(t) = \frac{|L_T|}{n(n-1} \sum_{i=1}^{n}$$

Network cross-K function can be write as

$$\breve{k}^{ba}(t) = \frac{|L_T|}{n_a n_b} \sum_{i=1}^{nb}$$

In both these cases the significance of departure of the networks $\breve{k}(t)$ and $\breve{k}^{ba}(t)$ from CSR can be estimated through Monte Carlo Simulation.

In spatial statistics, Ripley's K function (Ripley (1977)) is a classical tool to analyse spatial point patterns.

Yet, it faces two major limits: it is only pertinent for homogeneous point processes and it does not allow the weighting of points. Therefore, this tool seems to be inappropriate to analyse obviously non-stationary point sets or objects whose size matters, like the study of the spatial structure of manufacturing plants (Marcon and Puech (2003a).

A first solution was given by Cuzick and Edwards (1990) who developed a non-parametric test able to detect clustering in a non-homogeneous point set, followed by Diggle and Chetwynd (1991) who introduced the *D* function, defined as the difference between the *K* function for studied points (called *cases*) and the *K* function for the others (called *controls*).

This is not completely satisfactory yet because, since both *K* functions are computed separately, all the data contained in the relative position of cases and control is lost.

One demerits of nearest neighbor analysis is associated with its one scale interaction approach at time. Most commonly this technique detects clustering at short distances. Ripley's K function can be computed over a range of distances and be used to identify the scales over which clustering occurs.

It is a tool for analyzing completely mapped spatial point process data and K function is also known as second order analysis which indicate that the focus is on the variance, or second moment, of pair of arbitrary distances. It also a test of the CSR hypothesis that considers all combinations of pairs of points.

It compares the numbers of observed pair with the expectation at all distances based on a random spatial distribution of points. With this tool, the point's density, borders and the size of the samples are taken into consideration and the expected value of $L\,(d)$ under CSR is d. The observed $L(d)$ can be calculated as follows

$$L\,(d) = \sqrt{\frac{A\ \sum_{i=1}^{N}\sum_{j=1, j\neq i}^{N} k(i,j)}{\pi N\,(N-1)}}$$

Where, A is the study area, N is the number of points and d is the distance $\sum_{i=1}^{N}\sum_{j=1, j\neq i}^{N} k(i,j)$ is the number of j points within distances d of all I points, and $k\,(i,j)$ can be calculate by following equation with e_1 and e_2 (distance to the nearest vertical and horizontal borders)

$$k\,(i,j) = \left[1 - \frac{\cos^{-1}\frac{e_1}{d\,(i,j)} + \cos^{-1}\frac{e_2}{d\,(i,j)} + \frac{\pi}{2}}{2\pi}\right]^{-1}$$

Kamu *et al.* (2015) have utilized this approach to quantify the distribution of infected oil palms with *Ganoderma* basal stem root disease. Gidoin *et al.* (2014) have used this approach in cooco agro-ecosystems for black pod disease, caused by *Phytophthora megakarya*.

4.6.4.15.1.4. *Kinhom generalization*

Kinhom is a generalization of Ripley's K to non stationary processes. Stationary is required for the second-order property: this property is understated as soon as interactions between points are evaluated at a given distance. *Kinhom* can be defined as an integration of the radial distribution function *g* on the circle of radius *r*. When the density is not a constant, $\lambda(i)$ is the process density at the point

$$\acute{K}_{in\,hom}(r) = \frac{1}{A}\sum_{i=1}^{N}\sum_{j=1,\#j}^{N}\frac{c(i,j,r)}{\lambda(i)\lambda(j)}$$

The function developed by Baddeley *et al.*(2000) constitutes a theoretical milestone in the effort for characterizing non homogeneous point processes. Yet, applications are not straightforward. Its fundamental issue is the great difficulty, both theoretical and practical, to estimate local densities.

Morcon *et al.* (2012) have generalized Ripley's K function to get a new function *M* to characterize the spatial structure of point pattern relatively to another one. They consider that their new approach pertinent in ecology when space is not homogeneous and the size of object matter. Their theoretical examples illustrate that *M*(*r*) is equivalent to $K(r)/\pi r2$ when applied to a homogeneous, non weighted point pattern.

Univariate M Function As

$$M(r) = \frac{\sum_{f=1}^{N_f}\left(\sum_{n=1,x_n\#x_f}^{N_f} 1(\|x_f - x_n\| \le r)w_n / \sum_{a=1,x_a\#x_f}^{N_a} 1(\|x_f - x_a\| \le r)w_a\right)}{\sum_{f=1}^{N_f}((W_f - w_f)/(W_a - w_f))}$$

The bivariate $M_{f,n}$ function is

$$M_{f,n}(r) = \frac{\sum_{f=1}^{N_f}\left(\sum_{n=1}^{N_f} 1(\|x_f - x_n\| \le r)w_n / \sum_{a=1,x_a\#x_f}^{N_a} 1(\|x_f - x_a\| \le r)w_a\right)}{\sum_{f=1}^{N_f}(W_f/(W_a - w_f))}$$

In their approaches, r = radius; x = point, n= neighbour, f = reference point; x_f = point of the reference point; x_n =point of the neighbours; N = total point around study area; N_f =number of reference point; N_n =number of neighbour; N_a =number of point whatever their type; A_f = area of circle of the radius r centred on the point x_f located inside the study area (window); w_i = weight of the point and

$$W_i = \sum_{i-1}^{N} w_i = \text{is the total weight of the this type of point}$$

Morcon *et al.* (2012) have simplified both above uni-variant and bi-variant equations, when point are not weighted

$$M(r) = \frac{N_a - 1}{N_f(N_f - 1)} \sum_{f=1}^{N_i} \frac{\sum_{n=1, x_a \# x_f}^{N_a} 1(\|x_f - x_n\| \leq r)}{\sum_{a=1, x_a \# x_f}^{N_a} 1(\|x_f - x_a\| \leq r)}$$

$$M_{f,n}(r) = \frac{N_a - 1}{N_f N_n} \sum_{f=1}^{N_i} \frac{\sum_{n=1}^{N_a} 1(\|x_f - x_n\| \leq r)}{\sum_{a=1, x_a \# x_f}^{N_a} 1(\|x_f - x_a\| \leq r)}$$

They further emphasises that particular attention must be paid to case-control designs. According to them, all points of interest (called *cases*) are carefully referenced, and the benchmark point set (called control) is just sampled.

This approach has been widely used for spatial clustering of disease (Kingham *et al.*, 1995; Diggle and Chetwynd, 1991; Diggle *et al.*, 2007; Gatrell *et al.*, 1996; Gatrell *et al.*, 1996) in which sick people are the *cases* and the rest of the population are treated as *controls*.

They further clarified that case control design is not limitative to geographical epidemiology and can easily be applied to ecology questions.

For this feature the M function that was defined previously, can be slightly modify. Since the controls are chosen to be representative samples of the population at every scale, neighbours of any kind are replaced by controls, indexed by a. Reference and neighbour points are N_c; their total weight is W_c. After simplification, M_{cases} can be written as follows:

$$M_{cases}(r) = \frac{\sum_{f1}^{N_c} \left(\sum_{n=1, x_a \# x_f}^{N_c} 1(\|x_f - x_n\| \leq r) w_n / \sum_{a=1}^{N_a} 1(\|x_f - x_n\| \leq r) w_a\right.}{W_c (N_c - 1)/W_a}$$

The confidence interval of the null hypothesis is then computed by Monte Carlo technique (Kenkel, 1988).

However, Loosmore and Ford (2006) stated that the test based on Monte Carlo simulations are actually not corrected because it repeated at each stop of the function.

Morcon *et al.* (2012) follow Loosmore and Ford's goodness of fit to obtain a corrected P value to reject the null hypothesis. For this they first compute the average value of M (r)

$$\acute{M}(r) = \frac{1}{s-1}\sum_{i=1}^{s} M_i(r)$$

Where s is the number of simulation and M_i (r) the value of M (r) I the ith simulation. Then, the statistic u_i is calculated for the *ith* simulation by summing on all values of r, where δr is the difference between the next value of r:

$$u_i = \sum_{r} [M_r(r) - \acute{M}(r)]^2 \, \delta r$$

To get *P*-value following equation can be used

$$P_u = \frac{\sum_{i=1}^{s} 1(u_1 > u)}{s}$$

If u is greater than all simulated values, the *P*- value to reject the null hypothesis erroneously is around 0. To avoid 0 or 1 *P*-value, it can assume that the another simulation would have given a value of u_i higher or lower than u and write $P_u < 1/s$ or $P_u > 1\text{-}1/s$.

In an another attempt Marcon *et al.* (2013) have utilized Ripley's K function in a different way

$$\acute{K}(r) = \frac{I_1 I_2}{N(N-1)} \sum_{\varepsilon_i \neq \varepsilon_j} I\{d(\varepsilon_i, \varepsilon_j) \le r\}$$

Where I_1 = width; I_2 are the length, *P* is the intensity of the underlying point process; estimated by the number of observed points N divided by the area of the quadrate or window. They use r to denote vectors of distances (r_1___r_2.........r_d). Points are denoted by $[^\wedge_i$, and $I\{d([^\wedge_i, [^\wedge_j) \le r\}$ is an indicator function equal to 1 when the distance between two points is less that or equal to r, 0 else.

Marcon *et al.* (2012) have pointed out several advantages over traditional K function as well as $K_{in\ hom}$ function and these includes:

1. M is more convenient to use than K because no edge-effect correction is required. More than this, the domain limits do not have to be known, the point locations are enough. Therefore, complete geographical data sets can be treated without simplifying the domain shape and eliminating many border points.

2. M does not require a good knowledge of the underlying point process and a pre-computation of local densities like K_{inhom} does.
3. M is more powerful than D because it does not ignore a part of the data.
4. Neither K nor D nor K_{inhom} take into account the points' weight.

4.6.4.15.1.5 *O-ring statistics*

The function $K(r)$ is the expected number of points in a circle of radius r centered at an arbitrary point (which is not counted), divided by the intensity λ of the pattern.

The alternative pair correlation function $g(r)$, which arises if the circles of Ripley's K-function are replaced by rings (Ripley, 1981; Stoyan and Stoyan, 1994; Stoyan and Penttinen, 2000; Dale *et al.*, 2002), gives the expected number of points at distance r from an arbitrary point, divided by the intensity λ of the pattern.

For a homogeneous and isotropic point pattern, the second-order characteristics depend only on distance r, but not on the direction or the location of points. An appropriate geometry is therefore to adopt circular shapes, such as the circles of Ripley's K-function, as a basis for the spatial statistics. Curiously, the alternative approach of using rings (or annulus) instead of circles, *i.e.*, the pair-correlation function $g(r)$ or the O-ring statistic $O(r) = \lambda g(r)$ (Wiegand *et al.*, 1999), has rarely been used in ecology (Galiano, 1982; Wiegand *et al.*, 2000; Condit *et al.*, 2000; Revilla *et al.*, 2002).

Using rings instead of circles (has the advantage that one can isolate specific distance classes, whereas the cumulative K-function confounds effects at larger distances with effects at shorter distances (Getis and Franklin, 1987; Penttinen *et al.*, 1992; Condit *et al.*, 2000). Note that the K-function and the O-ring statistic respond to slightly different biological questions.

The accumulative K-function can detect aggregation or dispersion *up to* a given distance r and is therefore appropriate if the process in question (*e.g.*, the negative effect of competition) may work only up to a certain distance, whereas the O-ring statistic can detect aggregation or dispersion at a given distance r. The O-ring statistic has the additional advantage that it is a probability density function (or a conditioned probability spectrum, Galiano, 1982) with the interpretation of a neighborhood density, which is more intuitive than an accumulative measure (Stoyan and Pettinen, 2000).

On the other hand, the *O*-ring statistic will lose the advantage that it can isolate specific distance classes if the rings are too wide. Again, the univariate *O*-ring statistic $O(r)$ is calculated by setting pattern 2 equal to pattern 1. For CSR, $O(r) = \lambda$, $O(r) > \lambda$ indicates aggregation of the pattern at distance r, and $O(r) < \lambda$ regularity.

The *O*-ring statistic is a probability density function with the straightforward biological interpretation of a local neighborhood density, which is more intuitive than an accumulative measure (Stoyan and Penttinen, 2000). Because it is a scale-dependent density function, the ring statistic is only marginally biased by virtual aggregation (caused by larger gaps in the study region that violate the assumption that the pattern is homogeneous).

4.6.4.16. *Kernel estimation of intensity (k mean)*

Kernel density estimation is an interpolation technique that generalizes individual point locations or events, $\boldsymbol{s}_i$, to an entire area and provides density estimates, $\lambda(\boldsymbol{s})$, at any location within the study region R. From a visual point of view it can be thought of a three-dimensional sliding kernel function $K(\bullet)$ that 'visits' every location $\boldsymbol{s}$ (Fig. 11).

Distances to each observed event ***si*** that lies within a specified distance b, referred to as the bandwidth, are measured and contribute to the intensity estimate at $\boldsymbol{s}$ according to how close they are to $\boldsymbol{s}$ (Bailey, 1996). This produces a more spatially smooth estimate of variations in $\lambda(\boldsymbol{s})$ than could be attained by using a fixed grid of quadrates. Formally, if $\boldsymbol{s}$ represents a vector location anywhere in R and $\boldsymbol{s}_1, ..., \boldsymbol{s}_n$ are the vector locations of the n observed events, then the intensity $\lambda(\boldsymbol{s})$, at $\boldsymbol{s}$ is estimated as follows (Burt and Barber, 1996):

$$F(s) = \frac{1}{nb}\sum_{i=1}^{n} K\left(\frac{s - s_i}{b}\right)$$

There are a number of different kernel functions. However, since the result of an analysis is not strongly influenced by the chosen function as long as the function is symmetric (Silverman, 1978; Burt and Barber, 1996) there are no binding rules concerning the choice of an appropriate function. The most common kernel function is the normal distribution function (Jansenberger and Steinnocher, 2004).

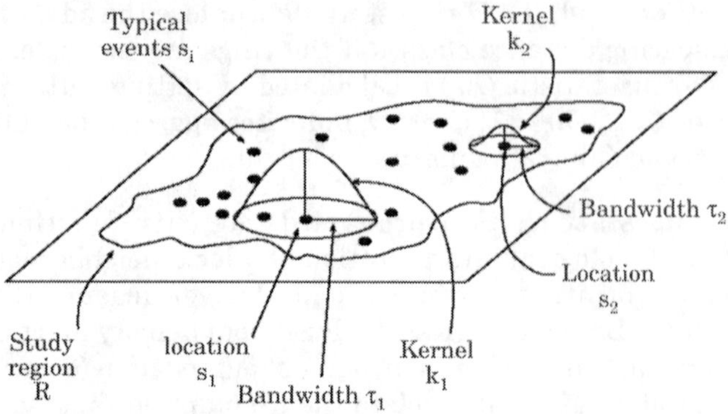

Fig. 11: Illustration of Kernel Estimation

$$K\left(\frac{d_i}{u}\right) = \frac{1}{2\pi}\exp\left(-\frac{1}{2}\left[\frac{d_i}{u}\right]^2\right)$$

With
$$u = \left(\frac{s - s_i}{b}\right)$$

Based on this function the density estimate is expressed as (Burt, 1996):

$$\lambda(s) = \frac{1}{b^2}\sum_{i=1}^{n} K\left(\frac{d_i}{b}\right)$$

where d_i is the distance between the point s and the observed event location s_i. Because the bandwidth b is the standard deviation of the normal distribution, this function extends to infinity in all directions, *i.e.*, it will be applied to each point in the region (Leitner, 2001). The quartic kernel function on the other hand, has a circumscribed radius b, which is also the and width. The quartic kernel function is therefore applied to a limited area around each event and has the following functional form (Bailey and Gatrell, 1995).

$$K(u) = \begin{cases} \frac{3}{4}(1 - u^2u)^2 \; for \; u^2u \leq 1 \\ 0 \; otherwise \end{cases}$$

The intensity at s is estimated as

$$\lambda(s) = \sum_{i=1 d_i \leq b}^{n} \frac{3}{\pi b^2}\left(1 - \frac{d^2}{b^2}\right)^2$$

Since $d_i \leq b$, the summation is only over values of ***di*** which do not exceed b. The region of influence within which the observed events contribute to λ^(***s***) therefore equals a circle of radius b centred on ***s***. In ***s*** the weight is $3/\pi b^2$ and drops smoothly to a value of zero at distance b. This function can be extended by a weighting variable (W_i) as is shown in next equation (Levine, 2002)

$$\lambda(u) = \sum_{i=1 d_i \leq b}^{n} \left[W_i \left(\frac{3}{\pi b^2}\right)\left(1 - \frac{d^2}{b^2}\right) \right]$$

More important than the kernel function is the choice of an appropriate bandwidth b (Silverman, 1986), which has considerable influence on the estimated distribution. The effect of increasing the bandwidth b is to stretch the region around ***s*** within which observed events influence the intensity estimate at ***s***. For very large b, λ^(***s***) will appear flat and local features will be obscured.

If b is small, then λ^(***s***) tends to a collection of spikes centred on ***si*** (Fischer, 2001). There are several methods which attempt to optimise the value of b given the observed pattern of event locations. Basically, the bandwidth can either be fixed (fixed kernel estimation) or adaptive (adaptive kernel estimation).

In adaptive smoothing sub-areas in which events are more densely packed than in others are visited by a kernel whose bandwidth is smaller than elsewhere in order to avoid smoothing out too much detail. In the fixed choice the bandwidth always stays the same. When using a fixed interval, the appropriate choice of the smoothing parameter will be influenced by the purpose for which the density estimate is to be used. If the KDE is applied to one variable, this is referred to as a single density estimate.

If it is applied to two variables, it is called a dual density estimate. In the latter case, a kernel density is estimated for each variable individually and then the two density estimates are related with each other through simple algebraic operations such as the sum, the difference, the quotient and the like. The most commonly used operation in this realm is the quotient.

The simplest way of exploring point pattern data is by examining a two-dimensional frequency distribution of counts of the number of events falling in each of the equal-area units imposed on the study area, giving an impression of how the intensity of the point process varies.

This method may give a global idea of sub-regions with high or low intensity, but much of the spatial details in the observed pattern are lost. As a result, the kernel estimation method is used to obtain smooth estimates of univariate (or multivariate) probability densities from an observed sample of observations (Silverman, 1986).

Using the kernel estimation method, the intensity estimation of a spatial point pattern is similar to estimating a bivariate probability density function. If s represents a vector location anywhere in R and $s_1, ..., sn$ are the locations of the n observed events, then the intensity, $\lambda(s)$, at s is estimated as (Bailey and Gatrell, 1995):

4.6.4.17. *F, G, K and J functions*

The statistical analysis of a point pattern usually begins with the computation of estimation of summary function F (empty space function), G (nearest –neighbour distance distribution function and K (reduced second moment function).

For a homogeneous Poisson process F, G and K take known functional forms, and deviations of estimates of F, G, K from these forms are taken as indications of clustered or inhibited alternatives (Baddeley and Gill, 1997).

The empty space function (also called the "*spherical contact distribution*" or the "*point-to-nearest-event*" distribution) of a stationary point process X is the cumulative distribution function F of the distance from a fixed point in space to the nearest point of X. An estimate of F derived from a spatial point pattern data set can be used in exploratory data analysis and formal inference about the pattern (Cressie, 1991; Diggle, 1983; Ripley, 1988).

In exploratory analyses, the estimate of F is a useful statistic summarising the sizes of gaps in the pattern. For inferential purposes, the estimate of F is usually compared to the true value of F for a completely random (Poisson) point process, which is

$$F(r) = 1 - exp(- lambda * pi * r^2)$$

Where *lambda* is the intensity (expected number of points per unit area). Deviations between the empirical and theoretical F curves may suggest spatial clustering or spatial regularity.

This algorithm estimates the empty space function F from the point pattern X. It assumes that X can be treated as a realisation of a stationary (spatially homogeneous) random spatial point process in the plane, observed through a bounded window.

The window (which is specified in X) may have arbitrary shape.

The estimation of F is hampered by edge effects arising from the unobservability of points of the are andom pattern outside the window.

The nearest neighbour distance distribution function (also called the "*event-to-event*" or" *inter-event*" distribution) of a point process X is the cumulative distribution function G of the distance from a typical random point of X to the nearest other point of X. An estimate of G derived from a spatial point pattern dataset can be used in exploratory data analysis and formal inference about the pattern (Cressie, 1991; Diggle, 1983; Ripley, 1988). In exploratory analyses, the estimate of G is a useful statistic summarising one aspect of the "clustering" of points. For inferential purposes, the estimate of G is usually compared to the true value of G for a completely random (Poisson) point process, which is:

$$G(r) = 1 - exp(- lambda * pi * r^2)$$

Where *lambda* is the intensity (expected number of points per unit area). Deviations between the empirical and theoretical G curves may suggest spatial clustering or spatial regularity.

The reduced second moment measure *Kappa* can be regarded as a generalisation of the more familiar K-function.

An estimate of *Kappa* derived from a spatial point pattern dataset can be useful in exploratory data analysis. Its advantage over the K-function is that it is also sensitive to anisotropy and directional effects.

In formal terms, the reduced second moment measure *Kappa* of a stationary point process X is a measure defined on the two-dimensional plane such that, for a 'typical' point x of the process, the expected number of other points y of the process such that the vector $y - x$ lies in a region A, equals *lambda * Kappa(A)*. Here *lambda* is the intensity of the process, *i.e.* the expected number of points of X per unit area.

The function value $K(t)$ is the value of the reduced second moment measure for the disc of radius t centred at the origin; that is, $K(t) = Kappa (b(0,t))$. J -function introduced by Van Lieshout and Baddeley (1996) to point processes with real-valued marks.

The underlying idea of this summary statistic is to compare distances seen from an arbitrary fixed point 0 to the nearest point of the pattern measured by the empty space function F—to those seen from a typical point of the pattern, as captured by the nearest neighbour distance function G. Thus, for (unmarked) point processes

$$J(t) = \frac{1 - G(t)}{1 - F(t)}$$

This is a nonparametric measure of the type of spatial interaction: the value 1 can be interpreted as indicating complete randomness or lack of interaction, while values less than 1 suggest clustered pattern and values greater than 1 suggest regular or inhibitory pattern. The power of the *J*-function in hypothesis testing was assessed in Baddeley *et al.* (2000), Chen *et al.* (2001), and Th½onnes onnes and Van Lieshout (1999). Extensions to multivariate point processes were proposed by Van Lieshout and Baddeley (1999), to germ grain models by Last and Holtmann (1999), and window based *J*-functions were suggested by Baddeley *et al.* (2000) and Chen *et al.* (2001).

4.6.4.18. *Nearest neighbour analysis*

Nearest-neighbour analysis provides an objective method for distinguishing among these spatial distributions. This is done by comparing the average observed distance between features and their nearest neighbor to an expected or theoretical average distance between nearest-neighbor points in a distribution generated by a random process.

Nearest neighbour analysis examines the distances between each point and the closest point to it. The Nearest neighbour is a method of exploring pattern in Locational data by comparing graphically the observed distribution functions of event-to-event or random point-to-event nearest neighbour distances, either with each other or with those that may be theoretically expected from various hypothesized models, in particular that of spatial, *i.e.*, it describes distribution of points according to their spacing.

4.6.4.18.1. *The nearest neighbour index*

The Nearest neighbour index measures the degree of spatial dispersion in the distribution based on the minimum of the inter-feature distances, *i.e.*, it is based on the distance between adjacent point features. Such that the distance between point features in a clustered pattern will be smaller than in a scattered (uniform) distribution with random falling between the two.

Methods for calculating Nearest Neighbor are as follows.

Ad = (å$_i$d$_i$)/n. d_i is the distance from point i to its nearest neighbour; Ad is the average of nearest neighbour distance of the point pattern; n is the total number of points in the chosen map area.

Ed=1/2sqr(A/n) Expresses the expected value of the average nearest distance; A denotes the map area

NNI=Ad/Ed Equation for the nearest neighbour index, it is defined as the ratio of Ad to Ed

The values of NNI range between two theoretical extremes, 0 and 2.1491. When all the points in a pattern fall at the same location, the pattern represents the theoretical extreme of spatial concentration, in this case, Ad = 0 and NNI = 0. The more closely the points are clustered together, the closer to 0 NNI will be, since the average nearest neighbour distance decreases. The closer NNI gets to 1, the more randomly spaced the points are. The value of NNI approaches 2.1491 for perfectly uniformly spaced points. Hence, the closer NNI is to 2.1491, the more uniformly spaced the data are.

4.6.4.18.2. *The logic of nearest neighbor analysis*

This analysis examines the distances between each point and the closest point to it. It then averages all these nearest neighbor distance.

If the average distance is less that the average for hypothetical random distribution or CSR pattern, the distribution of the features being analysed are considered clustered.

However, if the average distance is greater than a hypothetical random distribution, it is considered as dispersed. The sample mean of nearest neighbor distance can be calculated by following equation.

$$\bar{d} = \frac{\sum_{i=1}^{n} d_i}{N}$$

Where d_i is the nearest neighbor distance for point I, and N is the number of points. The expected average distance for a random pattern can be calculated through

$$E\left(\bar{d}_N\right) = \frac{1}{\sqrt[2]{\lambda}}$$

In which λ is the density and it the ration of N/A (size of study area = N). Variance and its standard deviation can be quantify through

$$Var\left(\bar{d}\right) = \frac{4-\pi}{4\lambda\pi} and\ \sigma\left(\bar{d}\right) = \sqrt{var\left(\bar{d}\right)} = \sqrt{\frac{(4-\pi)}{m4\pi\lambda}}$$

Where, π is the constant (3.14) and the null hypothesis of CSR is can be test using the *Z* statistic (standard normal variant).

$$Z = \frac{\bar{d} - E(\bar{d})}{\sigma\,(\bar{d})}$$

A negative *Z* score indicates clustering and a positive score indicates dispersion or evenness. Palacio-Bielsa *et al.* (2012) used this to quantify the spatial spread of *Erwinia amylovora* on pome fruit.

Kamu *et al.* (2015) have utilized this approach to quantify the distribution of infected oil palms with *Ganoderma* basal stem root disease. In their analysis this tool showed that spatial distribution of the infected palm in all the studied area was clustered. Gidoin *et al.* (2014) have used this approach in coco agro-ecosystems for black pod disease, caused by *Phytophthora megakarya*.

The analysis can be applied to behavioural phenomena which possess discrete spatial locations which may be mapped as points. Nearest neighbor analysis utilizes the fundamental concept of randomness.

A distribution is random when each spatial unit in the area containing the points has an equal opportunity of receiving a point. A nonrandom point pattern is either more clustered than random or more uniform than random.

Nearest neighbor analysis reduces the complexities of spatial distributions of observed points to a pattern description called random, more grouped than random, or more clustered than random. Relations between neighboring points are derived under the assumption that such points are randomly distributed in terms of the Poisson distribution.

$$R = r_a/r_e$$

R scale values range from 0 to 2.149. A value of 1.0 provides the standard of randomness, increasing R values are indicative of increasing dispersion leading to a limiting case of regularity, and decreasing R values provide evidence of increasing clustering.

4.6.4.18.3. *Conceptual and methodological problems*

The nearest neighbor statistic is a useful tool for dealing with certain spatial phenomena. Its successful application depends, of course, on an appreciation of its conceptual and technical limitations as well as the conditions under which its usefulness will be maximized..

4.6.4.18.4. *Distinguishing pattern from processes*

The R scale provides a pattern description. Patterns themselves are the product of underlying processes which develop over time and space.

Since patterns provide only static evidence of spacing, they must be approached with notions derived from a rationalization of processes thought to evolve over space.

In this sense patterns are no more than abstractions derived by artificially halting dynamic processes. Point patterns are therefore, a synthetic visual expression at a given point in time of processes which continuously operate over space. For this reason any observed point with a spatial distribution represents an event in time and space (Taylor, 1977).

Alternative values of R do not necessarily justify the conclusion that either a random or a systematic process is operating. A distribution of observed points provides no direct information about the underlying processes giving rise to the observed distribution (Amedeo and Golledge, 1975).

Inferences from patterns to processes can be defended only when processes are considered in forming hypotheses concerning changing R values which are observed over time (Dawson, 1975).

4.6.4.18.5. *Limiting values of the R scale*

As points in space tend to cluster interpoint distances, r_a, and hence R, would tend toward zero. The opposite tendency to clustering is dispersion. Taken to the extreme, dispersion results in a limiting regular pattern based on a triangular lattice and an R of 2.149. These theoretical limits to the R scale are rarely observed in practice.

The complex processes underlying spatial distributions tend to produce spatial distributions which are more complicated than those reflected by the theoretical limits of the R scale (Pinder and Witherick; 1972). Hence it has been observed (Taylor, 1977) that the empirical values of R are likely to range between 0.33 and 1.67.

4.6.4.18.6. *Processes generating nonrandom and random patterns*

Extreme cases of clustering and regularity are deterministic in the sense that they have singular R values associated with unique average interpoint distance's which satisfy each extreme case.

This determinism does not hold for a randomly produced distribution. A number of random patterns share are value of 1. In fact, we may purposely use a random procedure to generate a pattern whose R value differs from 1 (Dawson, 1975). In such a case, a known random process could be shown to produce a spatial pattern with a tendency toward clustering.

4.6.4.18.7. *Higher order analysis*

Considering the space or distance between all points and their first or closest neighbor can result in utilizing only a small portion of the information available in a spatial pattern. Imagine pairs of points which are evenly dispersed over space.

The distance between any given point and its nearest neighbor would be small. Yet obviously we are not dealing with a clustered pattern as would be indicated by a low R value. Unless each point is considered in relation to all others, the character of the spatial distribution will be obscured by nearest neighbors (Getis, 1964).

The solution is to identify and average the distance between each point and not only the original nearest neighbor but succeeding neighboring points (second, third .... nth nearest neighbors). At any order of analysis (so named to refer to the number of succeeding neighbors considered) Higher order analysis was made possible by Morrista's (1954) and Thompson's (1956) derivations of expected distance to nth order neighbors based on calculation of high order probability density functions.

In distributions lacking a pronounced spatial bias an order of analysis which is less than n can be employed. Similarly the absence of a pronounced spatial bias supports a direct selection of neighboring points while a bias may support the use of a sectoral decision rule. With this algorithm the space around each point is divided into the number of sectors equal to the order of analysis and distance from the reference point to the nearest neighbor in each sector is determined.

4.6.4.18.8. *Properties of observed and expected distances and the R scale*

The range and interpretation of the R scale depends on the type of analysis employed. The two principal types of more complex nearest neighbor analysis call for either a division of the space surrounding each observed point into k sectors (where $k > 1$) (Taylor, 1977) or measures of spacing between each observed point and its n nearest neighbors (where $n>1$) (Clark and Evans, 1954).

The sectoral method is more sensitive to randomness in a pattern biased toward uniform spacing while raising the order of analysis provides more power efficiency when points are biased toward grouping or clustering.

These types of analysis can be combined, *e.g.*, a space can be divided into k sectors and the distance to n neighbors in each sector can then be determined. For purposes of clarification we shall consider separately the effects of sectoral and nth order nearest neighbor analysis on expected average distances and R scale values. With first order nearest neighbor analysis in which both k and n equal 1, the R scale ranges from values of 0, indicative of maximum clustering through 1.0, indicative of the observed average distances equal to that expected under Poisson conditions ($r_a = r_e$), to a maximum of 2.149, which represents the case of maximum hexagonal regularity.

The interpretation of clustering is unaffected by altering the type of analysis in terms of either k or n. This follows from the fact that maximum clustering ($R = r_e/r_a = 0$) occurs when all points exist at one location and thus the value of observed average interpoint distance would be unaffected by increasing the number of sectors or increasing the number of neighboring distance measures.

By definition, an R value of 1.0 will define a situation in which the observed and expected (under random conditions) average distance are identical. This is obviously unchanged by altering the type of analysis.

4.6.4.18.9. *Refined nearest neighbor analysis*

Basically, this test is utilize to test the null hypothesis of CSR. In nearest neighbor test, a few very large nearest neighbor distances associated with isolated points could vague cluster pattern, while, refined nearest neighbor analysis overcomes this issue by examining the entire distribution (Fischer and Getis, 2010).

If for each distance, the observed nearest neighbor distance higher, $F(d_i \leq r)$, larger than the expected nearest neighbor distance for CSR, $p\ (d_i \leq r)$ a clustered pattern is indicated.

When, the observed nearest neighbor distances, $F(d_i \leq r)$, a regular pattern is indicated $F\ (d_i \leq r)$ can be quantify through taking the nearest neighbor distance (d_i) and the nearest distances to study the boundary (u_i) for each point (i). In this the value of d_i then be raked from smallest to the largest. According to Kamu *et al.* (2015) for every distance of interest (r), the number of points (n_1) for which $d_i \leq r$, and the number of points (n_2), for which $u_1<r<d_1$ are counted. The proportion of $F(d_i \leq r)$ and $P\ (d_i \leq r)$ are quantify through

$$f\ (d_i \leq r) = \frac{n_1}{N - n_2}\ and\ P(d_i \leq r) = 1 - e^{-\pi r^2 \lambda}$$

In above equation *e* is the constant (2.178) and *r* is the specified distance and λ is the density and calculated above. Kamu *et al.* (2015) have utilized this approach to quantify the distribution of infected oil palms with *Ganoderma* basal stem root disease.

4.7. SPATIAL AUTOCORRELATION

The spatial distribution of organisms may vary from aggregated (clustered) through a random pattern to the regular (uniform) case. Individuals are randomly positioned if the location of one individual is independent of the positions of all others.

Aggregation occurs when individuals tend to be close together, resulting in a high variance of density estimates across space (over-dispersion). On the other hand, regularity occurs when individuals tend to avoid each other (*e.g.*, spatial distributions of territorial organisms) so that density estimates are less variable (under-dispersion).

According to the type of spatial distribution, appropriate sampling should be designed to ensure proper estimates of abundances and their distributions (Andrew and Mapstone, 1987).

It follows that the spatial structures manifest themselves by the relationship (or lack of independence) between abundance values observed at neighboring sites in space. In many instances, individuals are aggregated so that sampling sites that are closer together tend to display abundance values that are more similar than sites that are further apart, resulting in positive spatial dependence. Regular distribution of individuals may produce the opposite effect, inducing negative spatial dependence.

Reynolds and Madden (1988) have provided the theory of spatio-temporal autocorrelation. According to them autocorrelation tool like the Moran I statistic and others conserve the spatial information content of a filed sampling scheme by making use of information on the location of each sample point.

Nicot *et al.* (1984) have provided the advantages of using the Moran I statistic to characterize the spatial pattern of disease, while Noe and Campbell (1985) used this technique to characterize the spatial pattern for nematode distribution.

Modjesk and Rawling (1983) and Madden *et al.* (1987) provided the information's about advantages of spatial autocorrelation coefficients for a series of sample time that can be used in conjugation with spatial pattern statistics such as Lloyd's mean crowding and patchiness indices to characterize changes in spatial aggregation over time.

Basically, models that are exercise to portray pattern evolution in time and space are known as spatio-temporal transfer function (STFs) and it may be composed by distributed inputs, system variables, and errors. The general form of the STF expresses the system variable (*e.g.*, disease incidence at time *t*), system inputs are variables that are external to the system and which are not influenced by the system (*e.g.,* weather variable)

This technique determine the spatial dependence of disease incidence between sampling units over various lags, thus measuring inter-cluster correlation in contrast to ρ. Typically, spatial autocorrelation decline exponentially over distance; the rate of this decline indicates the degree of disease gradient. This technique is related to semivariogram. Xu and Ridout (1998) found the following model to be useful in describing the relationship of spatial autocorrelaton with mean disease incidence and spatial lag for data generated from computer simulated epidemics

$$\gamma\,(d,\mu) = S\mu^{u}\,(1-\mu)^{v}\,\exp[-td\;x\;(1-\mu)]$$

Where S is a scaling parameter, u and v are the power law parameters, d is the distance and t determines the rate of decrease of correlation with distance at fixed incidence. Gent *et al.* (2006) had utilized first and second-order spatial autocorrelation to quantify the similarity or dissimilarity of disease incidence among cones of neighbouring plants with a transet by following Turechek and Madden (1999). For this analysis data were transformed using the logit transformation 1n[y/1-y)], where y = (x +0.5)/ (n+1).

When a plant population or community is sampled, the samples have a spatial relationship with each other. To a certain extent, samples that are close to each other are more likely to be similar (Dale, 1999).

For example if host is sampled using a transect of small contiguous quadrates, adjacent quadrates are likely to be more similar than those at greater spacing. The lack of independence is referred to as to as spatial autocorrelation because the correlation occurs within the data set itself and arises because of spatial relationships.

'Spatial autocorrelation' is the correlation among values of a single variable strictly attributable to their relatively close locational positions on a two-dimensional (2-D) surface, introducing a deviation from the independent observations assumption of classical statistics.

Spatial autocorrelation exists because real-world phenomena are typified by orderliness, (map) pattern, and systematic concentration, rather than randomness. Tobler's first law of geography encapsulates this situation: "everything is related to everything else, but near things are more related than distant things." To this maximum should be added the qualifier: "but not necessarily through the same mechanisms." In other words, spatial autocorrelation means a dependency exists between values of a variable in neighboring or proximal locations, or a systematic pattern in values of a variable across the locations on a map due to underlying common factors.

Some of the basic terminologies related with spatial autocorrelation are as follows

- ***Auto***: A prefix literally meaning self; spatial autocorrelation means self-correlation, or values within a given variable are correlated, resulting in the variable being correlated with itself.
- ***Correlation***: A description of the nature and degree of a relationship between a pair of quantitative variables
- ***Negative Spatial Autocorrelation***: For the geographic distribution of some variable across a map, high values tend to be geographic neighbors of low values, intermediate values tend to be geographic neighbors of intermediate values, and low values tend to be geographic neighbors of high values
- ***Positive Spatial Autocorrelation***: For the geographic distribution of some variable across a map, high values tend to be geographic neighbors of high values, intermediate values tend to be geographic neighbors of intermediate values, and low values tend to be geographic neighbors of low values.
- ***Redundant Information***: Information in data that is duplicated, and hence unneeded for spatial statistical analyses; for geo-referenced data, this duplication arises from locational closeness that results in mutually shared information, allowing an awareness of nearby values once the value for a given location is known.
- ***Moran Coefficient***: An index of spatial autocorrelation, involving the computation of cross-products of mean adjusted values that are geographic neighbors (*i.e.*, co-variations), that ranges from roughly (–1, –0.5) to nearly 0 for negative, and nearly 0 to approximately 1 for positive, spatial autocorrelation, with an expected value of $-1/(n-1)$ for zero spatial autocorrelation, where n denotes the number of areal units.
- ***Moran Scatter-plot***: A scatter-plot of standardized versus summed nearby standardized values whose associated bi-variate regression slope coefficient is the unstandardized Moran coefficient
- ***Geary Ratio***: An index of spatial autocorrelation, involving the computation of squared differences of values that are geographic neighbors (*i.e.*, paired comparisons), that ranges from 0 to 1 for negative,

and 1 to approximately 2 for positive, spatial autocorrelation, with an expected value of 1 for zero spatial autocorrelation

- ***Geographic Connectivity/Weights Matrix*:** An n-by-n matrix with the same sequence of row and column location labels, whose entries indicate which pairs of locations are neighbors
- ***Map Pattern*:** The systematic organization of values for some variable across a map resulting in visually conspicuous texture that consists of global, regional, and local trends, gradients, swaths, or mosaics.

At the community level, spatial dependence may be induced by the functional dependence of the response variables (species) on some explanatory variables (*e.g.*, environmental) that share some parts of their spatial structures (induced spatial dependence (Fortin and Dale, 2005; Legendre and Legendre, 2012).

Autocorrelation also designated as true autocorrelation, inherent autocorrelation, or autogenic autocorrelation (Fortin and Dale, 2005), or interaction model (meaning interaction among the sites) by Cliff and Ord (1981).

From a statistical standpoint, autocorrelation is the spatial structure found in the error component of a community–environment model, once the effect of all important spatially structured environmental variables has been accounted for (*i.e.*, included in the model). In practice, it is fairly difficult to know whether all important environmental drivers have been included (and with correct functional forms) in the analysis of a particular data set. Hence, if autocorrelation remains in residuals, it could be driven by any number of processes including unmeasured (small-scale) environmental variables, or population/community dynamics. Nevertheless, theoretically, a complete model can be written as (Dray *et al.*, 2012).

$$Y = f(E) + R = F + R = F + T + U \text{ (i.e)} = R + T + U$$

Where R is broken down into the spatial autocorrelation in the residuals, T, and a random error component, U. However, these two residual components are impossible to partition unless expectations from specific autocorrelated models (such as dispersal models for instance) can be obtained in the same metric as R, which no longer contains abundances, but signed deviations of the observed abundances from their fitted values as predicted by the environmental variables.

As a consequence, the presence of a spatial structure in residuals, which makes the species residuals in R has long been seen as a statistical nuisance for the environmental control model (Legendre, 1993, Griffith and Peres-Neto, 2006), often generating inflated type I error rates in the environmental model and spurious (apparent) species– environment concordance.

Spatial autocorrelation may be defined as the relationship among the values of a single variable that comes from the geographic arrangement of the areas in which these values occur. It measures the similarity of objects within an area, the degree to which a spatial phenomenon is correlated to itself in space (Cliff and Ord, 1981), the level of interdependence between the variables, the nature and strength of the interdependence, *i.e.* spatial autocorrelation is an assessment of the correlation of a variable in reference to spatial location of the variable.

Factors Causing Spatial Autocorrelation.

Millet *et al.* (2007) have enlisted the potential sources/underlying causes of spatial dependence in biogeographic data (plant and animals) and these includes

- Dispersal/growth
- Disturbance
- Predation/herbivory
- Environmental/physical barriers
- Historical biogeography
- Interaction/competition/mutualism
- Metapopulation dynamics
- Learned behaviors (foraging, fear)
- Spatial dependence in predictor variables
- Missing/mis-specified predictor variables

The manifestation of these factors and processes in bio-geographical data can be differentiated by the scale at which their influence is observed. At broad-scale spatial dependence, Haining's first order effect, refers to a predictable pattern or spatial trend that is deterministic and usually related to an unmeasured environmental variable.

Legendre and Legendre (1998) refer to this as a true gradient, where gradient implies a gradual change in values in geographic space, as would be observed in an environmental variable with a strong spatial structure.

In socio-economic models, Anselin (1992) describes the same phenomenon as 'nuisance' spatial dependence, as it does not characterize an observation that is truly jointly determined by its neighbours. This type of spatial dependence results in spatially correlated errors and would therefore be specified in the model by a lagged error term, and the mean of the response variable would not be affected (Anselin, 1992).

The second order effect described by Haining (1989) is fine scaled spatial dependence and occurs when the process(es) that produce the distribution of the response variable are inherently spatial, reflecting biotic interaction between sites. Legendre and Legendre (1998) call this same effect a false gradient. The spatial dependence that results from these complex spatial/behavioural/physiological processes comes closer to representing what Anselin (1992) terms 'substantive' spatial dependence, and would be specified in a model with a lagged response variable as an explanatory variable. Spatially, this type of pattern would generally be observed to be more patchy than gradual, as the processes that generate it are more stochastic than deterministic.

Legendre and Legendre (1998) note that distinguishing between false gradient and true gradient in ecological data is a "moot point". If the process resulting in the pattern is known, usually a determination as to whether it is a false gradient or true gradient can be made, but the pattern itself is not enough to enable that distinction.

Both types of spatial dependence can have potential predictive ability, but as they also both represent combinations of factors that are location-specific (possible environmental factors for the true gradient and biotic factors such as competition, dispersal, and disturbance for the false gradient), they should be considered indirect gradients.

It should be noted that Legendre and Legendre (1998) further differentiate true gradient as 'spatial dependence' and false gradient as 'spatial autocorrelation'.

The causes of spatial autocorrelation are manifold, but three factors are particularly common (Legendre and Fortin, 1989; Legendre, 1993; Legendre and Legendre, 1998): 1) biological processes such as speciation, extinction, dispersal or species interactions are distance-related; 2) non-linear relationships between environment and species are modeled erroneously as linear; 3) the statistical model fails to account for an important environmental determinant that in itself is spatially structured and thus causes spatial structuring in the response (Besag, 1974).

The second and third points are not always referred to as spatial autocorrelation, but rather spatial dependency (Legendre *et al.*, 2002). Since they also lead to autocorrelated residuals, these are equally problematic. A fourth source of spatial autocorrelation relates to spatial resolution, because coarse grains lead to a spatial smoothing of data. In all of these cases, spatial autocorrelation may confound the analysis of species distribution data.

Spatial autocorrelation may be seen as both an opportunity and a challenge for spatial analysis. It is an opportunity when it provides useful information for inference of process from pattern (Palma *et al.*, 1999), for example, increasing our understanding of contagious biotic processes such as population growth, geographic dispersal, differential mortality, social organization or competition dynamics (Griffith and Peres-Neto, 2006). In most cases, however, the presence of spatial autocorrelation is seen as posing a serious shortcoming for hypothesis testing and prediction (Lennon, 2000; Dormann, 2007), because it violates the assumption of independently and identically distributed (*i.e.*) errors of most standard statistical procedures (Anselin, 2002) and hence inflates type I errors, occasionally even inverting the slope of relationships from non-spatial analysis (Ku¨hn, 2007).

Spatial autocorrelation describes the correlation of the values of a single attribute related to the proximity of those values in geographic space (Suzuki, *et al.*, 2008 and Tobin, 2004). It relies on data interdependence and spatial statistics (Odland, 1988 and Ovoermars *et al.*, 2003).

Spatial autocorrelation exists when there is systematic spatial variation in the values of a given variable. This variation can exist in two forms called positive or negative spatial autocorrelation. In the positive case, the value of a variable at a given location tends to be similar to the values of that variable in nearby locations.

This means that if the value of some variable is low in a given location, the presence of spatial autocorrelation indicates that nearby values are also low. Conversely, negative spatial autocorrelation is characterized by dissimilar variant values in nearby locations. For example, a low variate value may be surrounded by high values in nearby locations when negative spatial autocorrelation exists.

A positive spatial autocorrelation refers to a map pattern where geographic features of similar value tend to cluster on a map, whereas a negative spatial autocorrelation indicates a map pattern in which geographic units of similar values scatter throughout the map. Positive spatial autocorrelation occurs when adjacent neighborhood locations have similar attribute values (Almeida-Neto and Lewinsohn, 2004; Griffth, 2003 and Mueller-Warrant *et al.*, 2008). When no statistically significant spatial autocorrelation exists, the pattern of spatial distribution is considered random (Chou, 1997).

Positive spatial autocorrelation over short distances is commonly observed and reflects phenomena that are contagious (clumped) over short distance, such a limited seed dispersal, clonal growth, nutrient-rich patches and positions on an environmental gradient. By contrast, negative spatial autocorrelation over short distances would suggest avoidance or competition among neighbors.

Spatial autocorrelation is a widespread phenomenon in ecology (*e.g.*, Fortin *et al.*, 1989, Henebry 1995; Torgersen *et al.*, 1995; Koenig 1998; Radeloff *et al.*, 2000). Plant species respond individualistically to their environment, both directly and indirectly through species interactions. Autocorrelation can be used to compare the pattern of the same species in different environments, and the patterns of different species in the same environment.

For plant analysis the spatial autocorrelations and its various uses were discussed by Heikkinen *et al.* (1996), Hawkins *et al.* (2003) and Dorman (2006). Goslee (2006) have discussed the behaviors of different vegetation sampling methods in the presence of spatial autocorrelation. He concluded that transect methods especially responded poorly to spatial autocorrelation and randomly-located quadrats were more efficient in highly patchy environments, but large numbers of quadrats are needed for either method if rare species must be located. He emphasized that multi-scale methods are more robust to spatial autocorrelation, and more effective at identifying rare species because of the larger proportion of the total area sampled. Cover and especially frequency estimates are highly variable, and depended on both the level of autocorrelation and the sampling method used.

4.8. METHODS DEALING WITH SPATIAL AUTOCORRELATION

Spatial dependence in point and area data has been studied more extensively in socioeconomic and medical geography applications (Cliff and Ord, 1981; Haining, 1990; Cressie, 1991). A suite of statistics has been developed to test for the presence and magnitude of spatial dependence in data.

These include global distance – or contiguity – based measures such as Moran's I, and Geary's c, as well as more recent local statistics that can characterize spatial clustering in a portion or geographic subset of the dataset (Getis and Ord, 1996).

These spatial dependence statistics can also be used as a more objective means of quantifying landscape pattern in terms of patchiness (Aspinall, 1999). While some ecological studies have used these statistics to test for spatial dependence (Fortin, 1999; Dale *et al.*, 2002; Perry *et al.*, 2002), their use has typically preceded removal or justified avoidance of spatial dependence in a model, rather than explicitly incorporate spatial dependence (Klute *et al.*, 2002; Lichstein *et al.*, 2002).

Another way in which spatial dependence has been quantified follows work done by Borcard *et al.* (1992) and Legendre (1993), where the variation of the response variable is separated into environmental (*e.g.*, climate, topography) and spatial (represented by a trend surface of geographic coordinates) components (Lobo *et al.*, 2002, 2004; Nogu´es-Bravo and Martinez-Rica, 2004; Titeux *et al.*, 2004).

Partial regression analysis (Legendre and Legendre, 1998) using either GLMs or GAMs is used to decompose the explanatory effects of a group of variables into single variation fractions. Figure. 1 shows a conceptualization of this model: the outer circle represents the variation in the response variable: (a) represents unexplained variation (model error); (b) represents the (nonspatial) effect of an environmental variable; (c) represents the overlapping effects of environment and spatial dependence (the coincident environmental and geographic gradients); (d) represents the effect of spatial dependence. However, what is described as the 'pure effect of spatial variation' (Lobo *et al.*, 2002, 2004), and the 'pure spatial fraction' (Nogu´es-Bravo and Martinez-Rica, 2004) represented by (d) in Figure 12 should be interpreted with caution (Legendre and Legendre, 1998). Unless specified to be extraordinarily complex, the polynomial equations defined in the trend surface analysis are usually not capable of describing fine-scaled spatial variation (Fig. 12a and 12d), as would result from patchy biotic processes. This method is particularly useful for differentiating between the broad-scale correlation between environmental variables and the response variable (Fig. 12c), and the nonspatial usually finer-scaled correlation between the environmental variables and response variable (Fig. 12 b) (Legendre and Legendre, 1998).

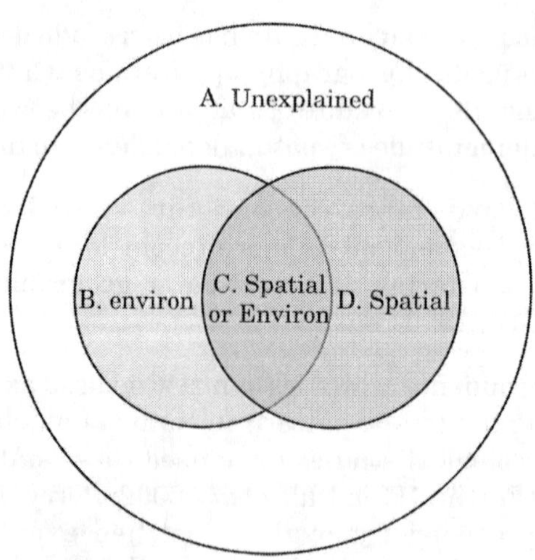

Fig. 12: A conceptualization of the variation in vegetation distribution; the outer circle represents the variation in the vegetation distribution: (a) represents unexplained variation (model error); (b) represents the variation explained by an environmental variable; (c) represents the variation explained by spatially structured environment; (d) represents the variation explained by spatial structure alone (Redraw from Miller *et al.*, 2007)

Spatial autocorrelation tools test whether the observed value of a variable at one locality is independent of values of the variable at neighboring localities. A variety of methods have consequently been developed to correct for the effects of spatial autocorrelation (partially reviewed by Miller *et al.*, 2007), but only a few have made it into the ecological literature.

Before considering the use of modeling methods that account for spatial autocorrelation, it is a sensible first step to check whether spatial autocorrelation is in fact likely to impact the planned analyses, *i.e.* if model residuals indeed display spatial autocorrelation. Checking for spatial autocorrelation (SAC) has become a commonplace exercise in geography and ecology (Sokal and Oden, 1978 and Fortin and Dale, 2005).

There are many indicators of spatial autocorrelation (Cliff and Ord, 1981; Goodchild, 1986; Haining, 1990; Chou, 1997; Isaaks and Shrivastava 1989, Perry *et al.*, 2002). Diniz-Filho *et al.* (2003) and Kuhn (2007) have argued that the presence of residual spatial autocorrelation should always be tested for in spatial ecology and appropriate methods should be used if there is shown to be significant spatial autocorrelation

Miller *et al.* (2007) have listed the various methods adopted by different workers for measuring the spatial dependency and they differentiate them into autoregressive method, geographically weighted method and parameter estimation method (Table 6).

Carl and Kuhn (2007) have visualized the methods deals with spatial autocorrelation into two group *i. e.*, methods for normally distributed data and methods related with non-normal distributed data.

For data with normally distributed responses, spatially autoregressive linear models are available and extensively used by ecologists (Lichstein *et al.*, 2002; Dark, 2004).

Augustin *et al.* (1996) applied an autologistic method to handle presence/ absence data which was later widely used by several authors (Wu and Huffer, 1997; Osborne *et al.*, 2001; Segurado and Araujo, 2004; Luoto *et al.*, 2005).

This method can also be extended to other distributions (Haining, 2003). Furthermore, binary spatial regression models for spatially aggregated data have been proposed using spatial Markov random fields (Pettitt *et al.*, 2002). Bayesian estimation methods are available using Markov chain Monte Carlo to approximate the posterior distributions (Besag *et al.*, 1991; Kuhn *et al.*, 2006, for an application in ecology). Carl and Kuhn (2007) have proposed an approach that was based on generalized estimating equations which was an extension of generalized linear models (GLM) (Liang and Zeger, 1986).

This method allows for a correlation structure for longitudinal (one-dimensional, *e.g.*, time series) data analysis.

This special data layout is necessary when responses are measured repeatedly on the same subject or unit across time. Therefore, the GEE method takes into account correlations within these units, while correlations between units can be assumed to be zero. Such units or clusters can also occur in a spatial context, *i.e.*, as independent regions.

Table 6: Methods utilized by various workers for measurement of spatial dependency (Source Miller *et al.*, 2007).

Study	***Modeling method(s)***	***Response variable***	***Spatial dependence incorporated as***	***Additional variables used***
1. Autoregressive methods				
Bullock and Burkhart (2005)	SAR	Host stem diameter	Distance-based weights matrix	Stem height
Dark (2004)	SAR	Number of invasive plant species	Spatial weights matrix	Elevation, native species richness, road density, population density
Ferrier *et al.* (2002)	GLM	Plant presence/ absence	Modified trend surface, modified autocovariate	Topography, climate, soil
He *et al.* (2003)	GLM	Plant presence/ absence	Autocovariate	Climate
Hoeting *et al.* (2000)	GLM	Simulated data	Autocovariate	Simulated data, search intensity
Hubbell *et al.* (2001)	GLM	Odds of host species survival	Autocovariate	Density, plant size, species richness
Huffer and Wu (1998)	GLM	Plant presence/ absence	Autocovariate	Climate
Silva *et al.* (2002)	GLM	Probability of species Colonization	Autocovariate	Altitude, solar radiation
Wu and Huffer (1997 and 2001)	GLM	Plant presence/ absence	Autocovariate	Climate
		Capacity	variogram	Yield, conductivity
Bolstad *et al.* (1998)	CK	Basal area of host Richness	Variogram	Elevation, terrain shape geographic coordinates

Table 6: *Contd.*

Table 6: *Contd.*

Study	*Modeling method(s)*	*Response variable*	*Spatial dependence incorporated as*	*Additional variables used*
Chong *et al.* (2001)	GLM, RK	Number and presence of plant species	Variogram, residual variogram	Location, elevation, slope, aspect, TM imagery
Ver Hoef (1993)	UK	Log-transformed plant species cover	Residual variogram	Area of woody plant species
2. Geographically weighted regression				
Wang *et al.* (2005)	GWR	Net primary production	Spatially varying coefficients	Climate, elevation, NDVI
Zhang *et al.* (2004)	GWR	Host height	Spatially varying coefficients	Host diameter at breast height
3. Parameter estimation methods				
Gotway and Stroup (1997)	GEE	Presence of plant damage, groundwater above sea level	Variogram, residual variogram	–
Hooten *et al.* (2003)	GLMM	Plant presence/ absence	Exponential correlation function based on distance	Aspect, elevation, land type, variable depth soil
Dirnbock and Dullinger (2004)	(Ordinal) GLM	Plant abundance	Dummy variable indicating neighborhood presence	Climate, geomorphology
Leathwick (1998)	GAM	Host species presence/ absence	Dummy variable indicating neighborhood presence	Climate, geology, plot size

Model abbreviations are—(C/S)AR: (conditional/simultaneous) autoregressive model; CK: co-kriging; DT: decision host; GAM: generalized additive model; GEE: generalized estimating equation; GLM: generalized linear model; GLMM: generalized linear mixed model; GS: Gaussian simulation; GWR: geographically weighted regression; MLR: maximum likelihood regression; RK: regression kriging; UK: universal kriging.

Spatial autocorrelation can be analyzed from two different perspectives. Global autocorrelation analysis involves the study of the entire map pattern and generally asks the question as to whether the pattern displays clustering or not.

Local autocorrelation, on the other hand, shifts the focus to explore within the global pattern to identify clusters or so called hot spots that may be either driving the overall clustering pattern, or that reflect heterogeneities that depart from global pattern. Global statistics answer the question: is there a spatial pattern (Y/N)? Local statistics, on the other hand, answer the question: where is there spatial pattern (map).

1. Global indicators of spatial autocorrelation includes methods like Gamma index of spatial autocorrelation, Join count statistics, Moran's I, Geary's C, and Getis and Ord's G
2. Local indicators of spatial autocorrelation includes local Moran's I, Local Geary and Local Gi and Gi* statistics.
3. The variogram approach to spatial association (geostatistical perspective) includes correlogram and semivariogram

Defining Contiguity

In spatial autocorrelation analysis some measure of contiguity is required. Contiguity has a rather broad definition depending on the research question, however, most analyses in spatial autocorrelation adhere to a common definition of neighbourhood relations. Namely, neighbourhood relations are defined as either rooks case, bishops case or queens (kings) case.

These are rather simple and intuitive as their names suggest (Fig. 13). Rooks case contiguity is by a neighbourhood of 4 locations adjacent to each cell, Bishops only considers the diagonals of the relation and queens or kings case considers a neighbourhood of eight cells. These are the most common forms of contiguity used in spatial autocorrelation when considering continuous data in a raster format. Of these three the rooks case is the most commonly used and most programmes only will compute this case.

Fig. 13: Type of Neighborhoods Relations

4.8.1. Global Autocorrelation

Global spatial autocorrelation is a measure of the overall clustering of the data. Global indices of spatial autocorrelation have been used to evaluate the degree to which similar observations tend to occur near each other (Rogerson, 1999; Walker and Gotway, 2004 and Jackson and Waller, 2005).

Six different tests for global spatial autocorrelation includes the Gamma index of spatial autocorrelation, join count statistics, Moran's I, Geary's C, General G statistics and Getis and Ord's G.

The common elements of the various spatial autocorrelation models are (1) a matrix of values representing the association between locations and (2) values representing a vector of the attributes of the various locations (Getis, 1991).

4.8.1.1. *Defining the neighborhood*

One important issue in spatial data analysis is the definition of locality. "Local" can be taken to mean near (in the sense of distance) or adjacent. Alternative definitions can lead to very different neighborhoods and hence different values for the statistics (Unwin and Unwin, 1998; Wilhalm and Steck 1998).

For point data (like mapped host data), distance is a clear cut basis for the definition of neighborhoods. The neighborhood is defined by a spatial proximity matrix, cij(d), for a given bandwidth in the computation of the spatial statistics. There are different ways to specify cij(d) (Lee and Wong 2001).

1. A binary matrix, in which the neighbors are assigned 1, and the nonneighbors are assigned 0. The cij denotes the value (either 1 or 0) in the ith row and jth column. A row represents how a subject host is spatially related to all the other hosts in the plot. Summing the values for that row across all columns, also known as the row sum indicates the total number of neighbors of the subject host.

$$C_i = \sum_j C_{ij}$$

2. A row-standardized matrix, in which a ratio (Cij /i) is calculated to indicate how much each neighbor contributes to the value of the subject host.
3. A distance matrix, in which the weight is measured by a function of the distance (dij) between the subject host and a neighbor. Several functions can be used, such as

$$\frac{I}{d_{ij}};\frac{1}{(d_{ij})^2} or\ exp(-(\frac{d_{ij}}{d})^2)$$

4.8.1.2. *Defining the Lag*

Lag refers to the spacing or interval between neighboring unit phenomena or neighboring sampling or analysis units. Unless clearly defined, there are a few possible interpretations of the word in either measurement or analysis.

Figure 14 shows that lag may refer to the distance between the centroids of the sampling units (Type IIIa) or the distance between the closest boundaries of the sampling units (Type IIIb), (Dungan *et al.*, 2002).

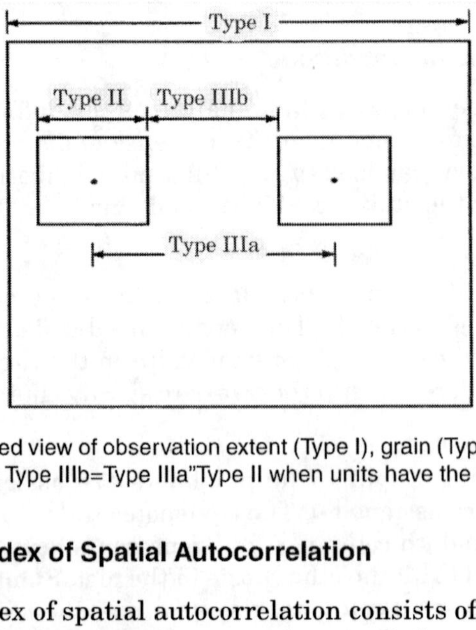

Fig. 14: A simplified view of observation extent (Type I), grain (Type II) and lag (Type III). Note that Type IIIb=Type IIIa"Type II when units have the same size and shape.

4.8.2. Gamma Index of Spatial Autocorrelation

The Gamma Index of spatial autocorrelation consists of the application of the principle behind a general cross-product statistic to measuring spatial autocorrelation Herbert *et al.* (1981).

The idea is to assess whether two similarity matrices for n objects, *i.e.*, n by n matrices A and B measure the same type of similarity. This is reflected in a so-called Gamma Index.

$$\Gamma = \sum_i \sum_j a_{ij}.b_{ij}.$$

In other words, the statistic consists of the sum over all cross-products of matching elements (i,j) in the two matrices.

The application of this principle to spatial autocorrelation consists of turning the first similarity matrix into a measure of attribute similarity and the second matrix into a measure of locational similarity.

Naturally, the second matrix is the a spatial weight matrix. The first matrix can be any reasonable measure of attribute similarity or dissimilarity, such as a cross-product, squared difference or absolute difference.

Formally, then, the Gamma index is:

$$\Gamma = \sum_i \sum_j a_{ij}.w_{ij}$$

Where w_{ij} are the elements of the weights matrix and a_{ij} are corresponding measures of attribute similarity. Inference for this statistic is based on a permutation approach in which the values are shuffled around among the locations and the statistic is recomputed each time.

This creates a reference distribution for the statistic under the null hypothesis of spatial randomness. The observed statistic is then compared to this reference distribution and a pseudo-significance computed as

$$p = (m + 1)/(n + 1)$$

Where m is the number of values from the reference distribution that are equal to or greater than the observed join count and n is the number of permutations. The Gamma test is a two-sided test in the sense that both extremely high values (*e.g.*, larger than any value in the reference distribution) and extremely low values (*e.g.*, smaller than any value in the reference distribution) can be considered to be significant.

Depending on how the measure of attribute similarity is defined, a high value will indicate positive or negative spatial autocorrelation, and *vice versa*. For example, for a cross-product measure of attribute similarity, high values indicate positive spatial autocorrelation and low values negative spatial autocorrelation. For a squared difference measure, it is the reverse. This is similar to the interpretation of the Moran's I statistic and Geary's C statistic respectively.

Many spatial autocorrelation test statistics can be shown to be special cases of the Gamma index. In most instances, the Gamma index is an unstandardized version of the commonly used statistics. As such, the Gamma index is scale dependent, since no normalization is carried out (such as deviations from the mean or rescaling by the variance).

4.8.2.1. *Joint count statistics*

Join-count statistics are the simplest measure of spatial autocorrelation which were put forward by Moran in 1948. They are used for a binary variable, "0" and "1", which are often used referred to as "white" and "black".

A join links two neighboring areas, the possible types of joins are limited to white-white (WW), black-white (BW), and black-black (BB).

Join-counts are frequencies or counts of the numbers of (0,0), (0,1), and (1,1) joins in the study area, and these numbers are compared to the expected numbers of them under the null hypothesis of no spatial autocorrelation. The observed number of (0,0), (0,1) and (1,1) joins are given by

$$(0,1) = \frac{\sum_{i=1}^{n}\sum_{j=1}^{n} C_{ij} \mid x_i - x_j}{2}$$

$$(0,0) = \frac{\sum_{i=1}^{n}\sum_{j=1}^{n} C_{ij} (1 - x_i)(1 - x_j)}{2}$$

$$(1,1) = \frac{\sum_{i=1}^{n}\sum_{j=1}^{n} C_{ij} x_i x_j)}{2}$$

Where xi is the binary attribute value related to region i, 1 or 0, Cij is a value assigned to region i and region j by a spatial weighting function. Join-count statistics have been widely applied to analysis spatial autocorrelation in remote sensing data (*e.g.*, Congalton, 1988; Chuang and Huang, 1992; Pugh and Congalton, 2001).

Most of these applications on join-count are restrained on binary variables, the values of each spatial unit are either 1 (black) or 0 (white).

4.8.2.2. *Moran's I*

Mihail (1989) used this tool for spatio-temporal dynamics related with *Macrophomina phaseolina.* Haas *et al.* (2011) used this for *Phytophthora ramorum* for forest species.

Moran (1950) introduced the first measure of spatial autocorrelation in order to study stochastic phenomena which are distributed in space in two or more dimensions. Moran's I has been subsequently used in almost all studies employing spatial autocorrelation (Upton and Fingleton, 1985).

Measuring the correlation (simultaneous change in value of two numerically valued random variables) among the neighboring observations in the pattern is done using the spatial autocorrelation statistic Moran's I.

Moran's I is defined as a measure of the correlation among neighboring observations in a pattern (Boots and Getis, 1988). Computation of Moran's I is achieved by division of the spatial covariation by the total variation. Resultant values are in the range from approximately ?1 to 1.

Positive signage represents positive spatial autocorrelation, while the converse is true for negative signage. With a Zero result representing no spatial autocorrelation.

$$I = \frac{N\sum_{i=1}^{n}\sum_{j=1}^{n} w_{ij}(x_i - \bar{x})(x_j - \bar{x})}{(\sum_{i=1}^{n}\sum_{j=1}^{n} w_{ij})\sum_{i=1}^{n}(x_i - \bar{x})^2}$$

Where N is the number of observation of the whole region

X is the mean of the variable

X_i is the variable value at a particular location

X_j is the variable value at another location

W_{ij} is a spatial weight between location of i relative to j

Moran's I formula is related to Pearson's correlation coefficient; its numerator is a covariance, comparing the values found at all pairs of points in turn, while its denominator is the maximum-likelihood estimator of the variance (i.e. division by n instead of n– 1); in Pearson's r, the denominator is the product of the standard deviations of the two variables, whereas in Moran's I there is only one variable involved.

Moran's I mainly differs from Pearson's r in that the sums in the numerator and denominator do not involve the same number of terms; only the terms corresponding to distances within the given class are considered in the numerator whereas all pairs are taken into account in the denominator.

4.8.2.2.1. *Statistical Test for Moran'sI*

Global Moran's I can be standardized to Z and this can be calculated as

$$Z = \frac{I - E(I)}{\sqrt{V(I)}}$$

$$E(I) = -1/(n-1)$$

$$Var\ (I) = \frac{1}{w_0^2\ (n^2 - 1)}\ (n^2 w_1 - n w_2 + 3w_0^2) - E^2 - (I)$$

$$w_o = \sum_{i=1}^{n}\sum_{j=1}^{n} w_{ij};\ w_1 = 0.5\sum_{i=1}^{n}\sum_{j=1}^{n}(w_{ij} + w_{ji})^2\ ;\ w_2 = \sum_{i=1}^{n}(w_{i*} + w_{*i})^2$$

Where w_{-j*} is the sum of all weights located in the row I, w_{*i} is the sum of all weights in the column i. The threshold of 1.96 can be applied to test the significance level of Z. If Z is greater than 1.96 or smaller than –1.96, it implies that the spatial autocorrelation was significant (Zhang and McGrath, 2004).

The tool computes a z-score and p-value indicating whether this difference is statistically significant or not. Index values cannot be interpreted directly; they can only be interpreted within the context of the null hypothesis.

The p-value is a probability. For the pattern analysis tools, it is the probability that the observed spatial pattern was created by some random process. When the p-value is very small, it means it is very unlikely (small probability) that the observed spatial pattern is the result of random processes, so we can reject the null hypothesis. Z-scores are simply standard deviations. Both z-scores and p-values are associated with the standard normal distribution (Table 7).

Very high or very low (negative) z-scores, associated with very small p-values, are found in the tails of the normal distribution. When we run a feature pattern analysis tool and it yields small p-values and either a very high or a very low z-score, this indicates it is unlikely that the observed spatial pattern reflects the theoretical random pattern represented by our null hypothesis (CSR).

A statistically significant positive z score means that similar values cluster spatially. High values are found closer together, and low values are found closer together, than we would expect from an underlying random spatial process. A statistically significant negative Z score means that similar values are spatially dispersed. High values are found far away from other high values, and low values are found far away from other low values, and this dispersion is more pronounced than we would expect from an underlying random spatial process (Table 8).

Table 7: Statistical relationships between z and p values

z-score (Standard Deviations)	*p-value (Probability)*	*Confidence level*
< -1.65 or > +1.65	< 0.10	90%
< -1.96 or > +1.96	< 0.05	95%
< -2.58 or > +2.58	< 0.01	99%

The Spatial Autocorrelation (Global Moran's I) tool is an inferential statistic, which means that the results of the analysis are always interpreted within the context of its null hypothesis. For the Global Moran's I statistic, the null hypothesis states that the attribute being analyzed is randomly distributed among the features in study area.

Table 8: Interpretation of the z and p values

The p-value is not statistically significant.	We cannot reject the null hypothesis. It is quite possible that the spatial distribution of feature values is the result of random spatial processes. The observed spatial pattern of feature values could very well be one of many, many possible versions of complete spatial randomness (CSR).
The p-value is statistically significant, and the z-score is positive.	We may reject the null hypothesis. The spatial distribution of high values and/or low values in the dataset is more spatially clustered than would be expected if underlying spatial processes were random.
The p-value is statistically significant, and the z-score is negative.	We may reject the null hypothesis. The spatial distribution of high values and low values in the dataset is more spatially dispersed than would be expected if underlying spatial processes were random. A dispersed spatial pattern often reflects some type of competitive process—a feature with a high value repels other features with high values; similarly, a feature with a low value repels other features with low values.

Dray *et al.* (2008) have introduced multivariate spatial analysis based on Moran;s I (MULTISPATI) by introducing a row-sum standardized spatial weight matrix. Their analysis implies a compromise between relations among variables (multivariate analysis) and their spatial structure (autocorrelation).

Chen *et al.* (2008) have quantified the strength of autocorrelation by using the Moran's I. They have reported that for both α and β host diversities the distance of the spatial autocorrelation increased slightly at the scale of 10–15 and 200 km and decreased around 20–40 km.

4.8.2.2.2. *Modified version of Moran's I*

Traditional calculation of Moran's I for heterogeneous populations are not working well (Jackson *et al.*, 2010). Therefore, several alternative version of Moran's I have been proposed to account for heterogeneous population, for example Oden (1995); Waldhor (1996); Assuncao and Reis (1999) and Waller *et al.* (2006).

Oden (1995) proposed the modified Moran's I function (I*pop) he noted that symmetry is not required for I^*_{pop} and $w_{ij\#}$ 0 (but can be fix at any specified value). In order to capture the variability present in the region, Oden includes the first term in the numerator which is used to model the spatial variation in a manner similar to conventional chi-squared for heterogeneity rates. The odens's I^*_{pop} can be written as

$$I^*_{pop} = \left[n^2 \sum_i \sum_j w^*_{ij} (e_i - v_i)((e_i - v_i) - n(1 - 2b') \sum_i w^*_{ij} e_i - nb' \sum_i w^*_{ij} v_i \right] X \frac{1}{b'(1 - b,)[n^2 + \sum_i \sum_j v_i v_j w^*_{ij} - x \sum_i w^*_{ii}]}$$

Jackson *et al.* (2010) improves the original version of Moran's I by incorporation of (a) a weight function in the variance computation (b) introducing the population density weight function and (c) conducting Monte Carlo simulation.

Their weight function is not only included in the differences of the geographic unit's cases from the overall mean, but also in the calculation of the variance. They also expand the definition of neighbors to a broader concept in the construction of Moran's I (*e.g.*, all geographic units included in a pre-specified geographic range will be considered to be neighbors of the geographic unit in the center. They have concluded that modified Moran's I achieved higher power than Moran's I and I^*_{pop} for evaluating global and local clustering patterns on geographic data with homogenous population.

4.8.2.3. *Geary's C (contiguity) ratio*

Computation of Geary's C results in a value within the range of 0 to +2. With zero being a strong positive spatial autocorrelation, through to 2, which represents a strong negative spatial autocorrelation.

Calculation is similar to Moran's I, For Moran, the cross-product is based on the deviations from the mean for the two location values while for Geary, and the cross-product uses the actual values themselves at each location. Geary's C statistic (Geary, 1954) is based on the deviations in responses of each observation with one another:

$$C = \frac{n-1}{2S_0} \frac{\sum_i \sum_j w_{ij}(x_i - x_j)^2}{\sum_i (x_i - \bar{x})^2}$$

The variance is:

$$\begin{aligned}\mathrm{Var}(c) \;=\; & \frac{1}{n(n-2)(n-3)S_0^2} \\ & \{S_0^2[(n^2-3) - k(n-1)^2] + S_1(n-1)[n^2 - 3n + 3 - k(n-1)] \\ & + \frac{1}{4}S_2(n-1)[k(n^2 - n + 2) - (n^2 + 3n - 6)]\}\end{aligned}$$

where S_0, S_1, and S_2 are the same as in Moran's I. However, interpretation of these values is very different, essentially the opposite. Geary's C varies on a scale from 0 to 2, C of approximately 1 indicates no autocorrelation/ random, C of 0 indicates perfect positive autocorrelation/clustered, C of 2 indicates perfect negative autocorrelation/dispersed. Can convert to a -/+1 scale by: calculating C* = 1 – C.

Geary's C coefficient is a distance-type function; it varies from 0 to some unspecified value larger than 1. Its numerator sums the squared differences between values found at the various pairs of sites being compared.

A Geary's C correlogram varies as the reverse of a Moran's I correlogram; strong autocorrelation produces high values of I and low values of C. Positive autocorrelation translates in values of c between 0 and 1 whereas negative autocorrelation produces values larger than 1. Hence, the reference 'no correlation' value is C= 1 in Geary's correlograms.

4.8.2.3.1. *Statistical significance tests for Geary's C*

Similar to Moran which based on the normal frequency distribution with

$$Z = \frac{C - E(C)}{S_{error}\,(C)}$$

$$E(C) = 1$$

A statistically significant positive Z score means high/larger values cluster spatially larger values are found closer together than we would expect if the underlying spatial process was random. A statistically significant negative Z score means that low/smaller values cluster spatially smaller values are found closer together than we would expect if the underlying spatial process was random.

Moran's I and Geary's C coefficients are used to measure the degree of spatial autocorrelation displayed by a quantitative variable, and to test the null hypothesis (Ho) that there is no significant spatial autocorrelation (positive: aggregation, segregation). Since these coefficients compare values for pairs of points, the set of available point pairs is divided into a number of distance classes.

This number of classes is left to the user. Like Pearson's correlation coefficient, Moran's I is based on the computation of cross products of centered data. Geary's C is a distance type coefficient, summing squared differences between adjacent pairs of values.

Spatial autocorrelation analysis should not be performed with fewer than 30 localities, because the number of pairs of localities in each distance class would then become too small to produce significant results (Cliff and Ord, 1981; Legendre and Fortin, 1989). Formulas for computing the coefficients as well as the standard error of the estimated statistics can be found in Cliff and Ord (1981), in Sokal and Oden (1978) and in Legendre and Legendre (1984).

Both Moran's I and Geary's C are calculated at various distances, d. Both test the null hypothesis that there is no significance spatial autocorrelation among the mapped points with respect to the measured variable (Gibson, 2002 and Dormann *et al.*, 2007).

Moran's I is produced by standardizing the spatial auto covariance by the variance of the data using a measure of the connectivity of the data. Geary's C uses the sum of squared differences between pairs of data values as its measure of covariation.

Moran's I is a more global measurement and sensitive to extreme values of x, whereas Geary's C is more sensitive to differences in small neighborhoods. In general, Moran's I and Geary's C result in similar conclusions. However, Moran's I is preferred because its more powerful than Geary's C. (Cliff and Ord, 1981).

4.8.2.3.2. *General G-statistic*

Moran's I and Geary's C will indicate clustering or positive spatial autocorrelation if high values (*e.g.,* host with high density) cluster together (often called hot spots) and/or if low values cluster together (host with low density), but they cannot distinguish between these situations. The General G statistic distinguishes between hot spots and cold spots. It identifies spatial concentrations. G is relatively large if high values cluster together while G is relatively low if low values cluster together.

The General G statistic is interpreted relative to its expected value (value for which there is no spatial association). Larger than expected value indicates potential "hot spot" while smaller than expected value indicates potential "cold spot". Z test statistic is used to test if the difference is sufficient to be statistically significant.

Calculation of G must begin by identifying a neighborhood distance within which cluster is expected to occur.

$$G\ (d) = \frac{\Sigma \Sigma W_{ij}(d) x_i x_j}{\Sigma \Sigma x_i x_j}$$

Where

d is neighborhood distance

W_{ij} weights matrix has only 1 or 0

1 if j is within d distance of i

0 if its beyond that distance

Statistical Significance Tests

$$Z_G = \frac{E\ (G)}{\sqrt{V\ (G)}}$$

Expected value (if no concentration) for G is given by:

$$E(G) = \frac{W}{n\ (n-1)} \text{ where } W = \sum_i \sum_j W_{ij}\ (d)$$

$$V_G = (E\ (G^2)\text{-}E\ (G)^2$$

For the General G, the terms in the numerator (top) are calculated "within a distance bound (d)," and are then expressed relative to totals for the entire region under study. As with all of these measures, if adjacent **x** terms are both large with the same sign (indicating positive spatial association), the numerator (top) will be large.

4.8.2.3.3. *Getis and Ord's G*

Method developed by Getis and Ord (1992; 1996) not only provide hypothesis testing to determine whether clustering has occurred within a dataset, but also provide information on the extent to which above and below average values cluster more strongly and identify local concentration of clustering (Laffan, 2006; Mueller-Warrant *et al.*, 2008).

The High/Low Clustering (Getis-Ord General G) tool is an inferential statistic, which means that the results of the analysis are interpreted within the context of the null hypothesis.

The null hypothesis for the High/Low Clustering (General G) statistic states that there is no spatial clustering of feature values. When the p-value returned by this tool is small and statistically significant, the null hypothesis can be rejected. If the null hypothesis is rejected, then the sign of the z-score becomes important.

The High/Low Clustering (Getis-Ord General G) tool is most appropriate when we have a fairly even distribution of values and are looking for unexpected spatial spikes of high values. Unfortunately, when both the high and low values cluster, they tend to cancel each other out.

The null hypothesis for both the high/low Clustering (Getis-Ord General G) and the Spatial autocoorelation (Global Moran's I) tool is complete spatial randomness (CSR); values are randomly distributed among the features in the dataset, reflecting random spatial processes at work. However, the interpretation of z-scores for the High/Low Clustering tool is very different from the interpretation of z-scores for the spatial autocorrelation (Global Moran's I) tool: It measures the degree of clustering for either high values or low values. Where y_i represent the value of feature i, y_j represent the vlaue of feature j, and w_{ij} is the weight assigned to each pair of features.

Fox *et al.* (2012) have use global and local Moran's I spatial autocorrelation statistics, Getis-Ord General G global high/low clustering and Getis-Ord* test statistic techniques to evaluate the spread patterns of pine beetle infestation over time.

Result	***High/Low Clustering***	***Spatial Autocorrelation***
The p-value is not statistically significant.	We cannot reject the null hypothesis. It is quite possible that the spatial distribution of feature attribute values is the result of random spatial processes. Said another way, the observed spatial pattern of values could well be one of many, many possible versions of complete spatial randomness.	

Table (*Contd...*)

Table (*Contd...*)

Result	*High/Low Clustering*	*Spatial Autocorrelation*
The p-value is statistically significant, and the z-score is positive.	We may reject the null hypothesis. The spatial distribution of high values in the dataset is more spatially clustered than would be expected if underlying spatial processes were truly random.	We may reject the null hypothesis. The spatial distribution of high values and/or low values in the dataset is more spatially clustered than would be expected if underlying spatial processes were truly random.
The p-value is statistically significant, and the z-score is negative.	We may reject the null hypothesis. The spatial distribution of low values in the dataset is more spatially clustered than would be expected if underlying spatial processes were truly random.	We may reject the null hypothesis. The spatial distribution of high values and low values in the dataset is more spatially dispersed than would be expected if underlying spatial processes were truly random. A dispersed spatial pattern often reflects some type of competitive process: a feature with a high value repels other features with high values; similarly, a feature with a low value repels other features with low values.

4.9. LOCAL AUTOCORRELATION

Spatial autocorrelation can also be measured at the local level to evaluate the extent of autocorrelation within local neighbourhoods.

Measures captures the many local spatial variation and spatial dependency while global measurements provide only one set of values that represent the extent of spatial autocorrelation across the entire study area (Mueller-Warrant *et al.*, 2008).

As an operational definition, Anselin, (1995) has defined the local indicator of spatial association (LISA) is as any statistic that satisfies the following two requirements: 1. the LISA for each observation gives an indication of the extent of significant spatial clustering of similar values around that observation; 2. the sum of LISAs for all observations is proportional to a global indicator of spatial association. the general LISA can be used as the basis for a test on the null hypothesis of no local spatial association. For each location, LISA values allow for the computation of its similarity with its neighbours and also to test its significance. Five scenarios may emerge:

1. Locations with high values with similar neighbours: high-high. Also known as hot spots
2. Locations with low values with similar neighbours: low-low. Also known as cold spots

3. Locations with high values with low-value neighbours: high-low. Potential "spatial outliers.
4. Locations with low values with high-value neighbours: low-high. Potential "spatial outliers.
5. Locations with no significant local autocorrelation.

These specific configurations can be first identified from a scatter plot showing observed values against the averaged value of their neighbours. This so-called Moran scatter plot is a useful exploratory tool.

A significance level is set, values can also be plotted on a map to display the specific locations of hot spots and potential outliers. Recently, LISA has been successfully used in various research areas outside geography: to identify and test "hot spots" (positive autocorrelation or similarity) and "cold spots" (negative autocorrelation or dissimilarity) in biological data sets (Sokal *et al.*, 1998a, 1998b), to characterize remote sensing and Landsat TM imagery (Wulder and Boots, 1998 and 2001), to detect spatial clustering of disease (*i.e.*, spatial scan statistic) in epidemiological studies (Hjalmars *et al.*, 1996; Kulldorff, 1997; Gangnon and Clayton, 2001), and to study the spatial point patterns in a minefield and in a plot of pine saplings (Cressie and Collins, 2001a and 2001b).

LISA is commonly formulated as a numerical measure of a random variable relating each individual datum to the values observed at neighboring locations. It is similar or compatible, at least conceptually, to traditional host competition indices. Therefore, LISA would be an appropriate indicator for local spatial heterogeneity and useful for investigating the effects of neighboring hosts of different species, sizes, and separation distances on the subject host's performance.

Furthermore, LISA can be readily incorporated into visualization tools, such as a geographic information system (GIS), because it provides geo-referenced information at a local level (Unwin, 1996 and Lee and Wong, 2001). Shi and Zhang (2003) have utilized various LISA approaches to explore the relationship between LISA and traditional host competition indices, to explore the possibility that LISA can be used as a predictor variable for predicting host growth, and to demonstrate the usefulness of LISA for identifying and testing clusters of hosts with similar or dissimilar sizes. They shows that like most of the competition indices, LISA had moderate correlations with host basal area growth.

Unlike the local Moran, the local Geary is a measure of the weighted sum of square differences between the observed values at location i and those of its surrounding locations. The calculation of the pseudo-significant level P-value is similar to that of the local Moran after the standardization.

A small P-value (*e.g.*, $P < 0.05$) indicates a large GR_i in extremes, which suggests a positive spatial association (*i.e.*, similarity) of observation i with its surrounding observations, whereas a large P-value (*e.g.*, $P > 0.95$) indicates a small GR_i in extremes, which suggests a negative spatial association (*i.e.*, dissimilarity) of observation i with its surrounding observations. From their previous work (Getis and Ord, 1992), Ord and Getis (1995) defined local G statistics, which give an index of spatial clustering of a set of observations over a defined neighborhood

$$G_i = \frac{\sum_j c_{ij}(d)X_j}{\sum_j X_j} \quad i \# j$$

$$G_i^* = \frac{\sum_j c_{ij}(d)X_j}{\sum_j X_j} \quad i = j$$

The two equations differ depending on whether the location i around which the clustering is measured is included in the calculation. There is no theory to guide which one to use in any particular situation, but the difference between the two is typically small. The Gi or Gi * compares the local (weighted) average with the global average.

If high values of X (*i.e.*, the variable under study) tend to cluster around i, the G_i or G_i* will be high; if low values of X tend to cluster around i, then the Gi or Gi * will be low; and no distinct clustering of high or low values of X around i will produce intermediate values of the G_i or G_i* Fotheringham *et al.* (2000).

After the standardization, a significant and positive G_i or G_i * indicates that the location i is surrounded by relatively large values, whereas a significant and negative Gi or Gi * indicates that the location i is surrounded by relatively small values.

The G_i or G_i * statistics can be used to identify spatial agglomerative patterns with high-value clusters or low-value clusters.

In summary, the local MC_i measures joint covariance of neighboring locations. In other words, it measures the correlation of each Xi with all X_{js} within the distance d of the subject location i. If these locations deviate significantly from the average in a given area and have similarly high or low values with the subject point, the local spatial autocorrelation is positive. If the subject location is widely different from the mean and has values unlike those of its neighbors, the local spatial autocorrelation is negative. The local GR_i measures (squared) differences of the values between the subject and the neighboring locations.

High values of GR_i indicate negative spatial autocorrelation. Data values at the subject location and its neighbors that are all close to the mean are shown as a positive spatial autocorrelation by the GR_i, but as weak or 0 autocorrelation by the MC_i. The local G_i and G_i * measure overall concentration or lack of concentration of all pairs of (Xi, Xj) such that i and j are within the distance of each other. The G_i and G_i * tend to agree more with the MC_i than with the GR_i (Getis and Ord, 1996, Anselin, 1995, Sokal *et al.*, 1998a and 1998b).

4.10. VARIOGRAM APPROACH

This methodology has been used in the study of area with probability of occurrence of *Armillaria* spp in roots of ponderosa pine by Kallas *et al.* (2003), Jamie-Garcia *et al.* (2001) used this to study spatial and temporal progress of *Phytophthora infestans* genotypes in tomatoes and potatoes and Zas *et al.* (2007) used variogram and kriging for screening *Pinus pinaster* resistant to *Armillaria ostoyae* in field conditions. Calonnec *et al.* (2009) used variogram for spatiotemporal spread of powdery mildew epidemics in the vineyard.

Alves and Pozza (2010) studied indicator kriging modeling epidemiology of common bean anthracnose. Azahar *et al.* (2011) used semivariogram for temporal analysis of basal stem rot disease in oil palm plantations.

Geostatistical methods deal with the question of how variance and covariance depend on the instance between observations (*i.e.*, quadrats).

Spatial autocorrelation, or distance dependence, is commonly modelled by fitting a variogram function to an empirical variogram (Isaaks and Srivastava, 1989; Cressie, 1991; Haining, 1997; Burrough and McDonnell, 1998).

An empirical variogram is a plot of half the squared difference between two observations (the semivariance) against their distance in space, averaged for a series of distance classes. A simple variogram model is defined by the model family and the parameters sill (the average half squared difference of two independent observations), range (the maximum distance at which pairs of observations will influence each other), and nugget (the variance within the sampling unit).

A correlogram is a graph in which autocorrelation values are plotted, on the ordinate, against distance classes among sites on the abscissa. Correlograms provide evidence for the autocorrelation intensity, the size of the zone of influence and the type of spatial pattern of the variable under study.

The shape of a correlogram gives indications about the spatial pattern of the variable, as well as about the underlying generating process (Sokal 1986; Legendre and Fortin, 1989). Inference about the underlying generating process can be made from the shape of the correlogram only when the correlogram is globally significant; Oden (1984) and Legendre and Fortin (1989) show how to compute such a global test, whose aim is to correct for simultaneous multiple testing.

Correlograms (Cliff and Ord, 1981) can be computed for single variables (Moran's I or Geary's c autocorrelation coefficients) or for multivariate data (Mantel correlogram). In all cases, a test of significance is available for each individual autocorrelation coefficient plotted in a correlogram.

Similarly, a variogram is a graph in which semi-variance is plotted, on the ordinate, against distance classes among sites on the abscissa (Fig.15). In the geo-statistical tradition, semi-variance statistics are not tested for significance, although they could be through the test developed for Geary's C, when the condition of second order stationary is satisfied. Statistical models may be fitted to variograms (linear, exponential, spherical, Gaussian, etc.); they allow the investigator to relate the observed structure to hypothesized generating processes or to produce interpolated maps by kriging.

Because they measure the relationship between pairs of observation points located a certain distance apart, correlograms and variograms may be computed either for preferred geographic directions or, when the phenomenon is assumed to be isotropic in space, in an all-directional way.

Huo *et al.* (2012) have plotted spatial correlogram with Moran's I. They have used this tool for mapping of soil heavy metals in Beijing, China. According to Legendre and Fortin (1989), the spatial correlogram can be standardized into a standardized correlogram, in which the ordinate is standardized Moran's I and Z.

The shape of the standardized correlogram provides inductions about the spatial pattern (spatial clusters and spatial outliers) and spatial correlation distance of a variable (Zhang *et al.*, 1995). However, the standardized correlaogram often has one or more positive correlation ranges. Zhang *et al.* (1998) have explained that the closer positive correlation range represents the average size of the zone of spatial clusters, that is, the spatial correlation distance.

4.10.1. Semivariogram

Calculating semivariogram (called variograms for simplicity) is a basic geo-statistical measure to determine the rate of change of a regionalized variable along a specific orientation (usually distances).

Semivariogram values are defined as the sum of squared differences between pairs of points separated by a certain distance divided by two times the number of points in a distance class.

Semivariance is a measure of the degree of spatial dependence between sampled locations and is computed as the sum of squared differences between all pairs of observations that belong to a given distance class (Biondi *et al.,* 1994).

Calculating the semivariance for different distance classes produces the experimental semivariogram, which is computed after sorting all possible pairs of locations into classes by distance.

Like correlograms, semivariograms (decompose the spatial (or temporal) variability of observed variables among distance classes.

The structure function plotted as the ordinate, called semivariance, is the numerator of equation given for Geary's C function.

$$\gamma(d) = \frac{1}{2W}\sum_{h=1}^{n-1}\sum_{i=h+1}^{n} W_{hi}\,(y_h - y_i)^2$$

γ(d) is thus a non-standardized form of Geary's c coefficient. γ may be seen as a measure of the error mean square of the estimate of yi using a value yh distant from it by d. The calculation is repeated for different values of d. This provides the sample variogram, which is a plot of the empirical values of variance g(d) as a function of distance d.

Generally, variograms tend to level off at a sill which is equal to the variance of the variable the presence of a sill implies that the data are second-order stationary. The distance at which the variance levels off is referred to as the range (parameter a); beyond that distance, the sampling units are not spatially correlated.

The discontinuity at the origin (non–zero intercept) is called the nugget effect; the geo-statistical origin of the method transpires in that name. It corresponds to the local variation occurring at scales finer than the sampling interval, such as sampling error, fine-scale spatial variability, and measurement error.

The nugget effect is represented by the error term *ij in spatial structure model. It describes a portion of variation which is not autocorrelated, or is autocorrelated at a scale finer than can be detected by the sampling design. The parameter for the nugget effect is C_0 and the spatially structured component is represented by C_1; the sill, C, is equal to $C_0 + C_1$. The relative nugget effect is $C_0/(C_0 + C_1)$.

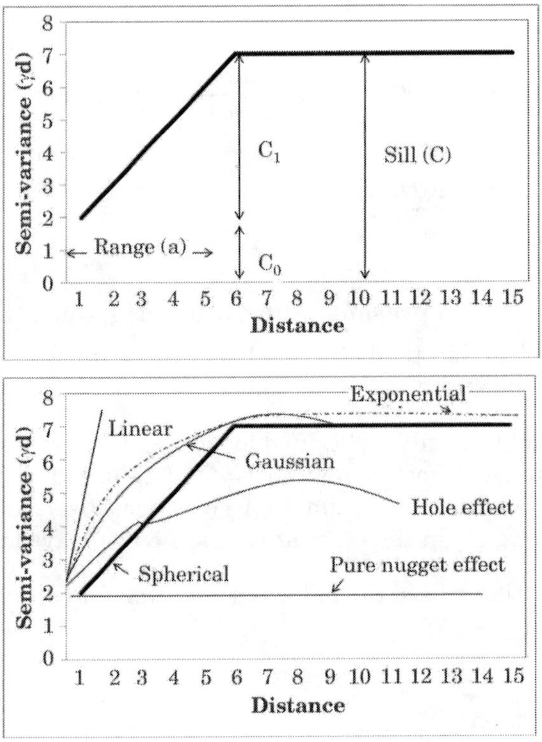

Fig. 15: Types of Semi-variance plots

4.10.2. Commonly Used Variogram Models

Although a sample variogram is a good descriptive summary of the spatial contiguity of a variable, it does not provide all the semivariance values needed for kriging.

A model must be fitted to the sample variogram; the model will provide values of semivariance for all the intermediate distances. The most commonly used models are the following

$$\text{Spherical Model: } \gamma(d) = C_0 + C_1\left[1.5\,\frac{d}{a} - 0.5\left(\frac{d}{a}\right)^3\right] \text{ if } d \le a;\ \gamma(d) = C \text{ if } d > a$$

$$\text{Exponential Model: } \gamma(d) = C_0 + C_1\left[1 - \exp\left(-3\frac{d}{a}\right)\right]$$

$$Gaussian\ Model: \gamma(d) = C_0 + C_1\left[1 - exp\left(-3\frac{d^2}{a^2}\right)\right]$$

$$Hole\ effect\ Model: \gamma(d) = C_0 + C_1\left[1 - \frac{\sin(ad)}{ad}\right]. An\ equivalent\ form\ is$$

$$\gamma(d) = C_0 + C_1\left[1 - \frac{a'\sin(d/a')}{d}\right]$$

$where\ a' = \frac{1}{a}.(C_0 + C_1)$ represents the value of g towards which the dampening sine function tends to stabilize. Linear model: $\gamma(d) = C_0 + bd$ where b is the slope of the variogram model.

A linear model with sill is obtained by adding the specification: $\gamma(d) = C$ if d ³ a. Pure nugget effect model: $\gamma(d) = C_0$ if d> o; $\gamma(d) = 0$ if d = 0. The second part applies to a point estimate. In practice, observations have the size of the sampling grain the error at that scale is always larger than 0.

Li *et al.* (2010) have used the following equation for estimation of semi-variogram

$$r(h) = \frac{1}{2N(h)}\sum_{i=1}^{n(h)}[Z(x_i) - z(x_i + h)]^2$$

where r(h) is the semivariance at a given distance h; $Z(x_i)$ is the value of the variable Z at the xi location, and N(h) is the number of pairs of sample points separated by the lag distance h. The parameters (1) "range" (A_0), (the separation distance over which the spatial dependence is apparent), (2) the "nugget" (C_0), (the variation found at a scale finer than the field sampling), and (3) the "sill" (C_0+C) (reflecting the total sample variation of the semivariogram levels for the patterned data) were calculated.

An especially important parameter ($C_0/(C_0$+C)), nugget-to-sill ratio, was calculated by the proportion of nugget to sill. In this case, a strong spatial dependence appears if the ratio is less than 25%, a moderate spatial dependence if the ratio is between 25% and 75%, and a weak spatial dependence if the ratio is more than 75%. Spherical, exponential, linear, and Gaussian relationships that best fitted the model were presented in the semivariogram model.

4.10.3. Kriging

Kriging is a method of optimal unbiased estimation of variables at unsampled locations based on parameters of the semi-variogram and initial data values (Wallace and Hawkins, 1994).

Predictions are weighted averages of the actual measurements, with values for the weights derived from solution of a set of equations determined by the semivariogram and the location and orientation of the sample points relative to each other and to the point or area being predicted. Weights are chosen to give unbiased estimates and to minimize estimation variance.

Kriging supplements predictions with standard errors, permitting quantification of the uncertainty in the predicted value. Because the kriging technique is robust, minor variations in semivariogram parameter estimates generally make little difference in the interpolated values or their standard errors.

Once the semivariogram is constructed and its parameters determined, it can be used in kriging.

The semivariogram's form defines the relationship between the regionalized variable at different lags and also leads to the optimal weights of the kriging estimators. There are two types of kriging: point kriging and block kriging (Cressie, 1991).

Point kriging is used to predict the value of a single measurement at a (generally unsampled) location. Block kriging is used to predict the average of the regionalized variable in some larger "support." The "support" of a sample is the amount of physical material it encompasses, for example, the topsoil in a 1 m^2 portion of the field centered at some location.

For prediction at arbitrary locations, both types of kriging problem require an explicit mathematical function for the semivariogram. Both point and block kriging can be used to produce contour maps, although the maps differ in purpose and appearance (Cressie, 1991). Particularly where there is a substantial nugget effect, the contour maps produced by block kriging will be much smoother than those produced by point kriging.

Wallace and Hawkins (1994) have used ordinary block kriging and they used following equation $T^* = w_1g_1 + w_2g_2 + w_3g_3 + w_4g_4 + \dots\dots\dots w_ng_n.$

Where T^* is the estimator; w_1, w_2, w_3.. are the krignig weight and g_1, g_2, g_{-3} are the sample value.

Zhang *et al.* (2012) have evaluated the performance of univariate kriging (ordinary kriging), hybrid kriging (co- kriging) and regression kriging.

According to them in land resource inventories, univariate geostatistical methods (ordinary kriging (OK) and simple kriging) have been widely used for spatial interpolation.

However, the predictive capability of univariate kriging is limited in complex terrain regions impacted by intense cultivation and land use changes on the other side, the application of hybrid geostatistical methods (*e.g.*, co-kriging (CK), regression kriging (RK) and residual maximum likelihood (REML)) can provide secondary information to improve the prediction accuracy.

RK uses traditional regression methods that model the trend or drift in conjunction with geostatistical methods that model the random residuals from the regression (Hengl *et al.*, 2004 and Rivero *et al.*, 2007).

In RK the trend and the residuals from. However, there are two disadvantages of RK. First, the trend is generally estimated by ordinary least square method, which does not yield estimates of minimum variance. Second, the estimates of semivariances obtained from residuals from the trend are biased (Webster and Oliver, 2001).

CK is an extension of OK that takes into account the spatial cross-correlation from two or more variables. For ordinary CK Zhang *et al.* (2012) have fitted a linear model of co-regionalisation as double spherical models to accommodate two distinct ranges a_1 and a_2:

$$\gamma(h) = c_0 + c\left\{\frac{3h}{2a_1} - \frac{1}{2}\left[\frac{4}{a_1}\right]^{-3}\right\} + c_2\left\{\frac{3h}{2a_2} - \frac{1}{2}\left[\frac{4}{a_2}\right]^{-3}\right\} \text{ for } 0 < h \leq a_1$$

$$= c_0 + c_1 + c_2\left\{\frac{3h}{2a_2} - \frac{1}{2}\left[\frac{4}{a_2}\right]^{-3}\right\} \text{ for } a_1 < h \leq a_2 = c_0 + c_1 + c_2 \text{ for } h > a_2$$

Where c_1 and c_2 are the sill variance associated with the two ranges.

Zhang *et al.* (2011) have reported that the prediction accuracy for soil organic carbon based on root mean sum error (RMSE) was in the following descending order RK (0.226) < CK (0.247) < OK (0.256).

The RMSE for OK was the greatest, suggesting that the prediction accuracy could be improved by incorporating ancillary variables into kriging prediction.

Zas (2006) have presented the iterative kriging procedure as a promising tool to account spatial heterogeneity and according to him semivariograms should be routinely inspected to test if residuals of the models are spatially correlated and in that case, the iterative kriging is strongly recommended to remove spatial heterogeneity .

4.11. SOFTWARE'S DEALS WITH SPATIAL AUTOCORRELATION APPROACHES

Software's name	*Web address*	*Details*
GeoDa (WIN) - Freeware	https://www.geoda.uiuc.edu/default.php	Spatial Autocorrelation: spatial weights creation (rook, queen, distance, k-nearest), higher order spatial weights, spatial weights characteristics (connectedness histogram).Moran scatterplot with inference, bivariate Moran scatterplot with inference, Moran scatterplot for rates (Empirical Bayes EB- standardization) Local Moran significance map, Local Moran cluster map, bivariate Local Moran, Local Moran for rates (EB standardization)
PASSAGE (WIN) - Freeware	http://www.passagesoftware.net/	Scattered Data Analyses. Correlograms (Moran's I, Geary's c and Mantel's correlation) Bearing, Angular Correlation.
Rookcase (An Excel 97/2000 Visual Basic (VB) Add-in) - Freeware	http://www.lpc.uottawa.ca/ or http://www.lpc.uottawa.ca/data/scripts/index.html	Regular dataset (lattice): Moran's I, Geary's c, Join-Count Statistics.Irregular dataset (lattice): moran's I, Geary's c, LISA functions (local Moran's I, Geary's c and Getis-Ord Gi and G_i . Monte-Carlo simulations: create random distributions of Moran's I, Geary's c or Joint-Count Statistics
SpPack (An Excel Visual Basic (VB) Add-in) - Freeware	Available by request to the author: george.perry@kcl.ac.uk	Grid-based and Spatial Autocorrelation: Simple Quadrat Based Mean (variance analysis), Global Moran's I and Geary's c, Dispersion Indices.
CrimeStat, version 3.0 (WIN) - Freeware	http://www.icpsr.umich.edu/NACJD/crimestat.html	Spatial distribution: statistics for describing the spatial distribution of incidents, such as the meancenter, center of minimum distance, standard deviational ellipse. Moran's I spatial autocorrelationindex, or directional mean, Moran correlogram that calculates Moran's I for different distance
Sspdep: Spatial Dependence (for use with R - WIN, MacOS, Linux) - Freeware	http://cran.r-project.org/src/contrib/Descriptions/spdep.html or http://spatial.nhh.no/R/spdep/	A collection of functions that run in the R language to:Create spatial weights matrix objects from polygon contiguities, from point patterns by distanceand tesselations, for summarising these objects, and for permitting their use in spatial data analysis.Tests for spatial autocorrelation, including global Moran's I, Geary's c, Hubert/Mantel generalcross

Table: (*Contd...*)

Table (*Contd...*)

Software's name	*Web Address*	*Details*
		product statistic, Empirical Bayes estimates and Assun¸c□ao/Reis Index, Getis/Ord G and multicoloured join count statistics, local Moran's I and Getis/Ord G, saddlepoint approximations for global and local Moran's I. Estimate spatial simultaneous autoregressive (SAR) models.
SAGE for Sun work-station only (running Solaris), for use with ARC/INFO - Freeware	ftp://ftp.shef.ac.uk/pub/uni /academic/D-H/g/sage/ sagehtm/sage.htm or http://www.shef.ac.uk/_ scgisa/newscgisa/research. htm	Computing distribution statistics of attributes. Correlation analysis for every selected pair of attributes and cross-tabulated analysis on categorical data. Local Getis and Moran I indicators and their significant values, and saving these results as new attributes.
IDRISI (Win) - Licensed	http://www.clarklabs.org/	Spatial Autocorrelation of an image (Moran's I).Variograms using Gstat (a program for geostatistical modeling, prediction and simulation).
Spacestat, version 1.91 and ArcView 3.x extension (WIN) - Licensed	http://www.terraseer.com/ products/spacestat.html	Includes techniques to describe and visualize spatial distributions, identify a typical locations (spatialoutliers), discover patterns of spatial association (spatial clusters) and suggest different spatial regimes and other forms of spatial non-stationarity.
S-Plus (S+ Spatial Stats and ArcView GIS extension), version 7 (WIN, Linux, UNIX) - Licensed	http://www.insightful.com/ products/splus/ or http://www.insightful. com/products/spatial/ default.asp or http://www.insightful. com/products/arcview/ default.asp	Geostatistical data: Contour plots, 3D point clouds, Variogram plots and box plots Directional variograms and correlograms, Empirical variogram estimation. Variogram models including spherical and exponential, Ordinary and universal kriging, Block and Point Kriging prediction at arbitrary locations with standard errors.Parametric and nonparametric trend surface.

4.12. CONCLUSION

The predictability of the physical arrangement of plant at whatever scale its viewed is reffered to their spatial pattern.

Various approaches are available to detect the spatial pattern. However, before starting the using various indices identification of natural randomness is prerequisite and for this CSR and its associated tests are the most essential features. On the other hand for quantifying the departures from randomness two categorize are available *i.e.,* natural sampling and arbitrary sampling.

In conclusion, commonly used approach to determine spatial patterns is to compare the observed frequency distribution with theoretical frequency distributions such as Poisson, binomial, negative binomial, Neyman type A, and beta-binomial distributions. It is generally accepted that for count data, a good fit to the Poisson distribution suggests random distribution, while a good fit to the negative binomial distribution implies heterogeneity.

Similarly, for a binary variable a good fit to the binomial distribution indicates homogeneity while a good fit to the beta-binomial distribution suggests heterogeneity. The frequency distribution approach has also been extended to calculate indices of aggregation or dispersion.

Blocked quadrat variances (BQV) method determines cluster size by identifying the peak of mean "local variance" (mean squared difference between adjacent blocks) as quadrat sizes increase by combining two contiguous quadrats to form a new quadrat.

The two major drawbacks with the BQV method are that the block sizes must be some power of 2, and that the starting position of blocking can affect the results. The two-term local quadrat variance (TTLQV) method offers more flexibility to the block size, and reduces the effects of the starting position. Reply's K function is a frequently used method for studying the spatial pattern of mapped point data in ecology.

The advantage of K-function analysis is that it uses all point-point distances, not just the nearest neighbour distances, to show spatial clustering at various scales of pattern, and the distance where clustering or over-dispersal becomes significance. Ripley's K-function is a cumulative distribution equation that calculates point densities within an entire circle of a given radius

Spatial autocorrelation is a widespread phenomenon in ecology. The spread of plants by the processes of growth, vegetative reproduction, and limited seed dispersal produces spatial autocorrelation etc are associated with spatial autocorrelation of plants. Spatial autocorrelation can be analyzed from two different perspectives.

Global autocorrelation analysis involves the study of the entire map pattern and generally asks the question as to whether the pattern displays clustering or not. Local autocorrelation, on the other hand, shifts the focus to explore within the global pattern to identify clusters or so called hot spots that may be either driving the overall clustering pattern, or that reflect heterogeneities that depart from global pattern.

Both types of methods have been extensively utilize in biological data sets, to characterize remote sensing and Landsat TM imagery, to detect spatial clustering of disease in epidemiological studies etc.

The study revealed the dominance of Moran's I approach for their extensive use compared to other methods. Our review of literature suggested the lesser efforts of using this technique for plant population in the Indian subcontinent; therefore there is an ample scope to utilize these techniques to quantify the spatial relationships between different plant variables located on different habitats.

There are many environmental and ecological processes that can result in spatial autocorrelation. These can include slow, large-scale processes such as uplift and erosion that form the geophysical template in which ecological interaction occur, mesoscale processes such as fire, local climate gradients and forest gap dynamics and fast, fine-scale interactions between individuals such as territoriality and conspecific attraction and repulsion.

Distance-limited dispersal and movement also generate autocorrelations in abundance, and these can oftentimes interact with broad-scale landscape. Some of the more interesting and esoteric (albeit largely theoretical) sources of spatial autocorrelation stem from species interactions coupled with dispersal.

Models of predator-prey and host-parasitoid dynamics have been shown to produce a wide range of autocorrelated spatiotemporal dynamics, including traveling waves, periodic checkerboard oscillations and spatial chaos (*e.g.* Bascompte and Sole, 1998).

4.13. REFERENCES

Alby, T., Ferris, J.M. and Ferris, V.R. 1983. Dispersion and distribution of *Pratylenchus scribneri* and *Hoplolaimus galeatus* in soybean fields. *Journal of Nematology*, 15: 418–426.

Almeida-Neto, M. and Lewinsohn, T.M. 2004. Small-scale spatial autocorrelation and the interpretation of relationships between phenological parameters. *Journal of Vegetation Science*, 15(4): 561–568.

Alston, R. 1994. Statistical analysis of animal populations. Ph.D thesis, University of Kent

Alston, D.G. and Schmitt, D.P. 1987. Population density and spatial pattern of *Heterodera glycines* in relation to soybean phenology. *Journal of Nematology*, 19: 336–345.

Alves, M.C. and Pozza, E.A. 2010. Indicator kriging modeling epidemiology of common bean anthracnose. *Applied Geomatrics*, 2: 65–72.

Amedeo, D. and Golledge, R.G. 1975. An introduction to scientific reasoning in geography New York: John Wiley.

Andersen, M. 1992. Spatial analysis of two-species interactions. *Oecologia*, 91: 134–140.

Andrew, N.L. and Mapstone, B.D. 1987. Sampling and the description of spatial pattern in marine ecology. Oceanography and Marine Biology. *Annual Review*, 25: 39–90.

Anscombe, F.J. 1949. The statistical analysis of insect counts based on the negative binomial distributions. *Biometrics*, 5: 165–173.

Anselin, L. 1992. Spatial Data Analysis with GIS: An introduction to application in the social sciences. NCGIA, Santa Barbara, CA, pp. 10–92.

Anselin, L. 1995. Local indicators of spatial association-LISA. Geographical analysis

Anselin, L. 2002. Under the hood: Issues in the specification and interpretation of spatial regression models. *Agriculture Economics*, 17: 247–267.

Armstrong, R. 2006. Measuring the spatial arrangement pattern of pathological lesion in histological sections of brain tissue. *Folia Neuropathologica*, 44(4): 229–237.

Askari, Y., Saei, E.K., Delpasand, S. and Rezaei, D. 2013. Evaluation of *Crategus* sp. spatial pattern in the central zagros forest. *International Journal of Advanced Biological and Biomedical Research*, 1(2): 179–185.

Aspinall, R.J. 1999. GIS for landscape conservation. *In*: Longley, P., Maguire, D.M., Goodchild, M.F., Rhind, D.R. (*Eds*.). GIS: Principles, Techniques, Applications and Management. Wiley, New York, pp. 967–980.

Assuncao, R.M. and Reis, E.A. 1999. A new proposal to adjust Moran's I for population density. *Statistics in Medicine*, 18(16): 2147–2162.

Augustin, N.H., Mugglestone, M.A. and Buckland, S.T. 1996. An autologistic model for the spatial distribution of wildlife. *Journal of Applied Ecology*, 33: 339–347.

Aylor, D.E. 2003. Spread of plant disease on a continental scale: Role of aerial dispersal of pathogens. *Ecology*, 84: 1989–1997.

Azahar, T. M., Mustapha, J. C, Mazliham, S. and Boursier, P. 2011. Temporal analysis of basal stem rot disease in oil palm plantations: An analysis on peat soil. *International Journal of Engineering and Technology*, 11(3): 96–102.

Baddeley, A.J., Moller J. and Waagespetersen, R. 2000. Non- and semi-parametric estimation of interaction in inhomogeneous point patterns. *Statistica Neerlandica*, 54: 329–350.

Bailey, T.C. and Gatrell, A.C. 1995. *Interactive Spatial Data Analysis*. Longman Higher Education, Harlow, UK.

Balanda, M. 2012. Spatio-temporal structure of natural forest: A structural index approach. *Beskydy*, 5(2): 163–172.

Barbour, M.G., Burk, J.H. and Pitts, W.D. 1987 Terrestrial Plant Ecology. Chapter 9: Method of sampling the plant community. Menlo Park, CA: Benjamin/Cummings Publishing Co.

Barot, S., Gignoux, J. and Menaut, J.C. 1999. Demography of a savanna palm host: Predictions from comprehensive spatial pattern analyses. *Ecology*, 80: 1987–2005.

Bartlett, M.S. 1971. Two-dimensional nearest neighbor systems and their ecological applications" in Patil, G.P., Pielon, E.C. and Waters, W.E. *Eds*., Statistical Ecology, Vol. 1 (University Park, Pennsylvania: The Pennsylvania University Press).

Bassanezi, R.B., Bergamin Filho, A., Amorim, L., Gimenes-Fernandes, N., Gottwald, T.R. and Bové, J.M. 2003 Spatial and temporal analyses of citrus sudden death as a tool to generate hypotheses concerning its etiology. *Phytopathology*, 93: 502–512.

Bassanezi, R.B., Bergamin Filho, A., Amorim, L. and Gottwald, T.R. (2005) Spatial and temporal analyses of citrus sudden death in Brazil. Proceedings, 16th Conference of the IOCV. Monterrey, México. pp. 217–229.

Bassanezi, R.B. and Laranjeira, F.F. 2007. Spatial patterns of leprosies and its mite vector in commercial citrus groves in Brazil. *Plant Pathology*, 56: 97–106.

Batista, L., Velαzquez, K., Estévez, I., Peρa, I., Lσpez, D., Reyes, M.L., Rodríguez, D. and Laranjeira, F.F. 2008a. Spatial temporal dynamics of citrus tristeza virus (CTV) in Cuba. *Plant Pathology*, 57: 427–437.

Batista, J.L.F. and Maguire, D.A., 1998. Modeling the spatial structure of tropical forests. *Forest Ecology Management*, 110: 293–314.

Batista, L., Bassanezi, R.B. and Laranjeira, F.F. 2008b. Comparative epidemiology of citrus tristeza in Cuba and citrus sudden death in Brazil. *Tropical Plant Pathology*, 33(5): 348–355.

Besag, J.E. 1977. Comments on Ripley's paper. *Journal of the Royal Statistical Society B*, 39(2): 193–195.

Besag, J., York, J. and Mollie, A. 1991. Bayesian image restoration with two applications in spatial statistics. *Ann. Inst. Stat. Math.*, 43: 1–59.

Besag, J.E. and Gleaves, J.T. 1973. On the detection of spatial pattern in plant communities. *Bulletin of the International Statistical Institute*, 45: 153–158.

Bilek, L., Remes, J. and Zahradnik, D. 2011. Mamaged Vs unmanged structure of beech forest stands (*Fagus sylvatica*) after 50 years of development, Central Behenia. *Forest System*, 20(13): 122–138.

Bliss, C.I. and Fisher, R.A. 1953. Fitting the negative binomial distribution to biological data. *Biometrics*, 9: 176– 200.

Boots, B.N. and Getis, A. 1988. Point Pattern Analysis, Newburry Park, CA, Sage Publication

Bossdorf, O., Schurr, F. and Schumacher, J. 2000. Spatial patterns of plant association in grazed and ungrazed shrublands in the semi-arid Karoo, South Africa. *Journal of Vegetation Science*, 11: 253–258.

Burie, J.B., Langlais, M. and Calonnec, A. 2011. Switching from a mechanistic model to a continuous model to study at different scales the effect of vine growth on the dynamic of a powdery mildew epidemic. *Annals of Botany*, 107: 885–895.

Burrough, P.A. and McDonnell, R.A. 1998. Principles of geographical information systems. Spatial information systems and geostatistics. Oxford university press, Carlon, D.B. 2002. New York, p. 333.

Burt, J.E. and Barber, G.M., 1996, *Elementary Statistics for Geographers*. New York: The Guilford Press.

Calonnec, A., Cartolaro, P. and Chadoeuf, J. 2009. Highlighting features of spatiotemporal spread of powdery mildew epidemics in the vineyard using statistical modeling on field experimental data. *Phytopathology*, 99: 411–422.

Camarero, J.J., Gutierrez, E., Fortin, M.J. and Ribbens, E. 2005. Spatial pattern of host recruitment in a relict population of *Pinus uncinata*: Forest expansion through stratified diffusion. *Journal of Biogeography*, 32: 1979–1992.

Camargo, J.A. 1993. Must dominance increase with the number of subordinate species in competitive interactions? *Journal of Theoretical Biology*, 161: 537–542.

Campbell, C.L. and Madden, L.V. 1990. Introduction to plant disease epidemiology. John Wiley and Sons, New York.

Campbell, C.L. and Noe, J.P. 1985. The spatial analysis of soilborne pathogens and root diseases. *Annual Review of Phytopathology*, 23: 129–148.

Carisse, O., McCartney, H.A., Gagnon, J.A. and Brodeur, L. 2005. Quantification of airborne inoculum as an aid in the management of Botrytis leaf blight of onion caused by *Botrytis squamosa*. *Plant Diseases*, 89: 726–733.

Carisse, O., Savary, S. and Willocquet, L. 2007. Spatiotemoral relationships between disease development and airborne inoculums in unmanaged and managed *Botrytis* leaf blight epidemics. *Phytopathology*, 98(1): 38–44.

Carl, G. and Kuhn, I. 2007. Analyzing spatial autocorrelation in species distribution using gaussian and logit models. *Ecological Modelling*, 270: 159–170.

Carneiro, F.S., Degen, B., Kanashiro, M., Lacerda, A.E.B. and Sebbenn, A.M. 2009. High level of pollen dispersal detected through paternity analysis from a continuous

Symphonia globulifera population in the Brazilian Amazon. *Forest Ecology and Management*, 258: 1260–1266.

Carpenter, S.R. and Chaney, J.E. 1983. Scale of spatial pattern: Four methods compared. *Vegetatio*, 53: 153–160.

Cassie, R.M. 1962. Frequency distribution models in the ecology of plankton and other organisms. *Journal of Animal Ecology*, 31: 65–92.

Chalcraft, D.R., Brian, J.W., Bowles, C. and Willing, M.R. 2009. The relationship between productivity and multiple aspects of biodiversity in six grassland communities. *Biodiversity Conservation*, 18: 91–104.

Chellemi, D.O., Rohrbach, K.G., Yost, R.S. and Sonoda, R.M. 1988 Analysis of the spatial pattern of plant pathogen and diseased plants using geostatistics. *Phytopathology*, 78: 221–226.

Chen, J. and Bradshaw, G.A. 1999. Forest structure in space: A case study of an old growth spruce-forest in Changbaishan natural reserve, PR. China. *Forest Ecology and Management*, 120: 219–233.

Chen, J., Hori, Y., Yamamura, Y., Shiyomi M. and Huang, D. 2008b. Spatial heterogenieity and diversity analysis of macrovegetation in the Xilingol region, Inner Mongolia, China, using the beta distribution. *Journal of Arid Environment*, 72: 1110–1119.

Chen, J., Shiyomi, M., Yamamura, Y. and Hori, Y. 2006. Distribution model and spatial variation of cover in grassland vegetation. *Grassland Science*, 52: 167–173.

Chen, J., Shyomi, M., Hori, Y. and Yamamura, Y. 2008a. Frequency distribution models for spatial pattern of vegetation abundance. *Ecological Modeling*, 211: 403–410.

Cheng, J., Liu, J., Yang, X., Wang, Y and Yu, X. 2011. The structure and spatial pattern of three desert shrub communities in the western ordos platru: Implication for biodiversity conservation. *Journal of Food, Agricultural and Environment*, 9(3-4): 714–722.

Chou, Y.H. 1997. Exploring spatial analysis in geographic information systems. Santa Fe: Onward Press.

Chuang, K.S. and Huang, H.K. 1992. Assessment of noise in a digital image using the joint-count statistic and the Moran test. *Physics in Medicine and Biology*, 37(2): 357–369.

Clark I. 1979. Practical Geostatistics. London: Applied Science Publishers.

Clark, P.J. and Evans, F.C. 1954. Distance to Nearest Neighbor as a Measure of Spatial Relationships in Populations," *Ecology*, 35: 445–453.

Cliff, A.D. and Ord, J.K. 1981. Spatial Processes, Models and Applications. London, Pion, pp. 34–41.

Cliff, A.D. and Ord J.K. 1973. Spatial autocorrelation. Pion Ltd., London. pp. 178

Cole, I.C. 1946. A theory for analyzing contagiously distributed populations. *Ecology*, 27: 329–341.

Cole, R.G. and Syms, C. 1999. Using spatial pattern analysis to distinguish causes of mortality: An example from kelp in north-eastern New Zealand. *Jounral of Ecology*, 87: 963–972.

Condes, S. and Martinez, M. 1998. A comparison between the most important forest spatial pattern index. *Forest System*, 7(1-2): 17–28.

Condit R, Ashton, P.S., Baker, P., Bunyavejchewin, S., Gunatilleke, S., Gunatilleke, N., Hubbell, S.P., Foster, R.B., Itoh, A., La- Frankie, J.V., Seng, Lee, H., Losos, E., Manokaran, N., Sukumar, R. and Yamakura, T. 2000. Spatial patterns in the distribution of tropical host species. *Science*, 288: 1414–1418.

Congalton, R.G. 1988. Using spatial autocorrelation analysis to explore the errors in maps generated from remotely sensed data. *Photogrammetric Engineering and Remote Sensing*, 54: 587–592.

Coomes, D.A., Rees, M. and Turnbull, L. 1999. Identifying aggregation and association in fully mapped spatial data. *Ecology*, 80(2): 554–565.

Costa, M.G., Barbosa, J.C., Yamamoto, P.T. and Leal, R.M. 2010. Spatial distribution of Diphorina citri Kuwagyama (Hemiptera: Psyllidae) in citrus orchids. *Scientica Agricola*, 67: 546–554.

Counts, S., and Martinez-Millan, J. 1998. Comparison of indices spatial distribution of hosts more used in forest. Investigacion agraria. *Sistemas Recursos Forestales,* 7(1-2): 173–187.

Cressie, N. 1991. Statistics for spatial data. New York: John Wiley.

Cressie, N.A. 1993. Statistics for spatial data. Revised edition. John Wiley and Sons, New York, New York, USA.

Cressie, N. and Collins, L.B. 2001b. Analysis of spatial point patterns using boundless of product density LISA functions. *Journal of Agricultural Biological and Environmental*, 6: 118–135.

Cressie, N., and Collins L.B. 2001a. Patterns in spatial point locations: Local indicators of spatial association in a minefield with clutter. *Naval Research Logistic*, 48: 333–347.

Cuzick, J. and Edwards, R. 1990. Spatial Clustering for Inhomogeneous Populations. *Anthyllis cytisoides* shrubland on abandoned land in southeastern Spain. *Journal of Vegetation Science*, 8: 627–634.

Dale, M.R.T. and Powell, R.D. 2001. A new method for characterizing point patterns in plant ecology. *Journal of vegetation Science*, 12(5): 597–605.

Dale, M.R.T. and MacIsaac, D.A. 1989. New method for the analysis of spatial pattern in vegetation. *Journal of Ecology*, 77: 78–91.

Dale, M.R.T., Dixon, P., Fortin, M.-J., Legendre, P., Myers, D.E., Rosenberg, M.S., 2002. Conceptual and mathematical relationships among methods for spatial analysis. *Ecography*. 25: 558-577

Dale, M.T. 1999. Spatial pattern analysis in plant ecology. Cambridge University Press, Campridge, UK

Dallot, S., Gottwald, T., Labonne, G., and Quiot, J.-B. 2003. Spatial pattern analysis of Sharka disease (*Plum pox virus* strain M) in peach orchards of southern France. *Phytopathology*, 93: 1543–1552.

Damgaard, C. 2009. On the distribution of plant abundance data. *Ecological Informatics*, 4: 76–82.

Danielson, B.J. and Gaines, M.S. 1987. Spatial pattern in two synteopic species of microtines: *Microtus ochrogaster* and *Synaptomys cooperi*. *Journal of Mammalogy*, 68(2): 313–322.

Dark, S.J. 2004. The biogeography of invasive alien plants in California: An application of gis and spatial regression analysis. *Diversity and Distribution*, 10: 1–9.

David, A.M., David, C.S. and Rose, G.A. 2003. Spatial pattern and patchiness during ontogeny: Post-settled *Gadus morhua* from coastal Newfoundland. *ICES Journal of Marine Science*, 60: 38–51.

David, F.N. and Moore P.G. 1954. Notes on contagious distributions in plant population. *Annals of Botany*, 18(1): 47–53.

Davis, J.H., Howe, R.W. and Davis, G. J. 2000. A multi-scale spatial analysis method for point data. *Landscape Ecology*, 15: 99–114.

Dawson, A.H. 1975. Are geographers indulging in a landscape lottery? *Area*, 7: 42–44.

Debski, I., Burslem, D.F.R.P. and Lamb, D. 2000. Ecolgical process maintaining differential host species distribution in an Australian subtropical rain forest: Implication for models of species coexistence. *Journal of Tropical Ecology*, 16: 387–415.

Diggle, P.J. 1983. Statistical analysis of spatial point patterns Academic Press, London

Diggle, P.J. 1985. A kernel method for smoothing point process data. - *Journal of the Royal*

Diggle, P.J. and Chetwynd, A.G. 1991. Second-Order Analysis of spatial clustering for inhomogenous population, *Biometrics*, 47(3): 1155–63.

Diggle, P.J. 1979. On parameter estimation and goodness-of fit testing for spatial point pattern. *Biometrics*, 35: 87–101.

Diggle, P.J., Gomez-Rubio, V., Brown, P.E., Chetwynd, A.G. and Gooding, S. 2007. Second –order analysis of inhomogeneous spatial point processes using case control data. *Biometrica*, 63(2): 550–638.

Diniz-Filho, J.A.F., Bini, L.M. and Hawkins, B.A. 2003. Spatial autocorrelation and red herrings in geographical ecology. *Global Ecology and Biogeography*, 12: 53–64.

Dixon, P.M. 2002. Ripley's K function. - *Encyclopedia of Environmetrics* 3: 1796–1803.

Dormann, C.F. 2007. Effects of incorporating spatial autocorrelation into the analysis of species distribution data. *Global Ecology and Biogeography*, 16: 129–138.

Dormann, C.F., Jana, M.M., Miguel, B.A., Roger, B., Bolliger, J., Carl, G., Davies, R.G., Hirzel, A., Jets, W., Kissling, O., Kuhn, I., Ohlemuller, R., Neto-Peres, P.R., Reineking, B., Schroder, B., Schurr, F.M. and Wilson, R. 2007. Methods to account for spatial autocorrelation in the analysis of species distributional data: A review. *Ecography*, 30: 609–628.

Dray, S. 2011. A new perspective about Moran's coefficient: Spatial autocorrelation as a linear regression problem. *Geographical Analysis*, 43: 127–141.

Dray, S., Pelissier, R., Couteron, P., Fortin, M.J., Legendre, P., Neto-Peres, P.R., Bellier, E., Bivand, R., Blanchet, F.G., Caceres, M.D., Dufour, A.B., Heegaard, E., Jombart, T., Munoz, F., Oksanen, J., Thioulouse, J. and Wagner, H.H. 2012. Community ecology in the age of multivariate multiscale spatial analysis. *Ecological Monographs*, 82(3): 257–275.

Dray, S., Said, S. and Debias, F. 2008. Spatial ordination of vegetation data using a generalization of wartenberg's multivariate spatial correlation. *Journal of Vegetation Science*, 19(1): 45–56.

Dungan, J.L., Perry, J.N., Dale, M.R.T., Legendre, P., Citron-Pousty, S., Fortinn, M.J., Jakomulska, A., Miriti, M. and Roesenberg, M.S. 2002. A balanced view of scale in spatial statistical analysis. *Ecography*, 25: 626–640.

Edwards, G.R, Parsons, A.J., Newman, J.A. and Wright, I.A. 1996. The spatial pattern of vegetation in cut and grazed grass/white clover pastures. *Grass and Forage Science*, 51: 219–231.

Elizabeth, G., Juliana, S., Ueli, M. and Sebastian, R. 2011. Powdery scab effect on the potato *Solanum tuberosum* ssp. andigena growth and yield. *Tropical Plant Pathology*, 36(6): 350–355.

Elliot, J.M., 1971. Some methods for statistical analysis of sampling of benthic invertebrates. Freshwater Biological Associates. The Ferry House, Ambleside, UK. p. 144.

Elliott, J.M. 1979. Some Methods for the Statistical Analysis of Samples of Benthic Invertebrates. Freshwater Biological Association, Ambleside, Cumbria, England.

Engeman, R.M., Constantine, B., Nelson, M. Woolard, J., Bourassa, J. 2001. Monitoring changes in fera swine abundance and spatial distribution. USDA national wildlife research centre. P. 168.

Errington, J.C. 1973. The effect of regular and random distributions on the analysis of pattern. *Journal of Ecology*, 77: 78–91.

Everhart, S.E., Askew, A., Seymour, L. and Scherm, H. 2012. Spatio-temporal patterns of pre-harvest brown rot epidemics within individual peach host canopies. *European Journal of Plant Pathology*, pp. 2–10.

Everhart, S.E., Askew, A., Seymour, L., Holb, I.J and Scherm, H. 2011. Characterization of three dimensional spatial aggregation and association pattern of brown rot symptoms with in intensively mapped sour cheery hosts. *Annals of Botany*, 108: 1195-1202.

Fehmi, J.S. and Bartolome, J.W. 2001. A grid-based method for sampling and analyzing spatially ambiguous plants. *Journal of Vegetation Science*, 12: 467–472.

Ferrandino, F.J. 1996. Length scale of disease spread: Fact or artifact of experimental geometry. *Phytopathology*, 86: 806–811.

Ferrandino, F.J. 1998. Past non-randomness and aggregation to spatial correlation: 2DCORR, a new approach for discrete data. *Phytopathology*, 88: 84–91.

Ferrandino, F.J. 1996. Two-dimensional distance class analysis of disease-incidence data: Problems and possible solutions. *Phytopathology*, 86: 685–691.

Ferrandion, F.J. 1993. Dispersive epidemic wavers I. Focus expansion within a linear planting. *Phytopathology*, 83: 795–802.

Fischer, M.M., Leitner, M. and Staufer-Steinnocher, P., 2001. Spatial Point Pattern Analysis – Some Useful Tools for Analysing Locational Data. *In*: Wohlschlδgl, H. (*ed.*): *Geographischer Jahresbericht aus Österreich*, 58: 49–66. Vienna: Institute for Geography and Regional Research. Vienna University.

Fishcer, M.M. and Gests, A. 2010. Handbook of Applied Spatial Analysis: Software Tools, Methods and Applications. New York: Springer.

Fisher, R.A. 1925. Statistical methods for research workers. Hafner, New York. p. 362

Fortin, M.J. and Dale, M.R.T. 2005. Spatial analysis: A guide for ecologists. Cambridge University Press.

Fortin, M.J., Drapeau, P. and Legendre, P. 1989. Spatial autocorrelation and sampling design in plant ecology. *Vegetatio*, 83: 209–222.

Fortin, M.-J. 1999. Effects of quadrat size and data measurement on the detection of boundaries. *Journal of Vegetation Science*, 10: 43–50.

Fortin, M.-J., 1999. Spatial statistics in landscape ecology. *In*: Klopatek, J.M. and Gardner, R.H. (*eds.*). landscape ecological analysis: Issues and applications. Springer-Verlag, New York.

Fortin, M.J. and Dale, M.R.T. 2005. Spatial analysis: A guide for ecologists. Cambridge University Press, Cambridge, UK.

Fotheringham, A.S., Burnsdon, S. and Charlton, M. 2000. Quantitative geography: Perspectives on spatial data analysis. Sage Publications, Thousand Oaks, CA. p. 270

Fox, E., Balram, S., Dragicevis, S. and Roberts, A. 2012. Spatial analysis of high resolution aerial photographs to analyse the spread of mountain pine beetle infestation. *Journal of Sustainable Development*, 5(9): 106–129.

Freeman, E.A. and Ford, E.D. 2002. Effects of data quality on analysis of ecological pattern using the K (d) statistical function. *Ecology*, 83(1): 35–46.

Freita-Nunes, A.F. and Rocha, C.F.D. 2007. Spatial distribution by *Canistropsis microps* (E. Morren ex Mez) Leme (Bromeliaceae: Bromelioideae) in the Atlantic rain forest in IIha Grande south-eastern Brazil. *Brazilian Journal of Biology*, 67(3): 467–474.

Gabriel, K.R. and Sokal, R.R. 1969. A new statistical approach to geographic analysis. *Systematic Biology*, 18(3): 259–278.

Gadow, K.V., Hui, G.U. and Albert, M. 1998. The neighbourhood pattern a new parameter for describing forest structure. Centralbl, *Gesamte Forstwes*, 115: 1–10.

Galiano, E.F. 1982. Pattern detection in plant populations through the analysis of plant-to-all plants distances. *Vegetatio*, 49: 39–43.

Galiano, E.F., Castro, I. and Sterling, A. 1987. A test for spatial pattern in vegetation using a Monte-Carlo simulation. *Journal of Ecology*, 75: 915–924.

Gangnon, R.E. and Clayton, M.K. 2001. A weighted average likelihood ratio test for spatial clustering of disease. *Statistics in Medicine*, 20: 2977–2987.

Gao, J., Queenborough, S.A. and Chai, J.P. 2012. Flowering sex rations and spatial distribution of dioecious hosts in a south-east Asian seasonal tropical forest. *Journal of Tropical Forest Science*, 24(4): 517–527.

Garrett, K.A., Madden, L.V., Hughes, G. and Pfender, W.F. 2004. New application of statistical tools in plant pathology. *Phytopathology*, 94: 999-1003.

Gatrell, A.C., Bailey, T.C., Diggle, P.J. and Rowlingson, B.S., 1996, Spatial Point Pattern Analysis and its Application in Geographical Epidemiology.

Geary, R. 1954. The contiguity ratio and statistical mapping. The Incorporated Statistician, 5: 115–45.

Gelman, A., Carlin, J.B., Stern, H.S. and D.B. Rubin. 2004. Bayesian data analysis. Second edition. Chapman and Hall/ CRC, Boca Raton, Florida, USA.

Gent, D.H., Mahaffee, W.F. and Turechek, W.W. 2006. Spatial heterogeneity of the incidence of powdery mildew on hop cones. *Plant Disease*, 90:1433–1440.

Gent, D.H., Turechek, W.W. and Mahaffee, W.F. 2007. Sequential sampling for estimation and classification of the incidence of hop powdery mildew I: Leaf sampling. *Plant Diseases*, 91:1002–1012.

Gent, D.H., Turechek, W.W. and Mahaffee, W.F. 2008. Spatial and temporal stability of the estimated parameters of the binary power law. *Phytopathology*, 98: 1107–1117.

Getis, A. and Franklin, J. 1987. Second-order neighbourhood analysis of mapped point data patterns. *Ecology*, 68: 473–477.

Getis, A. 1964. Temporal land-use analysis with the use of nearest neighbor and quadrant methods. *Annals, Association of American Geographers*, 54: 391–399.

Getis, A. 1991. Spatial interaction and spatial autocorrelation: Across-product approach. *Environment and planning A*, 23: 1269–1277.

Getis, A. and Ord, J.K. 1992. The analysis of spatial association by use of distance statistics. *Geographical Analysis*, 24(3): 189–206.

Getis, A. and Ord, J.K. 1996. Local spatial statistics: An overview. *In*: Longley, P. and Batty, M. (*Eds*.). Spatial Analysis: Modelling in a GIS environment. New York: Wiley, pp. 261–277.

Gibson, D.J. 2002. Methods in Comparative Plant Population Ecology, Oxford University, UK. p. 344.

Gibson, J.P. 2002. Genetic diversity and mating system comparisons between ray and disc achene seed pools of the heterocarpic species *Heterotheca subaxillaris* (Asteraceae). *International Journal of Plant Science*, 163: 1025–1034.

Gidoin, C., Babin, R., Bagny, B.L., Cilas, C., Ten Hoopen, G.M. *et al.* 2014 Host Spatial Structure, Host Composition and Resource Availability Influence Mirid Density or Black Pod Prevalence in Cacao Agro-forests in Cameroon. *PLoS ONE*, 9(10): e109405. doi:10.1371/journal.pone.0109405

Gignoux, J., Duby, C. and Barot, S. 1999. Comparing the performance of Diggle,s tests of spatial randomness for small samples with and without edge-effect correction: Application to ecological data. *Biometrics*, 55: 156–164.

Gonzalez-Andujar, J.I., Gonzalez-Diaz, I., and Navarrete, L. 2007. Characterizing spatial weed distribution with a distance index of dispersion and the effect of spatial scales. *World Journal of Agricultural Sciences*, 3(4): 558–561.

Goodall, D.W. 1974. *A new method for the analysis of spatial pattern by random pairing of* quadrats. *Vegetatio*, 29: 135–146.

Goodchild, M.F. 1986. Spatial Autocorrelation. Catmog 47, Geo Books.

Goreaud, F. and Pelissier, R. 1999. On explicit formulas of edge effect correction for Ripley's function. *Journal of Vegetation Science*, 3: 433–438.

Goreaud, F. and Pelissier, R. 2003. Avoiding misinterpretation of biotic interactions with the intertype K 12 function: Population independence *versus* random labelling hypothesis. *Journal of Vegetation Science*, 14(5): 681–692.

Goslee, S.C. 2006. Behavior of vegetation sampling methods in the presence of spatial autocorrelation. *Plant Ecology*, 187: 203–2012.

Gosme, M., Willocquet, L. and Lucas, P. 2007. Size, shape and intensity of aggregation of take-all disease during natural epidemics in second wheat crops. *Plant Pathology*, 56: 87–96.

Gottwald, T.R., Avinent, L., Llacer ,G., Hermoso de Mendoza and Cambra, M. 1995. Analysis of spatial spread of sharka (plum pox virus) in apricot and peach orchards in eastern Spain. *Plant Disease*, 79: 266–278.

Gottwald, T.R, Cambra, M., Moreno, P., Camarasa, E. and Piquer, J. 1996. Spatial and temporal analysis of citrus tristeza virus in eastern Spain. *Phytopathology*, 86: 45–55.

Gottwald, T.R. 2010. Current epidemiological understanding of citrus huanglongbing. *Annual Revies of Phytopathology*, 48: 119–139.

Gottwald, T.R., Bassanezi, R.B., Amorim, L. and Bergamin-Filho, A. 2007. Spatial pattern analysis of citrus canker-infected plantings in São Paulo, Brazil, and augmentation of infection elicited by the Asian leaf miner. *Phytopathology*, 97: 674–683.

Gottwald, T.R. 1995. Spatio-temporal analysis and isopath dynamics of citrus scab in nursery plots. *Phytopathology*, 85: 1082–1092.

Grau, H.R. 2000. Regeneration patterns of *Cedrela lilloi* (Meliaceae) in northwestern Argentina subtropical montane forests. *Journal of Tropical Ecology*, 16: 227–242.

Gray, S.M., Moyer, J.W. and Bloomfield, P. 1986. Two dimensional distance class model for quantitative description of virus-infected plant distribution lattices. *Phytopathology*, 76: 243–248.

Green, R.H. 1966. Measurement of non-randomness in spatial distributions. *Researches on Population Ecology*, 8: 1-7.

Greig-Smith, P. 1952. The use of random and contiguous quadrats in the study of the structure of plant community. *Annals of Botany*, 16(2): 293–316.

Greig-Smith, P. 1979. Pattern in vegetation. *Journal of Ecology*, 67: 755–779.

Greig-Smith, P. 1983. Quantitative Plant Ecology, Blackwell Scientific Publication, Oxford, pp. 1-19.

Greig-Smith, P., Kershaw, K.A. and Anderson, D.J. 1963. The analysis of pattern in vegetation: Comment on a paper by Goodall, D.W. *Journal of Ecology*, 51: 223–229.

Griffith, D.A. 2003. Spatial autocorrelation and spatial filtering: Gaining understanding through theory and scientific visualization. Berlin, Germany: Springer-Verlag.

Griffith, D.A. and Peres-Neto, P.R. 2006. Spatial modeling in ecology: The flexibility of eigen function spatial analyses in exploiting relative location information. *Ecology*, 87: 2603–2613.

Griffiths, D.A. 1973. Maximum likelihood estimation for beta-binomial distribution and an application to the household distribution of the total number of cases of a disease. *Biometrics*, 29: 637–648.

Guttorp, P. 1991. Spatial processes in ecology. *In*: Besag, J. and Simpson, J., *Eds*., Spatial Statistics and Digital Image Analysis, National Academy Press, Washington, pp. 129–146.

Haas, S.H., Hooten, M.B., Rizzo, D.M. and Meentemeyer, R.K. 2011. Forest species diversity reduces disease risk in a generalist plant pathogen invasion. *Ecology Letters*, pp. 1–9.

Haase, P. 2001. Can isotropy *vs*. anisotropy in the spatial association of plant species reveal physical *vs*. biotic facilitation? *Journal of Vegetation Science*, 12: 127–136.

Haase, P. 1995. Spatial pattern in ecology based on Ripley's K-function: Introduction and methods of edge correction. *Journal of Vegetation Science*, 6(4): 575–582.

Haase, P., Pugnaire, F.I., Clark, S.C. and Incoll, L.D. 1997. Spatial pattern in *Anthyllis cytisoides* shrubland on abandoned land in southeastern Spain. *Journal of Vegetation Science*, 8(5): 627–634.

Haase, P., Pugnaire, F.L., Clark, S.C. and Incoll, L.D. 1996. Spatial patterns in a two-tiered semi arid shrubland in southern Spain. *Jounral of Vegetation Science*, 7: 527–534.

Haining, R., 1989. Geography and spatial statistics: Current positions, future developments. *In*: Macmillan, B. (*ed.*). Remodelling Geography. Basil Blackwell, Cambridge, MA, pp. 191–203.

Haining, R.P. 2003. Spatial data analysis: Theory and practice. Cambridge University Press, Cambridge.

Hamill, D.N. and Wright, S.J. 1986. Testing the dispersion of juveniles relative to adults: A new analytic method. *Ecology*, 67: 952–957.

Hawkins, B.A., Field, R., Cornell, H.V., Currie, D.J., Guegan, J.F., Kaufman, D.M., Kerr, J.T., Mittelbach, G.G., Oberdorff, T., O'Brien, E.M., Porter, E.E. and Turner, J.R.G. 2003. Energy, water, and broad-scale geographic patterns of species richness. Ecology, 84: 3105–3117.

Haworth, J. and Vincent, P. 1976. Maximizing the Nearest-Neighbor Statistic. *Area*, 8: 299–302.

He, F.L. and Duncan, R.P. 2000. Density-dependent effects on host survival in an old-growth Douglas fir forest. *Journal of Ecology*, 88: 676–688.

Heikkinen, R.K. 1996. Predicting patterns of vascular plant species richness with composite variables: A meso-scale study in Finnish Lapland. *Vegetatio*, 126: 151–165.

Hengl, T., Heuvelink, G.B.M. and Stein, A. 2004. A generic framework for spatial prediction of soil variables based on regression-kriging. *Geoderma*, 120: 75–93.

Hill, M.O. 1973. The intensity of spatial pattern in plant communities. *Journal of Ecology*, 61: 225–236.

Hines, W.G.S. and Hines, R.J.O. 1979. The Eberhardt index and the detection of non-randomness of spatial point distributions. *Biometrika*, 66: 73–80.

Hinrichesen, V.L., Klassen, C.A., Song, C. and Kulldorff, M. 2009. Evaluation of the performance of tests for spatial randomness on prostrate cancer data. International *Journal of Health Geographic*, 81: 41–50.

Hjalmars, U., kulldorff, M., Gustafsson G. and Nagarwalla, N. 1996. Childhood leukaemia in Sweden: Using GIS and a spatial scan statistic for cluster detection. *Statistics in Medicine*, 15: 707–715.

Holgate, P. 1964. The efficiency of nearest neighbor estimators. *Biometrics*, 20: 647–649.

Holgate, P. 1965. Some new tests of randomness. *Journal of Ecology*, 53: 261–266.

Hopkins, B. and Skellam, J.G. 1954. A new method for determining the type of distribution of plant individuals. *Annals of Botany*, 18: 213–227.

Hubbell, S.P., Foster, R.B., Itoh, A., LaFrankie, J.V., Lee, H.S., Losos, E., Manokaran, N., Sukamar, R. and Yamakura, T. 2000. Spatial patterns in the distribution of tropical host species. *Science*, 288: 1414–1418.

Hughes, G. and Madden, L.V. 1992. Aggregation and incidence of disease. *Plant Pathology*, 41: 657–660.

Hughes, G. and Madden, L.V. 1993. Using the beta-binomial distribution to describe aggregated pattern of disease incidence. *Phytopathology*, 83: 759–763.

Hughes, G. 1999. Sampling for decision making in crop loss assessment and pest management: Introduction. *Phytopathology*, 89: 1080–1083.

Hughes, G. and Gottwald, T.R. 1999. Survey methods for assessment of *Citrus tristeza* virus incidence when *Toxoptera citricida* is the predominant vector. *Phytopathology*, 89: 487–494.

Hughes, G. and Madden, L.V. 1998. Comment: Using spatial and temporal patterns of Armillaria root disease to formulate management recommendations for Ontario's black spruce (*Picea mariana*) seed orchards. *Canadian Jounral of Forest Research*, 28: 154–158.

Hughes, G., Madden, L.V. and Munkvold, G.P. 1996. Cluster sampling for disease incidence data. *Phytopathology*, 86: 132–137.

Hughes, G., McRoberts, N., Madden, L.V. and Gottwald, T.R. 1997. Relationships between disease incidence at two levels in a spatial hierarchy. *Phytopathology*, 87: 542–550.

Huo, X., Li, N., Sun, H., Zhou, L.D. and Li, B.G. 2012. Combining geostatistics with Moran's I analysis for mappings soil heavy metals in Beijing, China. *International Journal of Environmental Research and Public Health*, 9: 995–1017.

Hurlbert, S.H. 1990. Spatial distribution of the montane unicorn, *Oikos*, 58: 257–271.

Irwin, J.O. 1954. A distribution arising in the study of infectious diseases. *Biometrika*, 41: 266–268.

Isaaks, E.H. and Shrivastava, R.M. 1989. An introduction to applied geo-statistics. Oxford Univ. Press.

Iwao, S. and Kuno, E. 1971. An approach to the analysis of aggregation pattern in biological population. *In*: Patill, G.P., Pielou, E.C. and Water, W.E. (Eds.). *Statistical Ecology*, vol. 1: Spatial Patterns and Statistical Distribution. Pennsylvania State University Press, University Park, PA, pp. 3-156.

Jackson, M.C. and Waller, L.A 2005. Exploring Goodness-of-fit and spatial correlation using components of Tango's Index of spatial clustering. *Geographical Analysis*, 37(4): 371–382.

Jackson, M.C., Huang, L., Xie, Q. and Tiwari, R. 2010. A modified version of Moran's I. *International Journal of Health Geographics*, 9: 33–43.

Jaime-Garcia, R., Orum, T.V., Felix-Gastelum, R., Trinidad-Correa, R., Vanetten, H.D. and Nelson, M.R. 2001. Spatial analysis of *Phytophthora infestans* genotypes and late blight severity on tomato and potato in the Del Fuerte Valley using geostatistics and geographic information systems. *Phytopathology*, 91: 1156–1165.

Jansenberger, E.M. and Steinnocher, P.S. 2004. Dual Karnel density estimation as a method for describing spatio-temporal changes in upper Austrian food retailing market 7th AGILE conference on Geographic Information Science, 29 April-1 May, Heraklion, Greece. pp. 551–559.

Jeger, M.J., Van den Boasch, F. and Dutmer, M.Y. 2002. Modeling plant virus epidemics in a field nursery sytem. *IMA Journal of Mathematical Applied Medicine*, 19: 75–94.

Jeltsch, F., Moloney, K.A. and Milton, S.J. 1999. Detecting process from snap-shot pattern: Lesson from host spacing in southern Kalahari. *Oikos*, 85: 451–466.

Jimenez, J.J., Rossi, J.P. and Lavelle, P. 2001. Spatial distribution of earthworms in acid-soil savannas of the eastern plains of Colombia. *Applied Soil Ecology*, 17: 267–278.

Johnson, D.A., Alldredge, J.R. and Fisher, R. 1988. Spatial pattern of downy mildew in hop yards. *Phytopathology*, 78: 1389–1395.

Johnson,R.A. and Zimmer, W.J. 1985. A more powerful test for dispersion using distance measurement. *Ecology*, 66: 1669–1675.

Jonathan, C.E., Franklin, S.C., Gibson, D.J. and Newman, J. 1998. Permutation of two-term local quadrat variance analysis: General concepts for interpretation of peaks. *Journal of Vegetation Science*, 9: 41–44.

Jones, S.J., Gent, D.H., Pethybridge, S.J. and Hay, F.S. 2011. Spatial characteristic of white mould epidemics and the development of sequential sampling plan in Australian bean fields. *Plant Pathology*, 60: 1169–1182.

Kallas, A.M., Reich, R.M., Jacobi, W.R. and Lundquist, J. E. 2003. Modeling the probability of observing *Armillaria* root disease in the Black hill. *Forest Pathology*, 33: 241–252.

Kamu, A., Phin, C.K., Seman, I.A. and Munn, H.C. 2015. Distribution of infected oil palms with *Ganoderma* basal stem root disease. *Journal of Scientific Research and Development*, 2(10): 49–55.

Kelly, M. and Meentemeyer, R.K. 2002. Landscape dynamics of the spread of Sudden Oak Death. *Photogrammetric Eng. Remote Sensing*, 68(10): 1001–1009.

Kenkel, N.C. 1988. Pattern of self-thinning in jack pine: Testing the random mortality hypothesis. *Ecology*, 69(4): 1017–1024.

Khojasteh, F. and Chahouki, Z.M.A. 2013. Interaction between *Gundelia tournefortii* and *Psathyrostachys fragilis* in rangelands of Taleghan, Middle East. Journal *of Scientific Research*, 13(10): 1388–1391.

Kingham, S.P., Gatrell, A.C. and Rowlingson, B. 1995. Testing for clustering of health events within a geographical information system framework. *Environment and Planning*, 27(5): 809–821.

Kint, V., Lust, N., Ferris, R. and Olsthoorns, A.F.M. 2000. Quantification of forest stand structure applied to scote pine (*Pinus sylvestris*) forest. Investment Agricultural Sist. Recur, pp. 147-164.

Klute, D.S., Lovallo, M.J. and Tzilkowski, W.M. 2002. Autologistic regression modeling of American woodcock habitat use with spatially dependent data. *In*: Scott, J., Heglund, P., Morrison, M., Haufler, J., Raphael, M., Wall, W. and Samson, F. (*eds*.). Predicting species occurrences; Issues of accuracy and scale. Island press, Washington, pp. 335–344.

Koenig, W.D. 1998. Spatial autocorrelation in California land birds. *Conservation Biology,* 12: 612–620.

Koening, W.D. and Knops, M.H.J. 2000. Pattern of annual seed production by northern hemisphere hosts: A global perspective. *The American Naturalist*, 155(1): 59–69.

Krebs, C.J. 1999. Ecological methodology. Second edition. Benjamin Cummings, Menlo Park, California, USA.

Krebs, C.J. 2000. *Ecological methodology*. Bennjamin/Cummings, Menlo Park, 2nd.ed. p. 654

Kuhn, I., Bierman, S.M., Durka, W. and Klotz, S. 2006. Relating geographical variation in pollination types to environmental and spatial factors using novel statistical methods. *New Phytology*, 172: 127–139.

Kuhn, I. 2007. Incorporating spatial autocorrelation may invert observed patterns. *Diversity and Distribution*, 13(1): 66–69.

Kulldorff, M. 1997. A spatial scan statistic. *Community Statics Theory and Methods*, 26: 1481–1496.

Kuno, E. and Iwao, S. 1968. Use of regression of mean crowding on mean density for estimating sample size and the transformation of data for the analysis of variance. *Research in Population Ecology*,10: 210–214.

Kuuluvainen, T. and Rouvinen, S. 2000. Post-fire understorey regeneration in boreal *Pinus sylvestris* forest sites with different fire histories. *Journal of Vegetation Science*, 11(6): 801–812.

Kuusinen, M. and Penttinen, A. 1999. Spatial pattern of the threatened epiphytic bryophyte *Neckera pennata* at two scales in a fragmented boreal forest. *Ecography*, 22: 729–735.

Laffan, S.W. 2006. Assessing regional scale weed distributions, with an Australian example using *Nassella trichotoma*. *Weed Research*, 46(3): 194–206.

Larsen, D.R. and Bliss, L.C. 1998. An analysis of structure of host seedling populations on a Lahar. *Landscape Ecology*, 13: 307–322.

Lee, J. and Wong, D.W.S. 2001. Statistical analysis with ArcView GIS. Wiley, New York. p. 192

Legendre, L. and Legendre, P. 1984. l~cologie numrrique. 2irme ed. Tome 2: La structure des donnres 6cologiques. Masson, Paris et les Presses de l'University du Quebec.

Legendre, P. Dale, M.T.R., Fortin, M.J., Gurvitch, J., Hohn, M. and Myers, D. 2002. The consequences of spatial structure for the design and analysis of ecological field surveys. *Ecography,* 25: 601–615.

Legendre, P. 1993. Spatial autocorrelation: Trouble or new paradigm? *Ecology*, 74: 1659–1673.

Legendre, P. and Fortin, M.J. 1989. Spatial pattern and ecological analysis. *Vegetatio*, 80: 107–38.

Legendre, P. and L. Legendre. 2012. Numerical ecology. Third English edition. Elsevier Science, Amsterdam, The Netherlands.

Legendre, P., Legendre, L., 1998. Numerical Ecology, 2nd English ed. Elsevier, Amsterdam

Leitner, M. and Staufer-Steinnocher, P. 2001. Kernel Density Estimation - An innovative exploratory tool for geomarketing. *In*: Strobl, J., Blaschke, T. and Griesebner, G. (*eds.*). *Angewandte Geographische formationsverarbeitung XIII*, pp.122–129. Heidelberg: Herbert Wichmann Verlag

Lennon, J.J. 2000. Red-shifts and red herrings in geographical ecology. *Ecography*, 23: 101–113.

Leps, J. and Kindlmann, P. 1987. Model of the development of spatial pattern of an even aged plant population over time. *Ecological Modelling*, 39: 45–57.

Lescure, J.P., Emperarire, L. and Francison, C. 1992. Leopoldinia piassaba, Wallace (Arecaceae) a few biological and economic data from the Rio Negro region (Brazil). *Forest Ecology and Management*, 55: 83–86.

Levick, S.R., Asner, G.P. and Smit, I.P.J. 2012. Spatial pattern in the effect of fire on savanna vegetation host dimensional structure. *Ecological Application*, 22(8): 2110–2121.

Levine, N. 2002. *CrimeStat Manual – Chapter 8 Kernel Density Interpolation*. Available from http://www.icpsr.umich.edu/NACJD/crimestat/CrimeStatChapter.8.pdf, accessed 8.3.2003

Li, C.J., Li, Y., Jian, M.A., Fan, L.L. and Wang, Q.X. 2010. Spatial heterogeneity of soil chemical properties between *Haloxylon persicum* and *Haloxylon ammodendron* population. *Journal of Arid Land*, 2(4): 257–265.

Li, F. and Zhang, L. 2007. Comparison of point pattern analysis methods for classifying the spatial distribution of spruce-fir stands in the north-east USA. *Forestry*, 80(3): 337–349.

Liang, K.Y., Zeger, S.L., 1986. Longitudinal data analysis using generalized linear models. *Biometrica*, 73: 13–22.

Lichstein, J.W., Simons, T.R., Shriner and S.A., Franzreb, K.E. 2002. Spatial autocorrelation and autoregressive models in ecology. *Ecological Monograph*, 72: 445–463.

Lin, Y., Hunting, M.L. and Augsperger, C.K. 2004. Causes of spatial pattern of dead hosts in forest fragment in Illinois. *Plant Ecology*, 170: 15–27.

Linden, A. and Mantyniemi, S. 2011. Using the negative binomial distribution model overdisperssion in ecological count data. *Ecology*, 92(7): 1414–1421.

Liu, C. 2001. A comparison of five distance-based methods for spatial pattern analysis. *Journal of Vegetation Science*, 12(3): 411–416.

Liu, D., Kelly, M., Gong, P. and Guo, Q. 2007. Characterizing spatial temporal host mortality patterns associated with a new forest disease. *Forest Ecology and Management*, 253: 220–231.

Lloyd, M. 1967. Mean crowding. *Journal of Animal Ecology*, 36: 1–30.

Lobo, J.M., Jay-Robert, P. and Lumaret, J.P. 2004. Modelling the species richness distribution for French Aphodiidae (Coleoptera, Scarabaeoidea). *Ecography*, 27: 145–156.

Lobo, J.M., Lumaret, J.-P., Jay-Robert, P., 2002. Modelling the species richness of French dung beetles (Coleoptera, Scarabaeidae) and delimiting the predictive capacity of different groups of explanatory variables. *Global Ecology and Biogeography*, 11: 265–277.

Lookingbill, T.R. and Zavala, M.A. 2000. Spatial pattern of *Quercus ilex* and *Quercus pubescens* recruitment in *Pinus halepensis* dominated woodlands. *Journal of Vegetation Science*, 11: 607–612.

Loosmore, N.B. and Ford, E.D. 2006. Statistical inference using the G or K point pattern spatial statistics. *Ecology*, 87(8): 1925–1931.

Lσpez, F.A., Matilla, M., Mur, J. and Ruiz 2010. A Non-parametric spatial independence test using symbolic entropy, *Regional Science and Urban Economic*, 40:106–115.

Lotwick, H.W. and Silverman, B.W. 1982. Methods for analysing spatial processes of several types of points. *Journal of Royal Statistical Society, Series B*, 44: 406–413.

Ludwig, J.A. and Goodall, D.W. 1978. A comparison of paired with locked-quadrat variance methods for the analysis of spatial pattern. *Vegetatio*, 38: 49–59.

Ludwig, J.A. and Reynolds, J.F. 1988. Statistical ecology: A primer on methods and computing. John Wiley and Sons, New York, NY.

Luoto, M., Poyry, J., Heikkinen, R.K., Saarinen, K., 2005. Uncertainty of bioclimate envelope models based on the geographical distribution of species. *Global Ecology and Biogeography*, 14: 575–584.

Maanen, A. and Xu, X.M. 2003. Modelling plant disease epidemics. *European Journal of Plant Pathology*, 109: 669–682.

Madden, L.V. and Hughes, G. 1995. Plant disease incidence: Distribution, heterogeneity, and temporal analysis. *Annual Review of Phytopathology*, 33: 529–64.

Madden, L.V. and Hughes, G. 1999. An effective sample size for predicting plant disease incidence in a spatial hierarchy. *Phytopathology*, 89: 770–781.

Madden, L.V., Hughe, G. and Van den Bosch, F. 2007. The study of plant disease epidemics american phytopathological society, St Paul, MN.

Madden, L.V. and Hughes, G. 1999. Sampling for plant disease incidence. *Phytopathology*, 89: 1088–1113.

Madden, L.V., Louie, R., A.N. and Knoke, J.K. 1982. Evaluation of tests for randomness of infected plants. *Phytopathology*, 72: 195–198.

Madden, L.V., Prione, T.P. and Raccah, B. 1987. Analysis of spatial patterns of virus-disease tobacco plants. *Phytopathology*, 77: 14–21.

Madden, L.V., Turechek, W.W. and Nita, M. 2002. Evaluation of generalized linear mixed models for analyzing disease incidence data obtained in designed experiments. *Plant Diseases*, 86: 316–325.

Marcon, E., Puech, F. and Traissac, S. 2012. Characterising the relative spatial structure of point patterns. *International Journal of Ecology*, pp. 1-13.

Marcon, E., Traissac, S. and Lang, G. 2013. A statistical test for Ripley' K function rejection of Poisson null hypothesis. *Ecology*, 1(1): 1-9.

Martens, S.N., Breshears, D.D., Meyer, C.W. and Barnes, F.J. 1997. Scales of aboveground and below-ground competition in a semi- arid woodland detected from spatial pattern. *Journal of Vegetation Science*, 8: 655–664.

Mast, J.N. and Veblen, T.T. 1999. Host spatial patterns and stand development along the pine-grassland ecotone in the Colorado Front Range. *Canadian Journal of Forest Research*, 29: 575–584.

Mathur, M. 2015. Spatial autocorrelation analysis in plant population: An overview. *Journal of Applied and Natural Sciences*, 7(1): 501–513.

Mathur, M. and Sundaramoorthy, S. 2008. Distribution patter and growth assessment of *Corchorus depressus* in semi arid Indian desert. *Tropical Ecology*, 49(1): 69–71.

May, R. 1973. Stability on randomly fluctuating versus deterministic environment. *American Naturalist*, 107: 621–650.

McCartney, H.A. and Fitt, B.D.L. 1998. Dispersal of foliar fungal plant pathogens: Mechanisms, gradients and spatial pattern. *In*: The epidemiology of plant diseases. Jones, D.G. *Ed*. Kluwer Academic Publishers, London. p. 460

Mcgarigal, K. and Marks, B.J. 1994. FRAGSTATS, Spatial pattern analysis programme for quantifying landscape structure. Forest science department oregon state university, Corvallis.

McLaughlin, C.C. and Boscoe, F.P. 2007. Effects of randomization methods on statistical inference in disease cluster detection. *Health Place*, 13(1): 152–163.

McMillen, D.P. 2003. Spatial autocorrelation or model misspecification? *International Regional Science Review*, 26: 208–217.

Michereff, S.J., Martins, R.B., Noronha, M.A. and Machado, L.P. 2011. Sample size for quantification of *Cercospora* leafspot in sweet pepper. *Journal of Plant Pathology*, 93(1): 183–186.

Mihail, J.D. 1989. *Macrophomina phaseolina*: Spatio-temporal dynamics of inoculums and of disease in a highly susceptible crop. *Phytopathology*, 79: 848–855.

Miller, J.F.L. and Quesenberry, C.P. 1979. Power studies of tests for uniformity, II. Communications in Statistics - *Simulation and Computation B*, 8: 271–290.

Miller, J., Franklin, J. and Aspinall, R. 2007. Incorporating spatial dependence in predictive vegetation models. *Ecological Modeling*, 202: 225–242.

Minogue, K.P. 1986. Disease gradients and the spread of disease. *In*: Plant disease epidemiology, Lenoard, K.J. and Fry, W.E. (*Eds*.). Macmillan Publishing Co., Inc., New York. pp. 285–310.

Minogue, K.P. and Frym, W.E. 1983. Models for the spread of diseases: Model description. *Phytopathology*, 73: 1168–1173.

Modjesk, J.S. and Rawling, J.O. 1983. Spatial correlation analysis of uniformity data. *Biometrics*, 39: 373–384.

Moeur, M. 1997. Spatial models of competition and gap dynamics in old growth *Tsuga heterophylla / Thuja plicata* forest. *Forest ecology and Management*, 94: 175–186.

Mohebbia, Z., Jafary, M., Tavilic, A., Zare, M.A., Baghbania, S. and Nourikia, Z. 2012. Studying the effect of index species *Astragalus ammodendron* on associated plants *Agropyron tauri* and *Bromus tomentellus* from different aspects (Case Study: Arak Province-Shazand, Iran). *Journal of Rangeland Science*, 2(2): 513–520.

Molles, M.C. 1999. Ecology: Concepts and applications. McGraw-Hill, New York.

Mollison, D. 1977. Spatial contact models for ecological and epidemics spread. *Journal of Royal Statistical Society B*. 39: 283–326.

Montane´, F., Casals, P. and Dale, M.R.T. 2011. How Spatial Heterogeneity of Cover Affects patterns of Shrub Encroachment into Mesic Grasslands. PLoS ONE, 6(12): 1–9.

Moran, P.A.P. 1950. Notes on continuous stochastic phenomena. *Biometrics*, 37: 17–23.

Mordi-Vajargah, M., Golizadeh, A., Rafiee-Dastjerdi, H., Zalucki, M.P., Hassanpour, M. and Naseri, B. 2011. Population density and spatial distribution pattern of hypera postica (Coleoptera: Curculionidae) in Ardabil, iran. *Not Bot Horti Agrobo*, 39(2): 42–48.

Morisita, M. 1962. Iδ-index, a measure of dispersion of individuals. *Research in Population Ecology*, 4: 1–7.

Morista, M. 1954. Measuring of the dispersion of individuals and analysis of the distributional patterns," *Memoirs of the Faculty of Science, Series E*, 2: 254–73.

Motz, K. Sterba, H. and Pommerening, A. 2010. Sampling measure of host diversity. *Forest Ecology and Management*, 260: 1985.

Mueller-Warrant, G.W., Whittaker, G.W. and Young, W.C. 2008. GIS analysis of spatial clustering and temporal change in weeds of grass seed crops. *Weed Science*, 56(5): 647–669.

Murolo, S., Mancini, V. and Romanazzi, G. 2014. Spatial and temporal stolbur population structure in a cv. Chardonnay vindeyard according to vmp1 gene characterization. *Plant Pathology*, 63: 700–707.

Nelson, S.C. 1995. Spatiotemporal distance class analysis of plant disease epidemics. *Phytopathology*, 85: 37–43.

Nelson, S.C., Marsh, P.L. and Campbell C.L. 1992. 2DCLASS, a two dimensional distance class analysis software for the personal computer. *Plant Disease*, 76: 427–432.

Nicot, P.C., Touse, D.L. and Yandell, B.S. 1984. Comparison of statistical methods for studying spatial patterns of soilborne plant pathogens in the field. *Phytopathology*, 74: 1399–1402.

Nordmeyer, H. 2009. Spatial and temporal dynamics of *Apera spica-venti* seedling populations. *Crop Protection*, 28: 831–837.

Numahara, T., Tanemura, M., Numaharai, K., Moriu, J., Shirahige, Y., Yokoi, I. and Kubota, Y. 2009. Spatial statistics for epidermal langerhans cells effects of protopic ointment 0.1% on the spatial distribution. *Forma*, 24: 49–59.

Oden N. 1995. Adjusting moran's I for population density. *Stattistical Medicine*, 14(1): 17–26.

Oden N.L. 1984. Assessing the significance of a spatial correlogram. *Geographical Analysis*, 16: 1–16.

Odland, J. 1988. Spatial autocorrelation. *In*: Thrall, G.I. (ed.). Sage University Scientific Geography Series no. 9. Sage Publications, Beverly Hills. p. 87

Palacio-Bielsa, A., Lopez-Quilez, A., Llorente, I., Ruz, L., Lopez, M.M. and Cambra, M.A. 2012. Criteria for efficient prevention of dissemination and successful eradication of *Erwinia amylovora* (the cause of fire blight) in Aragon Spain. *Phytopathologia Mediterranea*, 51(3): 505–518.

Palma, L. Beja, P. and Rodrigues, M. 1999. The use of sighting data to analyse Iberian lynx habitat and distribution. *Journal of Applied Ecology*, 36:812–824.

Pancer-Koteja, E., Szwagrzyk, J. and Bodziarczyk, J. 1998. Small-scale spatial pattern.

Payne, L.X. 1997. Dispersion patterns of sandpipers migrating through the Eastern United States. Master's thesis. University of wisconsin, madison, Wisconsin, USA.

Payne, L.X., Schindler, D.E., Parrish, J.K. and Temple, S.A. 2005. Quantifying spatial pattern with evenness indices. *Ecological Applications,* 15(2): 507–520.

Pedigo, L.P., Buntin, G.D. 1994. Handbook of sampling methods for arthropods in agriculture. CRC Press, Florida.

Pelissier, R. and Goreaud, F. 2001. A practical approach to the study of spatial structure in simple case of heterogenous population. *Journal of Vegetation Science*, 12(1): 99–108.

Penttinen, A., Stoyan, D. and Henttonen, H.M. 1992. Marked point processes in forest statistics. *Forestry Science*, 38: 806–824.

Perry, G.L.W., Miller, B.P. and Enright, N.J. 2006. A comparison of method for the statistical analysis of spatial point pattern in plant ecology. *Plant Ecology*, 187: 59–82.

Perry, J.N and Hewitt, M. 1991. A new index of aggregation for animal counts. *Biometrics* 47: 1505–1518.

Perry, J.N., Liebhod, A.M., Rosenbery, M.S., Dungan, I., Miriti, M., Jakomulska, A. and Citron-Pousty, S. 2002. Illustrations and guidelines for selecting statistical methods for quantifying spatial paterrns in ecological data. *Ecography*, 25: 578–600.

Perry, J.N. 1995. Spatial analysis by distance indices. *Journal of Animal Ecology*, 64: 303–314.

Pettitt, A.N., Weir, I.S. and Hart, A.G., 2002. A conditional autoregressive Gaussian process for irregularly spaced multivariate data with application to modelling large sets of binary data. *Statistical Computation*, 12: 353–367.

Pielou E.C., 1959. The use of point-to-plant distances in the study of the pattern of plant populations. *Journal of Ecology*, 47: 607–613.

Pielou, E.C. 1975. Ecological diversity. John Wiley and Sons, New York, New York, USA.

Pielou, E.C. 1977. Mathematical Ecology. John Wiley and Sons, New York, pp. 128–131.

Pinder, D.A. and Witherick, M.E. 1972. The Principles, Practice and Pitfalls of Nearest Neighbour Analysis. *Geography*, 57: 277–8.

Podani, J. and Czαrαn, T. 1997. Individual-centered analysis of applied point patterns representing multi-species assemblages. *Journal of Vegetation Science*, 8: 259–270.

Podani, J., Czaran, T. and Scheuring, I. 1998. Individual-centred analysis of community pattern: Some case studies. *Abstracta Botanica*, 22: 101–112.

Pommerening, A. 2002. Approaches to quantifying forest structure. *Forestry*, 75(3): 1–84.

Poole, R.W. 1974. *An introduction to quantitative ecology*. McGraw-Hill Inc, New York, USA, p. 532

Pugh, S. and Congalton, R. 2001. Applying spatial autocorrelation analysis to evaluate error in New England forest cover type maps derived from Landsat Thematic Mapper Data. *Photogrammetric Engineering and Remote Sensing,* 67(5): 613–620.

Radeloff, V.C., Miller, T.F., He, H.S., and Mladenoff, D.J. 2000. Periodicity in landscape pattern and geostatistical models: Autocorrelation between patches. *Ecography*, 23: 81–91.

Real, L.A., McElhany, P. 1996. Spatial Pattern and process in plant-pathogen interactions. *Ecology*, pp. 1011-1025.

Reich, R.M. and Davis, R. 1998 Quantitative spatial analysis *(Course notes)*. Colorado State University.

Reich, R.M. and Lundquist, J.E. 2005. Use of spatial statistics in assessing forest diseases, pp. 127–143. *In*: Lundquist, J.E. and Hamelin, R.C. (*eds*.). Forest Pathology: From genes to landscapes. American Phytopathological Society Press, St. Paul, MN, p. 175

Revilla, E. and Palomares, F. 2002. Spatial organization, group living and ecological correlates in low-density populations of Eurasian badgers, *Meles meles*. *Journal of Animal Ecology*, 71: 497–512.

Reynolds, K.M. and Madden, L.V. 1988. Analysis of epidemics using spation-temporal autocorrelaton. *Phytopathology*, 87: 240–246.

Rieux, A, Soubeyrand, S, Bonnot, F, Klein, E.K. and Ngando, J.E. *et al.* 2014. Long-distance wind-dispersal of spores in a fungal plant pathogen: Estimation of anisotropic dispersal kernels from an extensive field experiment. *PLoS ONE*, 9(8): e103225. doi:10.1371/journal.pone.0103225

Ripley, B.D. 1977. Modelling Spatial Patterns. *Journal of the Royal Statistical Society B* 39(2): 172–212.

Ripley, B.D. 1981. Spatial statistics. Wiley, New York

Ripley, B.D. 1988. Statistical Inference for Spatial Process. Cambridge: Cambridge University Press

Rivero, R.G., Grunwald, S. and Bruland, G.L. 2007. Incorporation of spectral data into multivariate geostatistical models to map soil phosphorus variability in a Florida wetland. *Geoderma*, 140: 428–433.

Roche, B.M., Alexander, H.M. and Maltby, A.D. 1995. Dispersal and disease gradients of anther-smut infection of *Silene alba* at different life stages. *Ecology*, 76(6): 1863–1871.

Rosset, S.D. and Alcalde, L. 2004. Distribution of the burrows of *Bufo fernandezae* (Anura, Bufonidae) outside the breeding season. *Phyllomederma*, 3(92): 95–99.

Rouse, D.I., MacKenzie, D.R., Nelso, R.R. and Elliott, V.J. 1981. Distribution of wheat powdery mildew incidence in field plots and relationship to disease severity. *Phytopathlogy*, 71: 1015–1020.

Sackett, K.E. and Mundt, C.C. 2005. The effects of dispersal gradient and pathogen life cycle components on epidemic velocity in computer simulations. *Phytopathology*, 95: 992–1000.

Salas, C., LeMay, V., Nunez, P., Pacheco, P. and Espinosa, A. 2006. Spatial pattern in an old-growth Nothofagus oblique forest in south-central Chile. *Forest Ecology and Management*, 231: 38–46.

Saunders, M.R. and Wanger, R.G. 2008. Long-term spatial and structural dynamics in Acadian mixedwood stands managed under various silvicultural systems. *Canadian Journal of Forest research*, 38: 498–517.

Schabenberger, O. and Pierce, F.J. 2002. Contemporary statistical model. Taylor and Francis. London.

Schuh, W., Frederiksen, R.A. and Jeger, M.J. 1986. Analysis of spatial pattern with Morisita's index of dispersion in sorghum downy mildew. *Phytopathology*, 76: 446–450.

Segurado, P. and Araujo, M.B. 2004. An evaluation of methods for modelling species distributions. *Journal of Biogeography*, 31: 1555–1568.

Shannon, C.E. and Weaver, W. 1949. The mathematical theory of communication. University of Illinois Press, Urbana, Illinois, USA.

Shaukat, S.S, Aziz, S., Ahmed, W. and Shahzad, A. 2012. Population structure, spatial pattern and reprorcutive capacity, of two semi desert undershurbs *Seena holosericea* and *Fagonia indica* in southern Sind, Pakistan. *Pakistan Journal of Botany*, 44(1): 1–9.

Shaukat, S.S., Ahmed, W., Khan, M.A. and Shahzad, A. 2009. Inter-specific competition and aggregation in a population of *Solanum forskalii* Dunal in a semi arid habitat: Impact on reproductive output, growth and phenolic content. *Pakistan Journal of Botany*, 41(6): 2751–2763.

Shi, H. and Zhang, L. 2003. Local analysis of host competition and growth. *Forest Science*, 49(6): 938–946.

Shiyomi, M. and Yoshimura, J. 2000. Measures of spatial heterogeneity for species occurrence or disease incidence with finite-count. *Ecological Research*, 15: 13–20.

Shiyomi, M., Akiyama, T. and Takashashi, S. 1984. A spatial pattern model of plant biomass in grazing pasture. *Journal of Japanese Grassland Science*, 30: 40–48.

Shu, C. and Burn, D.H. 2003. Spatial pattern of homogeneous pooling groups for flood frequency analysis. Hydrological Science, *Journal de-Science Hydrologiques*, 48(4): 601–619.

Silverman, B.W. 1978. Choosing the Window Width When Estimating a Density. *Biometrika*, 65: 1- 11

Simpson, E.H. 1949. Measurement of diversity. *Nature*, 163: 688

Skellam, J.G. 1948. A probability distribution derived from the binomial distribution regarding the probability of success as variable between sets of trials. *Journal of Statistical Society of Serial, B*. 10: 257–261.

Smith, B. and Wilson, J.B. 1996. A consumer's guide to evenness measures. *Oikos* 76: 70–82.

Sokal, R.R. 1986. Spatial data analysis and historical processes. *In*: Diday, E. (*ed.*). Data analysis and informatics, IV. Proceedings of the fourth international symposium on data analysis and informatics, Versailles, France, North-Holland, Amsterdam. pp. 29–43.

Sokal, R.R. and Oden, N.L. 1978. Spatial autocorrelation in biology. 1. Methodology. *Biological Journal of Linnaeus Society Society*, 10: 199–228.

Sokal, R.R. Oden, N.L. and Thomson, B.A. 1998a. Local spatial autocorrelation in a biological model. *Geographical Analysis*, 30: 331–354.

Sokal, R.R. Oden, N.L. and Thomson, B.A. 1998b. Local spatial autocorrelation in a biological variables. *Biological Journal of Linnaeus Society*, 65: 41–62.

Soubeyrand, S., Held, L., Hohle, M. and Sache, I. 2008. Modelling the spread in space and time of an airborne plant disease. *Journal of the Royal Statistical Society Series C-Applied Statistics*, 57: 253–272.

Southwood, T.R.E. 1978. *Ecological Methods*. 2nd Edn. Methuen, London.

Spolti, P., Sanhueza-Valdebenito, R.M., Laranjeira, F.F. and Del Ponte, E.M. 2012. Comparative spatial analysis of sooty blotch/flyspeck disease complex, bull eye and bitter rots of apples. *Plant Pathology*, 61: 271–280.

Spooner, P.G., Lunt, I.D., Okabe, A. and Shiode. 2004. Spatial analysis of roadside *Acacia* populations on a road network using the network K-function. *Landscape Ecology,* 19: 491–499.

Stamatellos, G. and Panourgias, G. 2005. Simulating spatial distributions of forest hosts by using data from fixed area plots. *Forestry*, 78(3): 70–77.

Sterba, H. 2008. Diversity indices based on angle count sampling and their interlationshiop when used in forest inventories. *Forestry*, 31(5): 587–597.

Sterner, R.W., Ribic, C.A. and Schatz, G.E. 1986. Testing for life historical changes in spatial patterns of four tropical host species. *Journal of Ecology*, 74**:** 621–633.

Stoyan D, Kendall, W.S. and Mekce, J. 1995. Stochastic geometry and its applications, 2nd edn. John Wiley and Sons, Chichester, UK

Stoyan, D. and Penttinen, A. 2000. Recent application of point process methods in forestry statistics. *Statistical Science,* 15(1): 61–78.

Stoyan, D. and Stoyan, H. 1994. Fractals, Random Shapes and Point Fields. Methods of Geometrical Statistics. *ZAMM - Journal of Applied Mathematics and Mechanics / Zeitschrift für Angewandte Mathematik und Mechanik*, 75(8): 614.

Sule, H. Muhamad, R., Omar, D., Hee, A.K.W. and Zazali, C. 2012. Dispersion Pattern and Sampling of *Diaphorina citri* Kuwayama (Hemiptera: Psylidae) Populations on *Citrus suhuiensis* Hort. Ex Tanaka in Padang Ipoh Terengganu, Malaysia. *Pertanika Journal of Tropical Agricultural Science*, 35: 25–36.

Sutton, J.C., Swanton, C.J., and Gillespie, T.J. 1978. Relation of weather variables and host factors to incidence of airborne spores of *Botrytis squamosa. Can. J. Bot.* 56: 2460–2469.

Suzuki, S.N. Kachi, N. and Suzuki, J.I. 2008. Development of a local size hierarchy causes regular spacing of hosts in an even-aged Abies forest: Analyses using spatial autocorrelation and the mark correlation function. *Annals of Botany*, 102(3): 435–441.

Szmyt, J. 2010. Spatial pattern of hosts of different diameter classes in managed pine stands (*Pinus sylvestris* L.) of different age. *Acta Sci. Pol. Sil. Colendar. Rat. Ind. Lignar*. 9(3-4): 37–49.

Tango, T. 1995. A class of tests for detecting 'general' and 'focused' clustering of rare diseases. Statistical Methods, 14(21-22): 2323–2334.

Tanne, E., Marcus, R., Dubitzky, E., and Raccah, B. 1996. Analysis of progress and spatial pattern of corky bark in grapes. *Plant Dis*. 80: 34–38.

Taylor, L.R. 1961. Aggregation, variance and the mean. *Nature.* 189: 732–735.

Taylor, L.R. 1984. Assessing and interpreting the spatial distribution of insect populations. *Annual Review of Entomology*, 29: 321–357.

Taylor, L.R. and Tylor, R.A.J. 1978. The dynamic of spatial behaviour. *In*: Ebling, F.J. and Stoddart, D.M. (*eds.*). Population control by spatial behaviour. Institute of Biology. London.

Taylor, P.J. 1977. Quantitative Methods in Geography, (Boston: Houghton Mifflin).

Thal, W.M. and Campbell, C.L. 1986. Spatial pattern analysis of disease severity data for alfalfa leaf spot caused primarily by Leptosphaerulina brisiana, *Phytopathology*, 76: 190–194.

Thompson, H.R. 1956 Distribution of distance to nth neighbourin a population of randomly distributed individuals. *Ecology*, 37: 391–394.

Titeux, N., Dufrene, M., *et al.* 2004. Multivariate analysis of a fine-scale breeding bird atlas using a geographical information system and partial canonical correspondence analysis: Environmental and spatial effects. *Journal of Biogeagraphy*, 31: 1841–1856.

Tobin, P.C. 2004. Estimation of the spatial autocorrelation function: Consequences of sampling dynamic populations in space and time. *Ecography*, 27(6): 767–775.

Torgersen, C.E., Jones, J.A. Moldenke, A.R. and LeMaster, M.P. 1995. The spatial heterogeneity of soil invertebrates and edaphic properties in an old growth forest stand in western Oregon. *In*: Collins, H.P. Robertson, G.P. and Klug, M.J. (*eds.*). The significance and regulation of soil biodiversity. Kluwer Academic Publishers, Dordrecht, The Netherlands. p. 225–236.

Trifkovi, S. 2013. Spatial distribution and composition of host in an aging coppice forest in Japan. *Formath*, 12: 35–53.

Trumper, E.V. and Gorla, D.E. 1997. The sensitivity of the ordinary runs test for evaluating the spatial pattern of infected plants. *Ecologia Austral*, 7: 79–86.

Tsuji, K., Kasuya, E., 2001. What do the indices of reproductive skew measure? *American Naturalist*, 158: 155–165.

Tufto, J., Engen, S. and Hindar, K. 1997. Stochastic dispersal processes in plant populations. *Theoretical Population Biology*, 52: 16–26.

Turechek, W.W. and Mahaffee, W.F. 2004. Spatial pattern analysis of hop powdery mildew in the Pacific northwest: Implications for sampling. *Phytopathology*, 94: 1116–1128.

Turechek, W.W. and McRoberts, N. 2013. Considerations of scale in the analysis of spatial pattern of plant disease epidemics. *Annual Review Phytopathology*, 51: 453–472.

Turechek, W.W. and Madden, L.V. 1999. Spatial pattern analysis and sequential sampling for the incidence of leaf spot on strawberry in Ohio. *Plant Diseases*, 83: 992–1000.

Turechek, W.W., and Madden, L.V. 1999. Spatial pattern analysis of strawberry leaf blight in perennial production systems. *Phytopathology*, 89: 421–433.

Turechek, W.W., Ellis, M.A. and Madden, L.V. 2001. Sequential sampling for incidence of Phomopsis leaf blight of strawberry. *Phytopathology*, 91: 336–347.

Unwin, A. and Unwin. D. 1998. Exploratory spatial data analysis with local statistics. *Statistics* 47: 415–421.

Unwin, D.J. 1996. GIS, spatial analysis and spatial statistics. *Progress in Human Geography,* 20: 540–551.

Upton, G.J. and Fingleton, B. 1985. Spatial data analysis by example, volume 1: Point pattern and quantitative data. Wiley, Toronto Singapore, Brisbane, New York, Chichester

Usher, M.B. 1975. Analysis of pattern in real and artificial plant populations. *Journal of Ecology*, 63: 569–586.

Vacek, S., Noskova, I., Bilek, L., Vacek, Z. and Schwarz, O. 2010. Regeneration of forest stands on permanent research plot in Krkonose MTS. *Jounral of Forest Science*. 11: 541–554.

Van den Bosch, F., McRoberts, N., Van den Berg, F. and Madden, L.V. 2008. The basic reproduction number of plant pathogens: Matrix approaches to complex dynamics. *Phytopathology*, 98: 239–249.

Van der Plank, J.E. 1963. Plant Disease: Epidemics and Control. Academic Press, New York, p. 349

Van Leeuwen, G.C.M., Stein, A., Holb, I. and Jeger, M.J. 2000. Yield loss in apple cuased by *Monilinia fructigena* (Aderh, and Ruhl) Honey, and spatio-temporal dynamics of disease development. *European Journal of Plant Pathology*, 106: 519–528.

Ver Hoef, J.M., Glenn-Lewin, D.C. and Werger, M.J.A. 1989. Relationship between horizontal pattern and vertical structure in a chalk grassland. *Vegetatio*, 83: 147–155.

Vincelli, P.C. and Lorbeer, J.W. 1988. Relationship of precipitation probability to infection potential of *Botrytis squamosa* on onion. *Phytopathology*, 78: 1078–1982.

Vincent, P. 1976. The General Case: How not to measure spatial point patterns, *Area*, 8: 161–63.

Waldhor, T. 1996. The spatial autocorrelation coefficient Moran's I under heteroscedasticity. *Statistics in Medicine*, 15(7-9): 887–892.

Wallace, M.K. and Hawkins, D.M. 1994. Application of geostatistics in plant nematology. *Journal of Nematology*, 26(4): 626–634.

Waller, L.A. and Gotway, C.A. 2004. Applied Statistics for Public Health Data. New York: Wiley.

Waller, L.A., Hill, E.G. and Rudd, R.A. 2006. The geography of power: Statistical performance of tests of clusters and clustering in heterogeneous populations. *Statistics in Medicine*, 25(5): 853–865.

Ward, J.S, Parker, G.R. and Ferrandino, F.J. 1996. Long-term spatial dynamics in an old-growth deciduous forest. *Forest Ecology and Management*, 83: 189–202.

Ward, J.S. and Ferrandino, F.J. 1999. New derivation reduces bias and increases power of Ripley's L index. *Ecological Modelling*, 116(2-3): 225–236.

Ward, J.S., Parker, G.R. and Ferrandino, F.J. 1996. Long-term spatial dynamics in an oldgrowth deciduous forest. *Forest Ecology and Management*, 83: 189–202.

Webster, R. and Oliver, M.A. 2001. Geostatistics for Environmental Scientists. Wiley, Chichester

West, P.W. 1984. Inter-host competition and small-scale pattern in monoculture of *Eucalyptus obliqua* (L'Herit). *Australian Journal of Ecology*, 9: 405–411.

White, K.A. and Gilligan. 1998. Spatial heterogeneity in three-species, plant-parasite-hyperparasite, system. *Philosophical Trans Royal Society of London B*, 353: 543–557.

Wiegand, K., Jeltsch, F. and Ward, D. 2000. Do spatial effects play a role in the spatial distribution of desert-dwelling *Acacia raddiana*? *Journal of Vegetation Science*, 11: 473–484.

Wiegand, T. and Moloney, K.A. 2004. Rings, circles and null models for point pattern analysis in ecology. *Oikos*, 104: 209–229.

Wiegand, T., Moloney, K.A., Naves, J. and Knauer, F. 1999. Finding the missing link between landscape structure and population dynamics: A spatially explicit perspective. *American Naturalist*, 154: 605–627.

Williams, S.M. and Sais, J. 2010. Temporal and spatial distribution pattern of Echinoderma larvae in La Parguera, Puerto Rico. *Revista de Biologia Tropical*, 58(3): 81–88.

Williams, S.M., Garcia, J.R. and Capella, J. 2009. Temperoal variation of early larval stage of long-spined sea urchin Diatema antillarum in La Parguera, Puento Rico. *Carabian Journal of Science*, 45(1): 110–117.

Willocquet, L., Sombardier, A., Blancard, D., Jolivet, J. and Savary, S. 2008. Spore dispersal and disease gradients in strawberry powdery mildew. *Canadian Journal of Plant Pathology*, 30: 434–441.

Wu, H. and Huffer, F.W. 1997. Modeling the distribution of plant species using the autologistic regression model. *Environmental Ecololgy Statement*, 4: 49–64.

Wu, X.M. and ubbaroa, K.V. 2004. Analysis of spatial patterns in plant pathology. *Recent Research Developments in Plant Pathology*, 3: 167–187.

Wulder, M. and Boots, B. 2001. Local spatial autocorrelation characteristics of Landsat TM imagery of a managed forest area. *Canadian Journal of Remote Sensing*, 27: 67–75.

Wulder, M. and Boots, B. 1998. Local spatial autocorrelation characteristics of remotely sensed imagery assessed with the Getis statistic. *International Journal of Remote Sensing*, 19: 2223–2231.

Xiao, C.L., Hao, J.J. and Subbarao, K.V. 1997. Spatial patterns of microsclerotia of *Verticillium dahliae* in soil and Verticillium wilt of cauliflower. *Phytopathology*, 87: 325–331.

Xu, X.M. and Madden, L.V. 2002. Incidence and density relationships of powdery mildew on apple. *Phytopathology*, 92: 1005–1014.

Xu, X.M., Robinson, J.D., Berrie, A.M. and Harris, D.C. 2001. Spatio-temporal dynamics of brown rot (*Monilinia fructigena*) on apple and pear. *Plant Pathology*, 50: 569–578.

Xu, X.-M. and Ridout, M.S. 1998. Effects of initial epidemic conditions, sporulation rate, and spore dispersal gradient on the spatiotemporal dynamics of plant disease epidemics. *Phytopathology*, 88: 1000–1012.

Xu, X.-M. and Ridout, M.S. 2000. Effects of quadrat size and shape, initial epidemic conditions, and spore dispersal gradient on spatial statistics of plant disease epidemics. *Phytopathology,* 90: 738–750.

Yin, Z.Y., Guo, Q., Ren, H. and Peng, S.L. 2005. Seasonal changes in spatial patterns of two annual plants in the Chihuahuan Desert, USA. *Plant Ecology*, 178: 189–199.

Zahradnik, D., Vacek, S., Bilek, L., Nskova, I. and Vacek, Z. 2010. Horizontal structure of forest stands on permanent resources plot in the Krkonose Mts and its development. *Journal of Forest Science*, 11: 531–540.

Zas, R., Solla, A. and Sampedro, L. 2007. Variography and kriging allow screening *Pinus pinaster* resistant to *Armillaria ostoyae* in field conditions. *Forestry*, 80(2): 201–249.

Zhang, C.S. and McGrath, D. 2004. Geostatistical and GIS analyses on soil organic carbon concentrations in grassland of southeastern Ireland from two different periods. *Geoderma*, 119: 261–275.

Zhang, C.S., Tao, S., Yuan, G.P. and Liu, S. 1995. Spatial autocorrelation analysis of trace element contents of soil in Tianjin plain area (in Chinese, with English abstract). *Acta Pedology Sinica*, 32: 50–57.

Zhang, C.S., Zhang, S. and He, J.B. 1998. Spatial distribution characteristics of heavy metals in the sediments of Changjiang River system—Spatial autocorrelation and fractal methods (in Chinese, with English abstract). *Acta Geogaphy Sinica*, 53: 87–96.

Zhang, W., Wang, K., Chen, H., He, X. and Zhang, J. 2012. Ancillary information improves kriging on soil organic carbon data for a typical karst peak cluster depression landscape. *Journal of Science for Agriculture*, 92: 1094–1102.

CHAPTER

5 Plant Disease Forecasting and Warning System's

5.1. INTRODUCTION

Plant disease forecasting involves all the activities through which farmers/ growers are notify that particular conditions are favorable for particular diseases and by adoption of timely control measures the economic losses may be minimized. A disease-forecasting model can reduce uncertainty about control decisions by providing a quantitative description of disease pressure

- Disease forecasting is an applied epidemiology.
- This approach is most appropriate for polycyclic diseases.
- It provides timely management opportunity
- Weather data's are the common factor for most plant disease forecasting

For development of forecasting system of any plant disease, a developer must have the Information about:

- Information's about pathogen and its life cycle.
- Information about weather data.
- Timely identification and analysis of inoculums.
- Information's about host

Disease forecasting having applied values for disease that must fulfill following conditions:

- Plant disease that causes severe economic losses.
- Disease that having variable onset and distribution.
- Availability of environmental data set.
- Knowledge and availability of control measure and that must be economically feasible.

In general appropriateness of any forecasting system depends on:

- Amount or efficacy of initial inoculums: this approach is most appropriate for disease induced by most monocyclic pathogen. They are also appropriate for polycyclic diseases if there is large amount of initial inoculums or if the number of generation (secondary cycles) is small. Examples of such diseases includes Stewart's wilt of corn, root rot of peas and apple scab.
- Rapidity of the secondary cycles: this approach can be use for pathogen with rapid secondary cycles and for such pathogen the prediction is based on disease severity. This approach is appropriate for diseases that have small amount of initial inoculums and many secondary cycles. Examples are potato late blight and *Cercospora* leaf spot of peanut.
- If the amount of initial inoculums is large and secondary cycles are rapid and many, then forecasts should consider both and examples of such diseases includes, beet yellows and *Cercospora* blight of celery.

The methods used in disease forecasting mostly fall into 5 major groups which are rely on:

- Weather conditions during the intercrop month particularly those affecting the survival of inoculums
- Weather conditions during the crop.
- Amount of disease in the young crop
- Number of propagules of the pathogen in the air, soil or at planting material
- Changes in host physiology

Disease forecasting model should be based on economic, environmental, and political factors and ultimately aims to reduce uncertainty about control decision. The three main objectives for developing a disease forecasting model includes:

- First objective is to increase farm income by providing more efficient allocation of disease management resource.
- To decrease the risk of large economic losses in crop value from disease and
- To decrease the amount of pesticide applied to crops, and thereby decrease the potentially harmful effects of the pesticide on the environment and human and animal health.

Advantages of forecasting of plant disease include:

- Growers and their advisers can select the most suitable crops and varieties

- Provide time for adopting appropriate management practice
- Resistant varieties can be deployed most effectively
- Lesser use of chemicals can be approached
- Cultural practise and some of instance harvesting date can be changed to decrease the lossess
- Models that accurately simulate incidence and crop loss per unit preliminary evaluations of different control strategies and cost benefit analysis.

5.2. PREDICTIVE SYSTEM

Predictive system" is used as a general term for formalized algorithms that assess disease risk factors to inform the need for crop protection.

The term 'forecast' and 'prediction' can be used as synonyms according to Webster's and Oxford dictionaries.

Both are related with something future; they make a statement *ante factum.* With respect to pathogen infection and subsequent symptoms development, such disease-forecast models are actually *post* factum.

In other words, the disease- forecasting model typically identifies when disease symptoms will appear only after an infection event has already taken place.

Predictive disease- forecasting models can be designed to predict disease development before symptoms are present, but they require knowledge of when inoculum is released and the time required for the pathogen to cause symptom development after infection.

Gent *et al.* (2013) have reviewed the use and role of predictive systems in disease management and according to them predictive systems ranges from sampling schemes (*e.g.* treat after disease detection but before a forecasted rain event) to simple categorical rules (*e.g.*, treat after co-occurrence of >2.54 mm rain and temperature >10°C) to complex, multi-criteria optimization programs based on process sub models.

However, some of them included all phases of a cropping situation integrating mechanistic simulation models, databases of information, and decision analysis rules to predict possible outcomes or specify management actions. De Wolf and Isard (2007) reported that such predictive systems quantify attributes of disease cycle, implicitly or explicitly like weather, pathogen, host, cultural practices and other crop management actions and risk expectation.

The utility of predictive systems can be assessed on the basis of three potentially overlapping aspects: conceptual utility, developmental utility, and output utility.

Conceptual utility denotes to the utility of a model as a frame of reference for thought. This is the base for lucid mental models and communicating concepts. All predictive systems should have some measure of conceptual utility.

Developmental utility refers to training of the modeler, as distinct from the disease manager, in thinking about a patho-system. This utility can be exemplified by simulation models developed as research tools and is applied in clarifying epidemic development, in testing hypotheses, in making predictions, and in shaping research directions.

Output utility is the usefulness of a model in its application by practitioners, be they farmers, their advisers, or policymakers.

Predictive systems developed to justify inputs or satisfy regulatory mandates, to prioritize resource allocation, to reduce pesticide use, or to improve disease control are examples of output utility. Output utility obviously is desired in applied research and is the primary focus of this review. Gent *et al.* (2013) have further explained its adoption criteria, constraints, risk assessment and cost-benefit ratio.

For development of any predictive model, both mechanistic and empirical approaches are being used and their comparative statements are as follows

Mechanistic Approach	***Empirical Approach***
This approach typically uses a series of sub-models to characterize the various sub-stages of pathogen life various are often based on the results of experiments designed to quantify pathogen biology under controlled situations.	With this approach, model developed through using statistical methods that describe relationships among environmental variables and increases in disease intensity.
With this approach, if knowledge about the system is not fully known then sub-models integrate statements about host-pathogen-environment	Such types of models are often rely on experiments that simulate production conditions, and may use the results of controlled experiments to direct model construction.
This approach is generally consider to have more explanatory then empirical approach	This approach do not characterize cause and effect relationships, however they can be used to make presumption about the underlying biology of a system.

Ponter *et al.* (2006) have reviewed the models and applications for risk assessment and prediction of asian soybean rust epidemics (ASR). They suggested that ASR model can be classified under two groups, simulation and empirical models. Simulation (*i.e.* process driven) models are rely on concept derived from a pathosystem and help to improve the understanding of the structure and behaviour of such biological system, on the hand, empirical models are typically constructed through the statistical relationships of explanatory variables with experimental data. Further, simulation ASR models can be separated into two groups: epidemiological and aerobiological models. Approaches, epidemic components and main estimation of 12 different ASR are as follows (Ponte *et al.*, 2006).

Approach for Model Development	*Epidemic Component*	*Main Estimation*
Mechanistic, process-driven	Epidemic and loss potential	Disease severity progress (%)
Analytic, process- driven	Establish potential	Daily number of infectious units
CLIMX	Survival potential	Disease stress-free index (0-1)
HYSPLIT	Entry potential	Spore dispersal and concentration
HYSPLIT	Entry potential	Spore availability index (0-7)
HYSPLIT	Entry and establishment potential	Spore dispersal and qualitative disease risk
Aerobiological process-driven	Entry, establishment and epidemic potential	Spore dispersal and disease development
Non-linear regression	Infection potential	Daily values of probability for infection (0-4)
Non-linear regression	Infection potential	Daily infection efficiency (0-100)
Neural networks	Epidemic potential	Apparent infection rate and disease severity progress
Linear regression	Epidemic potential	Final or maximum disease severity (%)

Crops and their disease for which prediction models have been developed are summarized as follows

Crop and Diseases	*Casual Organism*	*Models Developed By*
Canola (Sclerontinia stem rot)	*Sclerotinia sclerotiorum*	Twengstrom *et al.* (1998)
Corn (Gray leaf spot)	*Cercospora zeae-maydis*	Paul and Munkvold (2004 and 2005)
Groundnut (Earl leaf spot)	*C. arachidicola*	Phipps *et al.* (1997)
(Sclerotinia blight)	*Sclerotinia minor*	Langston *et al.* (2002)
Sugarbeet (Leaf spot)	*C. beticola*	Wolf and Verreet (2002 and 2005)
(Powdery mildew)	*Erysiphe betae*	Wolf (2002)
Wheat		
Fusarium head blight	*Fusarium graminearum*	Del Ponte *et al.* (2005); Moschini *et al.* (2004)

Table (*Contd...*)

Table (*Contd...*)

Crop and Diseases	*Casual Organism*	*Models Developed By*
Leaf rust	*Puccinia triticina*	Audsley *et al.* (2005)
Powdery mildew	*Bulmeria graminis*	Audsley *et al.* (2005)
Stripe rust	*Septoria tritici*	Heger *et al.* (2003)
Take all	*Gaeumannomyces graminis*	Roget (2001)
Apple		
Apple scab	*Venturia inaequalis*	Turechek *et al.* (2005)
Fire blight	*Erwinia amylovara*	Shtienberg *et al.* (2003)
Powdery mildew	*Podosphaera leucotricha*	Berrie and Xu (2003)
Grapes (Downy mildew)	*Plasmopara viticola*	Orlandini *et al.* (2003)
Peach (Leaf curl)	*Taphrina deformans*	Giosue *et al.* (2000)
Pear (Brown spot)	*Stemphylium vesicarium*	Llorente *et al.* (2000)
Lettuce (Downy mildew)	*Bremia lactucae*	Wu *et al.* (2002)
Onion (Downy mildew)	*Peronospora destructor*	Gilles *et al.* (2004)
Potato (Late blight)	*Phytophthora infestans*	Andrade-Piedra *et al.* (2005a,b)
Tomato (Early blight)	*Alternaria solani*	Batista *et al.* (2006)
Rice (Blast)	*Magnaporthe grisea*	Kaundal *et al.* (2006
Rice (Blast) and scab and rust of pear	Various	Kang *et al.* (2010)

5.3. DISEASE PRESSURE INDEX

Disease Pressure Index (DPI) have four basic modules (Holtslag *et al.*, 2004)

- An assessment of inoculum potential and production
- A signal for inoculums release
- A calculation of disease development based on the interaction between leaf-wetness duration and temperature (*i.e.,* environmental parameters) and
- An estimation of host susceptibility

DPI can be quantified through $Y = A[BC]D$, where, A is an estimate of the amount of inoculums, B signifies whether the inoculums is being released, C represent a measure of infection success based on environmental parameter and D is an assessment of host susceptibility

5.4. INFORMATION'S NEEDED FOR DISEASE FORECASTING

Forecasting diseases is a part of applied epidemiology. Hence, knowledge of epidemiology (development of disease under the influence of factors associated with the host, pathogen) is necessary for accurate forecasting.

The factors of epidemic and its components should be known in advance before forecasting is done. A disease forecasting model should be a quantitative description of the relationship between host, pathogen and environment and should provide a foundation for calculating whether or not these applications of a particular pesticide is justifiable.

The information's required for forecasting are:

Host Factors

Prevalence of susceptible varieties in the given locality

- Response of host at different stages of the growth to the activity of pathogen *e.g.*, some diseases are found during seedling stages while others attack grown up plants and
- Density and distribution of the host in a given locality. Dense populations of susceptible variety invite quick spread of an epidemic. Growing susceptible varieties in scattered locations and that too in a limited area are less prone to epiphytotic.
- Hosts that have high levels of disease resistance may show few to no visible symptoms, while highly susceptible host may completely succumb to the pathogen.
- Young plant tissue is often more susceptible to disease than older fully expanded leaves.

Pathogen Factors

- Amount of primary (initial) inoculum in the air, soil or planting material
- Dispersal of inoculum
- Spore germination
- Infection
- Incubation period
- Sporulation on the infected host
- Re-dispersal / Dissemination of spores
- Inoculum potential and density in the seed, soil and air

Environmental Factors

- Temperature
- Humidity
- Light intensity
- Wind velocity

The number of secondary cycles a pathogen can produce in a growing season, and the environmental conditions that affect each cycle are important factors to consider in disease- forecasting dynamics. For example, a forecast model for the polycyclic pathogen *Uncinula necator* (Causes powdery mildew on grapes) should incorporate the rapidity of secondary cycles or the amount of secondary inoculums being produced.

5.5. ATTRIBUTES OF FORECASTING MODELS

Major attributes related with a successful forecasting model reflects with its simplicity, usefulness, reliability, availability, multi-purpose applicability, significance and cost effectiveness

There are many different types of forecasting models that can be categorized under the multivariate time series models, mechanistic model, phenology and visual models.

Models which are based on similar environmental parameters as required for plant growth, can be used to predict fungal inoculums dissemination, germination, infection success and disease development

A disease-forecast model can be empirical or mechanistic in nature. First one describe the relationships between variables without referring to any underlying biological or physical structure that may exist between the variables. Multiple-regression model is one of them.

On the other hand, mechanistic models attempt to clearly represent causality between all of the interacting environmental variable.

Computers and real-time weather stations that are located in the crop canopy are essential tools for constructing dynamic disease forecasting models *e.g.,* BLITECAST

Disease forecast model using disease symptoms as a basic component are rare, because in many cases it is too late to intervene when sign of the pathogen are already visible.

When crop phenology is used as the only input, data are used to suggest the start of fungicide treatment when a specific plant development stage is reached.

Forecasting models can rely on quantification of inoculums production, and its efficacy.

Model efficiency can be improved through incorporation of time when inoculums is released in relation to environmental factors. (*e.g.,* temperature, humidity, leaf wetness period and light.

Inoculum presence and its quantification are the key requisite for most disease forecasting, thus, most of these methods are based on some extent on the assessment of either primary or secondary inoculums. Based on disease cycles forecasting methods can be categories into source and amount inoculums, survival of inoculums and development of inoculums.

An effective forecast combines two or more of these methods. For example, in some areas spore traps are used to detect the initial availability of apple scab inoculums (ascospores). This establishes the starting point of the plant-host timing window. Then, of period of wetness and temperature are used to detect infection periods and determine how long control measures are needed.

5.6. PATHOGEN DISEASE CYCLE AND THEIR ROLE IN FORECASTING

Pathogen disease cycle indicates pathogen biology as a series of interconnected stages of development including dormancy, reproduction, dispersal, and pathogensis. Abstractly, these stages can be divided apparently into infinite numbers of events with each having a unique relationship with the physical environment and host.

De Wolf and Isard (2007) have reviewed components of the disease cycle that are included in disease prediction models. According to them dormancy stage often divided into colonization and survival components, dispersal partitioned into released, transport, survival and deposition and similarly pathogenesis divided into infection, incubation, latency and senescence sub-stages.

When modelling diseases caused by pathogen that survive locally within debris of a previous crop, it may be likely to assume that inoculums is not restrictive and characterize only the pathogenesis phase in the disease prediction model, in other conditions, it may be rational to presume that infection is seldom limited by the local environment. Thus, useful predictions of disease intensity may be based on circumstances that favour either pathogen reproduction or the dispersal process. However, in most cases, models that provide useful information for plant disease management include host-pathogen-environment relationships for multiple components of the disease cycle.

Basically, models for a specified disease system frequently addressed similar stages and sub-stage of the disease cycle. For example, timing of initial inoculums, production, spore survival during dispersal and host infection are generally incorporated for downy mildew of grape cause by *Plasmopara viticola* (Orlandini *et al.*, 2003), while Rossi *et al.* (2005) have incorporated rate of primary inoculums production and information about the dispersal components of the disease cycle.

On the hand, Rouzet and Jacquin (2003) have used a different approach with incorporating conditions that supports the colonization of host tissues at the end of the previous growing season and oospores maturation in spring.

Among other lifecycle events Roget (2001) developed prediction model for take-all of wheat with colonization and survival, his model predicts intensity of take-all in wheat based on detection of DNA of *Gaeumannomyces graminis* and quantifies the survival of pathogen as function of rainfall and management practices. This approach is more appropriate for pathogen that overwinter sporadically or if spatial heterogeneity of survival within a region influences the presence of inoculums.

Rossi *et al.* (2005) simulate the predictive model with help of maturation of oospores (*Plasmopara viticola*) as a function of temperature and provides a mechanism to limit inoculums development if the moisture of leaf litter is too low or if there is no rain.

Manter *et al.* (2005) predictive model on Swiss needle cast of Douglas fir caused by *Phaeocryptopus gaeumannii* used winter temperature to quantify ascocarp development of pathogen.

Environmental variables that triggers the spore release have also been utilize, for example Xu and Butt (1994) for *Nectria galligena* (causes European apple canker) in which pathogen activities like production of active sporodochia and conidal released related with threshold of rainfall.

Other models related with this event of disease cycle concluded that temperature and relative humidity effects on survival of sporangia (Powdery mildew of grape) and the reduced risk of infection associated with unfavourable conditions for spore survival.

Conclusively, the accuracy of plant disease prediction models can be evaluated by different ways depending on the approach to model development. Bayesian decision theory (described in chapter 3) offer unique means of quantifying the accuracy of prediction models and evaluating their use in plant disease management.

5.7. MATHEMATICAL MODELS ANDTHEIR ROLE IN DISEASE FORECASTING

Multiple regression model and non-linear model are the basic mathematic approaches for disease forecasting. Multiple linear regressions use multiple predictor variables such as mean leaf wetness period, temperature, humidity etc. While, in non-linear model potential predictors tend to be constant and stay within the disease-conducive range. The relationship among precipitation, temperature, and disease severity may take many form. Usually, the best option is to model the rate of change in disease severity or pathogen population as a function of weather variable. For such tool logistic growth model are utilize with following equation with parameters like D_t (percentage infection at time t and R (growth rate)

$$D_{t+1} = D_1 + R * D_t * [(1-D_t)/100]$$

5.8. METHODS OF PLANT DISEASE FORECASTING

Plant disease forecasting allows the prediction of probable outbreaks or increases in intensity of disease and, therefore, allows us to determine whether, when and where a particular management practice should be applied.

There are many ways to categorize various methods related with disease forecasting. It may classify weather based predictions, inoculums based prediction and host-physiology based prediction.

5.8.1. Weather-Based Predictions

Weather forecasting is application of science and technology to foresee the atmospheric conditions of a particular location. Forecasting based on weather data are utilize sophisticated automated weather station, web based internet system, wireless communication infrastructure, high resolution camera etc. (Tilva *et al.*, 2013).

Most forecasting based on this approach utilize temperature, relative humidity, precipitation rate, wind speed, solar radiation and leaf wetness duration (LWD) parameters (Magarey *et al.*, 2007).

LWD defines as the period of time during which free water is present on the outer surfaces of crop plants. It is reliant on the surfaces properties as well as the atmospheric conditions and its incidence is linked to with the amount of dew, rainfall, fog and irrigation.

Among many parameters LWD and air temperature are the most important micro-metrological parameters influencing the development of many foliar and fruit diseases (Gillespie and Senelhas, 2008).

LWD influences epidemiological episodes affecting infection and sporulation processes on several foliar plant pathogenic fungi and favors any plant pathogen whose spores or cells require a water film to germinate or divide.

Some fungal pathogens require continuous wetness episodes lasting several days to sporulate, although sporulation can occur during a succession of short wet periods interrupted by dry intervals. Bacteria require a water film to increase on the foliage portions of their host plants.

Relative humidity can be physically well-defined, whereas leaf wetness duration (LWD) is a parameter difficult to measure or approximation because different various portions of leaves and canopies wet and dry at different times (Huber and Gillespie, 1992). In contrast to air temperature, LWD is a hard variable to measure or to estimate because it is determined by both atmospheric conditions and their interactions with the structure and composition of the vegetative community.

In groundnut the severity and further spread of tikka disease *Cercospora personata* and *C. arachidicola* depend upon the prevalence of high humidity and low temperature during crop season.

Cercospora leaf spot of peanut rapidly develop when the foliage remain wet for a period greater than 10 hours or above and minimum temperature is 21°C or higher for next 24 hours.

For postbloom fruit drop, caused by *Colletotrichum acutatum* amount of inoculum, the total rainfall and LWD for the last 5 days (Peres *et al.*, 2009) are important predisposing features.

Padmanabhan (1965) used weather parameters for predicting rice blast caused by *Pyricularia oryzae*, he reported that night temperature ranging from 20-24°C with combination of 90% relative humidity or above lasting for a week during susceptible phase of plant accelerate blast epidemic.

According to Nagarajan *et al.* (1980) epidemic of leaf rust (*Puccinia recondita*) of wheat is likely to develop if (a) rust infections are observed in Uttar Pradesh and Bihar between 15th and 20th January and 5-6 different locations separated from each other by a minimum of 25 ± 5 km (b) the number of rainy days during January to mid April over northwest India is at least twice the normal that is 12 days and (c) the weekly mean maximum temperature during March to mid April over northwest India, is within 1^{0}C of the normal temperature (26°C). If the first two conditions exist fully, then a severe epidemic of leaf rust is likely to occur.

Saxena and Lal (1985) have developed a linear equation based on sunshine hours and rainfall for prediction of *Erwinia* stalk of rot of maize.

In central India, groundnut rust disease more severe during rainy days.

A forecasting system has also been developed for downy mildew (*Plasmopara viticola*) on grapes based on the 10: 10:24 rule. In other words, a temperature of at least 10°C while at least 10 mm of rain falls in a 24 hour period on wet, chemically unprotected foliage will result in disease development.

Outbreaks of the Downy mildew (*Peronospora destructor*) on onion are more from July to mid-September (winter to early spring) when pre-dawn relative humidity is greater than 95% and, most critically, dew remains on plants until 10 a.m.

An early forecasting system for late blight of potato in the Netherlands relied Upon the so-called 'Four Dutch Rules' (Everdingen method) which combined the following criteria for temperature, leaf wetness, cloudiness and rainfall: (A) Night temperatures must be no lower than 10ºC. (B) Leaves must remain wet for at least 4 hours on a day following night dew, (C) the day following a night dew must remain cloudy with no less than 80% of the sky covered by cloud and (D). During periods of the above weather patterns there must be at least 0.1 mm of rain. Forecasting schemes developed from this early Dutch system relied on the more detailed prediction of a 'critical period' of weather conducive to sporulation, dispersal and infection by the pathogen. The following weather variables were monitored: (A) the saturation deficit between 3 and 6 a.m., which influences the likelihood of dew, which in turn determines the amount of sporulation; (B) the duration of bright sunshine before noon, which determines the maturation of the newly formed spores and (C) the amount of precipitation (including rainfall, drizzle and fog) between 9 a.m. and 6 p.m., which determines the amount of infection of the crop. Thus, the weather pattern leading to infection was a leaf wetness period of at least 13 hours with a total duration of daily sunshine of less than 2 hours. Symptoms appeared 3 or 4 days after such a weather pattern.

Other systems for late blight of potato are known as Wallin and Hoymon's method, Cook System and Hyre and Bonde System. These systems rely on BLITECAST which was developed by Krause and Massie (1975).

BLITECAST represent the overall favorableness of environmental conditions and control applications usually are determined when the accumulation of severity value reaches some critical level. BLITECAST combined two older late blight forecasting techniques, the first moles in BLITECAST is the 'favorable day' which is used to indicate when first spray should be applied. A favorable day is one in which the mean temperatures of the previous 5 days was less than 256ºC and total rain for 10 predicting days was 3 centimetres. Disease is forecast to occur within 1-2 weeks after occurrence of 10 consecutive favorable days. The second models incorporated in to BLITECAS T is relay on humidity and temperature and in this model 'severity value' are dispensed to different combination of relative humidity (>90%) and temperature for different period of time. Using this model disease forecast to occur 1-2 weeks after 18 severity values have accrued. Subsequent sprays are recommended after each accumulation of 3 additional severity values.

Kang *et al.* (2010) have developed web-based information system for plant disease forecast based on weather data at high spatial resolution.

Hyre system	*Wallin and Hoymon method*	*Cook system*	*Hyre and Bonde method*
Based on • Daily rainfall • Maximum and minimum temperature • Data not collect from within canopy	Based on • Relative Humidity and temperature • Based on accumulation of severity value. • Severity value show relationships between humidity and temperature	Based on • Average rainfall, and • Temperature for 7 days	Based on • Total rainfall and • Temperature for 10 days

Their system consists of four sub-systems, such as weather data acquisition system (WDAS), job process system (JPS), data storage system (DSS), and web service system (WSS) (Fig. 1). Weather elements used by the forecasting system are hourly data on air temperature (^{o}C), relative humidity (%), and rainfall (mm) Fig. 2.

Weather parameters affects the crops differently during various phenophases of its development. Crop –weather models can be categorized into three basic sorts. (A) crop growth simulation models, (B) Crop-weather analysis and (C) empirical statistical models.

Crop-growth simulation model	*Crop-weather analysis models*	*Empirical statistical model*
This is a simplified depiction of the chemical, physical and physiological means underlying plant growth courses. Simulation can be generated through proper understanding of processes-production and modules biomass and their relationships with water factors. This approach provides a free hand from differentiation between climatic regions. In general humid climate with low temperature and radiation levels, the model will basically present the high response to yields to increase in total radiation received. On the other hand in an arid and hot climate, total precipitation is the most important factor	These are product of two or more factors, each one of them representing the functional relationship between a peculiar plant response and the variations in selected attributes at different phenophases. Such models not require a hypothesis generation about basic plant and environmental process, which indicates that input requirement are less but the output information is more dependent on the input data and less detailed than in the case of simulation models. Basic statistical tools like regression analysis, covariance test, correlation coefficient are utilized.	With this approach, one or many variables (weather or climate, soil or a time trend) are correlate with crop responses like yield. Weighting coefficient in these equations are by necessity obtained in an empirical manner using statistical approaches like multivariable regression analysis. The independent variables in these models include weather, agrometeorological, soil or some indices. Water Requirement Satisfaction Index (WRSI), Thermal Interception Rate Index (TIR) and Growing Degree Days (GDD) are some agro-climatic indices used in the models

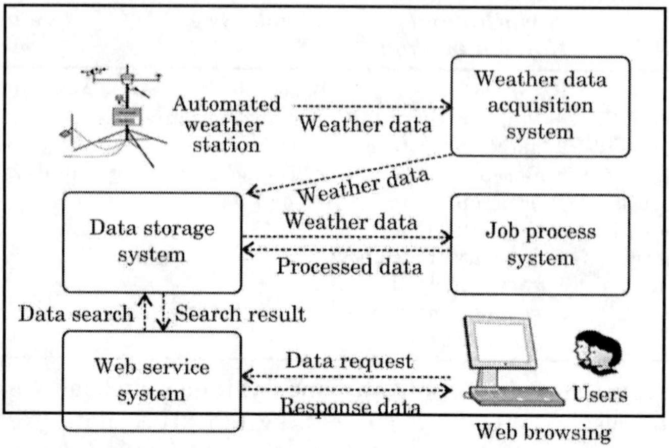

Fig. 1: Structure diagram of the overall disease forecasting system (Reproduce with kind permission of Wee Soo Kang (original author).

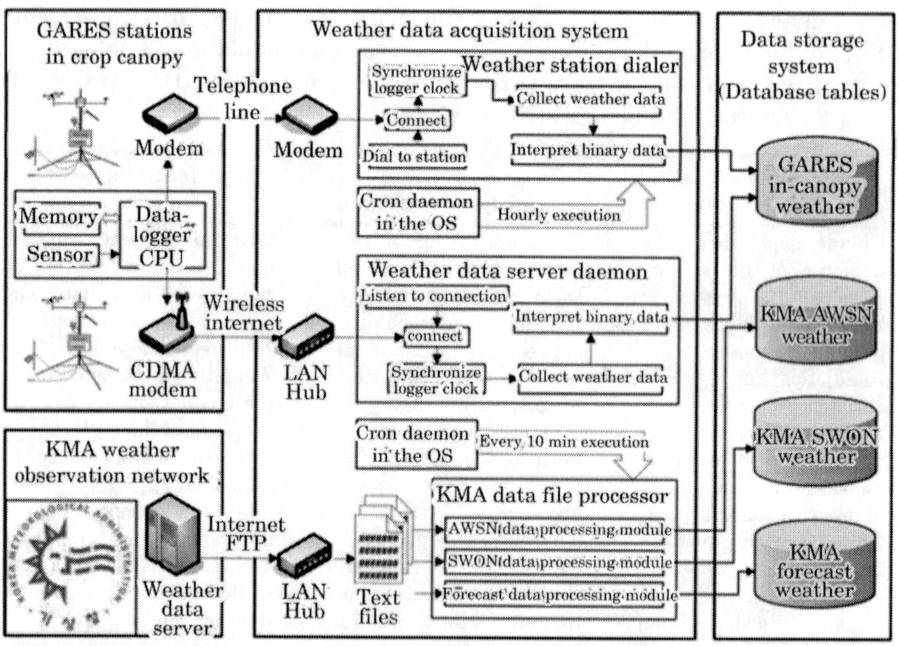

Fig. 2: Structure diagram of the weather data acquisition system (WDAS).

5.8.2. Forecasting Based on Inoculum

Inoculum may be any part of pathogen that come in contact with the host and this contact phenomena known as inoculation. Inoculum are either primary or secondary inoculums.

Primary Inoculum	*Secondary Inoculum*
These are the survival tools	They develop through primary inoculums
Survive during overwinter/over summer and caused primary infection	They generally spread the disease
Generally they are sexual spores	They are generally asexual spores

Inoculum potential refers to amount of energy available at host surface for infection, *i.e.,* Inoculum potential denotes inoculum density x inoculum capacity. This variable is largely used in epidemic and forecasting of plant disease. Basically non-specialized pathogen (saprophytes) produces high inoclums, while, specialized (obligate) pathogen produces low inoculums.

Some disease forecasting methods are based exclusively on monitoring the level of inoculum, often in the form of the amount of disease already present. This approach is improbable to be effective for diseases such as potato late blight and grape downy mildew that can develop explosively under favorable weather conditions.

By the time predictions based on disease progress are made, it is likely to be too late to act effectively. However, methods based on inoculums monitoring can be effective when disease is developing steadily under relatively uniform or expected weather conditions and the aim is to predict whether the disease is likely to progress above a certain threshold at which control measures become economic. For example, outbreaks of rice panicle blast caused by *Pyricularia oryzae* are predicted by monitoring the numbers of spikelets infected throughout inflorescence development. With powdery mildew (*Uncinula necator*) on grapevines, potential yield loss is related to when the disease first appears in vineyards. Where there is a high risk of disease carrying over from the previous season, three sprays early in the season (2, 4 and 6 weeks after budburst) are recommended, followed by further checks for disease and the application of additional fungicide sprays if necessary.

Inoculum for various types of pathogen can be quantifies through spore trap, soil sieving, fungal sclerotia, plate count on selective media for bacteria and fungi and by serological tests.

Verticillium dahliae can be quantified directly from the soil through dispersing a small amount of soil on non-nutrient media like water agar having low concentration of ethanol. This certain type of media restricts the amount of myecial growth but provide sufficient carbon to allow production of microsclerotia, thus, giving a direct measure of the inoculums potential in soil.

Spore trap are an effective measure tool for specialized structure spores like *Alternaria, Fusarium, Helminthosporium, Venturia,* smut and rust.

Trapping urediniospores of wheat stem rust on Vaseline-coated slides or glass rods is an useful approach. The Burkard spore trap, sticky slides, Petri-dish containing specific media and trapping of vectors of viruses are useful for predicting the occurrence of disease at a particular location.

Indexing is a useful method particularly for identification of viruses, in this approach a specific host plant is utilize for identification of specific virus and that viruses produces peculiar symptoms on that host. Examples of some indexing plants for their respective viruses are as follows

Index Host	***Virus***
Phalaris floridiana	TMV
Pisum sativum	NEPO (Nematode Transmitted Polyhedral virus)
Brassica pekinensis	Brassica virus
Citrus trieste	Kagzi lemon
Lycopersicon esculentum	PSTV

Forecasting method based on amount of inoculums can be partitioned into (A) based on amount of initial inoculums, (B) based on amount of secondary inoculums, and (C) based on both primary and secondary inoculums. This classification is elaborate in following Fig. 3.

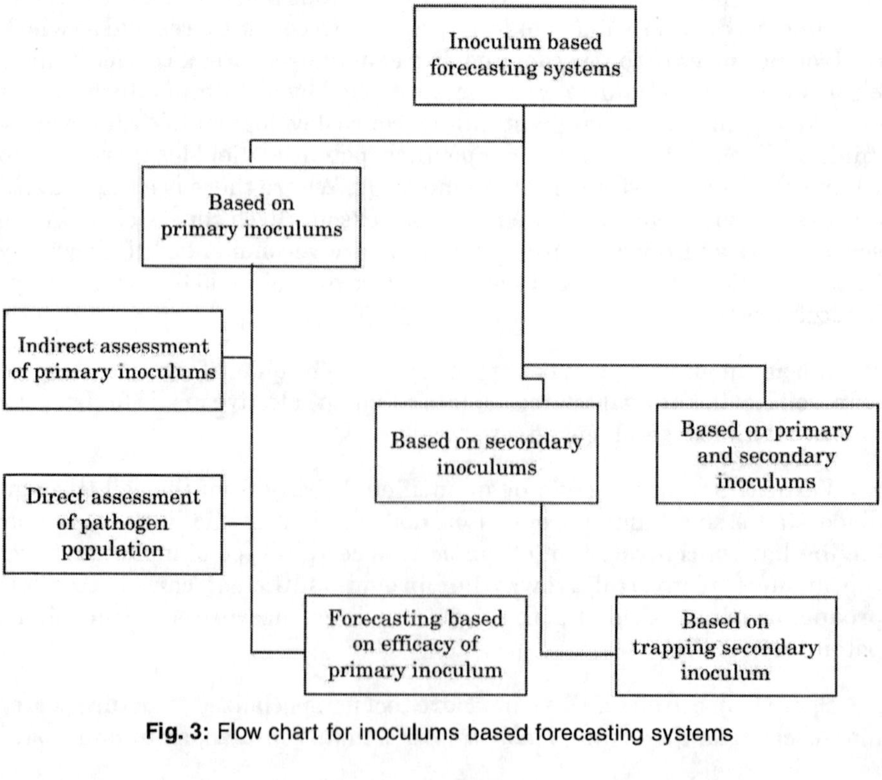

Fig. 3: Flow chart for inoculums based forecasting systems

5.8.2.1. *Forecasting based on primary inoculum*

Most monocyclic diseases like root rot of pea, stewart's wilt of corn and few polycyclic diseases that have large amount of initial inoculums like apple scab, disease development can be predicted by assessing the amount of initial inoculums.

Presence of primary inoculum, its density and viability are determined in the air, soil or planting material. Occurrence of viable spores or propagules in the air can be assessed by using different air trapping devices (spore traps). In the case of soil-borne diseases the primary inoculum in the soil can be determined by monoculture method.

Presence of loose smut of wheat, ergot of pearl millet and viral diseases of potato can be detected in the seed lots at random by different seed testing methods. Seed testing methods can be used to determine potential disease incidence and enable decision to be made on the need for chemical seed treatment. The extent of many virus diseases is dependent on the severity of the preceding winter which affects the size of vector population in the growing season. *e.g.*, Sugarbeet yellows virus.

5.8.2.1.1. *Indirect assessment of initial inoculum*

5.8.2.1.1.1. Stewart's Wilt of corn

This disease is caused by bacterium *Erwinia stewartii* which survives off season in the body of the vector (corn flea beetle). The bacterium is introduced into leaf tissue as the beetle feeds. Thus, knowing the number of vectors surviving previous seasons would give an idea of the density of initial inoculums available in the next season.

The disease would be so severe if the average monthly temperature for December-January-February greater than or equal to +7°C, however the disease would be mild if the average temperate is less than –1.1°C.

Basically such low temperature reduced the population of overwintering beetles

5.8.2.1.1.2. Pea root rot

Caused by *Aphanomyces euteiches* and for this disease initial inoculums determines severity because inoculums is generally not disseminated during the growing season.

Thus, quantification of initial inoculums is appropriate bases for disease forecasts.

For this, susceptible plants are planted in green house taking soil from infected fieled. Fieled with soils which permitted severe root rot in controlled conditions should not be planted to pea, whereas the field which permitted little root rot could use pea plantation.

Initial inoculums can also be accessed directly by isolating the sclerotia.

5.8.2.1.1.3. Fire blight of apple and pear

Caused by *Erwinia amylovora* and indirect assessment of pathogen population is useful for this disease.

Blossoms are the most susceptible tissue and this bacteria need daily average temperature more than 16.7°C in March, 15.6°C in April or 14.4°C in May

Below 15°C bacterium grows much less rapidly than it does at temperature 17°C or above.

Thus, the forecast reflects very closely the influence of temperature on pathogen growth

5.8.2.1.2. *Direct assessment of initial inoculum*

Forecasts for disease initiated by polycylic pathogen can also be developed by measuring pathogen population.

Many soil-born pathogen are described by these characterstics.

Assessment about these pathogen should be made before planting the crop.

This approach is most restricted for soil-borne plant pathogen.

5.8.2.1.2.1. Root and crown rot and damping off

Pathogen *Scleotium rolfsii* induces root rot and crown rot (Southern blight) and damping off many plants.

Pathogen produces sclerotia usually 1-3 mm in diameter which can be detect through soil sieving.

Viability of these sclerotia can be determined by allowing them to germinate on soil surface and also alternatively, most soil incubated in shallow trays stimulated growth and subsequent detection of mycelia from viable sclerotia.

The effects of *S. rolfsii* on yield of some crops are partially known and for example, under field conditions one viable sclerotium per 500 gram soil, can infect about 1% of the sugar beets

Assessment of population of this pathogen would enable farmers to detect a potentially damaging population. The grower could avoid planting sugar beets in heavily infested fields, and thus avoid severe losses.

5.8.2.1.3. *Forecasting based on efficacy of primary inoculum*

Some of initial pathogen populations are routinely very large but their further disease severity still varies from season to season.

Thus, there is no direct relation between the size of the initial population and subsequent disease.

Fluctuation in environmental conditions affects the pathogenesis, and forecast based on environmental measurement predict the efficacy of primary inoculums.

5.8.2.1.3.1. Apple scab

This disease is caused by fungal pathogen *Venturia inaequalis*

It is a polycyclic pathogen, but the initial inoculums is generally present in large amount and there are only few secondary cycle of pathogenesis.

Thus the amount of initial inoculums is very important for epidemic development and its reliable forecasting.

Inoculum consists of ascospores produce in perithecia in overwintered leaves on the orchard floor.

Mature ascospores are forcibly discharged after the perithecia have become wetted.

Leaves and fruts remain susceptible to this pathogen for only a limited period of time, thus only juvenile and recently mature tissue must be protected.

Perithecia mature during a rather long period; therefore, ascospores are available for 1-2 months during the period of tree growth.

Conidia from the primary lesions servers as inoculums for fruit infection

The apple scab forecast system identified periods, known as "Infection period", during which environmental conditions favour inoculation and penetration by the pathogen.

The forecast assumes that initial inoculums is abundant and then predicts, on the basis of leaf wetness and temperature, as slight, moderate or severe disease

Free water is necessary for germination of the ascospores and penetration of tissue, and the rate that these processes are accomplished is inclined by temperature

Thus, severity of apple scab is related to amount of perithecia surviving in dead leaves and the subsequent amount of ascospores liberated from them.

The discharge of ascospore takes place when perithecia are thoroughly wetted and the infection take place 30 hours after leaf wetness at 5°C, 14 hours at 10°C or 9.5. hours 15°C.

Relationships between minimum temperature for infection at different wetness durations are as follows

Wetness duration (hrs)	41	35	30	28	21	18	15	13	12	11	9	8	7	6	5
Temperature (^0C)	1	2	3	4	5	6	7	8	9	10	11	12-13	14-15	16-24	25

There is a method named potential ascospore dose or 'PAD', that estimates the quantity of ascospores based on a sampling of the leaves in late autumn, but before the leaves fall. The estimation predicts the next spring overall "scab-risk" as high or low.

- Randomly select the shoots within the orchard block, select both clusters and terminals shoots, all from parts of the tree canopy (high, low, exterior, and interior). If sucker shoots are present, make sure to select some sucker shoots.
- If the number of scabbed leaves is below the threshold (green zone), a full fungicide program schedule may not be necessary the following spring. Low risk infections (short wetness period, low amount of ascospores or new leaves) may be left uncontrolled especially at the beginning and the end of the pi-mary infection period.
- If the number of scabbed leaves is above the threshold (red zone), a full fungicide program will be needed the following spring. All primary infections should be controlled
- If the number of scabbed leaves is in the intermediate zone (orange zone), sampling must continue with another series of shoots before a decision can be made.

Mills and La Plante (1954) developed a Mills' Chart according to which the primary infection could occur within a temperature range of 0 to 28°C and leaf wetness for 9 to 48 hours. A Mills' period was defined as a period of leaf wetness required at a particular temperature for causing light, medium or heavy scab infection on the leaves.

By combining the temperature and duration of leaf wetness for varying periods, they suggested that combination of 30, 20, and 14 and 10 hours of leaf wetness, respectively with a mean temperature of that period of 5.6, 7.2, 10 and 15°C satisfied the requirements for successful infection among others. It also indicated the expected severity of disease along with incubation period. Later, Smith (1961) utilized a threshold level of 90% relative humidity following rain in place of leaf wetness record.

Jones *et al.* (1981) and Ellis *et al.* (1984) devised computer based apple scab predictor, which utilized special purpose micro-processor for analyzing detailed information on various relevant components.

5.8.2.2. *Forecasting based on secondary inoculum*

Forecasting based on secondary inoculums amount are useful specifically for those pathogen that have low level of primary inoculums, but they have the potential for many secondary cycle.

Examples of such types of disease include late blight of potato, *Cercospora* leaf spot of peanuts, early blight of celery and downy mildew diseases.

Specifically for late blight of potato and for *Cercospora* leaf spot weather conditions that favours the secondary cycles is generally being utilized.

5.8.2.2.1. *Leaf spot of peanuts*

Leaf spot of peanuts caused by two species of *Cercospora.*

Rapid epidemics are associated with periods of high relative humidity (>95%) for more than 10 hours.

This relationship of temperature and duration of humidity can be estimated through series of graphs.

Collected information's processed through computer programme which generate a daily peanut leaf-spot spray advisory.

5.8.2.2.2. *Early blight of celery*

For this disease spore trapping of *Cercospora apii* has been useful in timing of fungicide sprays to control it.

The forecast related the number of spores detected in a spore trap to probable disease intensity and spray frequency is recommended.

0-100 spores requires one fungicide spray, 100-300 requires two, 300-500 requires 3, and > 500 spore count/day requires 3-7 spray.

5.8.2.3. *Forecasting based on primary and secondary inoculums*

Forecasting of polycyclic disease would be more meaningful if they are based on primary inoculums as well as on the spread of their subsequent secondary cycles of pathogenesis.

Example of disease for which this approach has been utilized are apple scab, black rot of grape, blister blight of tea, cereal rust and sugar beet yellow.

5.8.2.3.1. *Blister blight of tea*

Disease caused by *Exobasidium vexans*.

Number of basidiospores in air at a given time related to percentage of infected shoot.

The duration of sunshine is negatively correlated with the duration of surface wetness, therefore duration of sunshine used as a measure of surface wetness indirectly.

Moderate disease will appear with 83% relative humidity for 10-14 days

While, severe disease will appear with 83% relative humidity for 20-24 days.

5.8.2.3.2. *Sugar beet yellow*

Caused by Beet Yellow Virus (BYV) and Beet Mild Yellowing Virus (BMYV)

Disease is sporadic in nature and severity is strongly influenced by vector population.

Vectors *Myzus persicae* and *Aphis fabae* transmit BYV virus into semi-persistent manner, while, *Myzus persicae* transmit BMYV in persistent manner

Yellow incidence is usually higher in those years when aphid infestations of the crop develop early than when infestations develop late.

The first part of the forecasting relay on the knowledge of secondary cycles that resulted from transmission by aphids.

In such case aphids population during May to July period is very important, while vector population during month of August have little effect of yield.

Next winter and early spring temperatures are monitored. Above average temperatures during late winter and early spring is generally associated with high incidence of this disease. Low temperature reduces the disease intensity

It is believed that low temperature reduces overwintering population of vector.

Low temperatures describes in to two measurements, first during month of January to March when minimum temperatures <–0.3°C and the second measurement was the mean temperature of April.

If there were relatively few frost days and April was relatively warm, then the overwintering vector population was likely higher, than normal and consequently disease will also more pronounce.

Farmers can use the winter and April temperatures to get an early prognosis of aphid population and yellows severity later in the year.

5.9. FORECASTING WORKS IN INDIA

Nagarajan and Singh (1975) provided the "Indian Stem Rust Rules" to explain the south to north transportation and deposition of stem rust spores along with rain and consequently formulated set of three upper air rules. These rules explain satisfactorily whether the primary inoculums from South India would travel tens of hundreds of kilometres before its deposition in central India. The stem rust prediction methods proposed by them consist of the following steps. A. Check if there were any rains during November in central India together with southerly wind. B. Check the urediniospore content of the rain. C. Check if the Indian stem rules are satisfied. If a tropical cyclone occurred during November disease would appear near the area where cyclone dissipated. If the inoculums arrive early the disease can develop into severe proportions. If rainfall satisfying the weather conditions of rule b or c occurred during January or February, only traces of the disease would appear. D. Check the weather over the area of urediniospore deposition. E. Check the viability of the transported spores, that is if the trajectory for 700-850 k bar passed near the south foci within 120 hours, spores would be viable and F. Check if the ground level conditions following the deposition by rain, are favourable for infection or not. If all conditions are satisfied, the first prominent infection could occur in 25-30 days at location closed to where the cyclone dissipated. They further suggested that once infections have occurred in central India, their further spread can be predicted following the same rules with the removal of steps A and C.

He also developed forecasting system for leaf rust and according to Nagarajan *et al.* (1980) epidemic of leaf rust (*Puccinia recondite*) of wheat is likely to develop if (a) rust infections are observed in Uttar Pradesh and Bihar between 15th and 20th January and 5-6 different locations separated from each other by a minimum of 25 ± 5 km (b) the number of rainy days during January to mid April over northwest India is at least twice the normal that is 12 days and (c) the weekly mean maximum temperature during March to mid April over northwest India, is within 1°C of the normal temperature (26°C). If the first two conditions exist fully, then a severe epidemic of leaf rust is likely to occur.

He along with Singh (1973) and with Joshi (1980 and 1985) monitored the satellite data particularly, related with tropical cyclone and their studies revealed that Nilgiris and Palani hills are the main source for the spread of stem rust urediniospores. They further summarized that though stem rust can survive in the northern hills during all the summer months, however, due to prevalence of low temperature Himalayas have little role to play in the recurrence of stem rust epiphytotics in northern India. The stem rust urediniospores that are available in South Indian hills get carried to north by tropical cyclones and are deposited along with rains over the central India where the infections are established on one months old wheat crop. They provided that subsequent spread of stem rust from central India to other parts of Northern India is favoured by 'Western Disturbances' linking both area. The associated winter rainfall permits nationwide appearance and built up of stem rust. Set of these three weather situation (cyclone from the Bay of Bengal, western disturbance and winter rainfall) which favours north-west movement of stem rust urdinospore also known as the "Puccinia Path".

Bhattacharya *et al.* (1982) had developed methods/rules for late blight of potato under the Indian conditions. According to them, under hill conditions, rainfall is the primary factor for the pathogen generation. They correlate the weather data and disease establishment at Shimla, Ootacamund and Shillong area and reported that (A) if the day moving precipitation (30 mm for Shimla, 28.9 mm for Ootacamund and 38.5 for Shillong) associated with a mean of temperature 23.9°C or less continues for 7 consecutive days the disease will accumulate within three weeks and (B) If hourly temperature within 18-20°C and are associated with high relative humidity (80% or more) for 18 continuous hours for 2 consecutive days, the disease will appear within a week. According to their criteria's a warning service for spraying of fungicides on potato crop against late blight of potato was initiated n the Northern India hills and Nilgiris.

Murlidharan and Rao (1981) had developed a rice blast warming with the help of trap plants.

Manibushanroa and Krishnan (1991) developed simulation model EPIBLA for forecasting rice blast in india.

Shukla and Anjaneyulu (1992) correlate the rice tungro development with vector population.

Bedi and Saini (1979) showed that the severity of the early blight of tomato (*Alternaria solani*) positively correlate with concentration of inoculums sprayed on the crop.

Padmanabhan (1963) studied the meteorological factors associated with the Bengal famine

Adiver *et al.* (1998) developed forecasting system for foliar disease (leaf spot and rust) of groundnut. They suggested that foliar disease increased with lowering of maximum temperature and increase of relative humidity.

Singh *et al.* (2000) developed a computerized forecast model (JHULSACAST) for western Uttar Pradesh using hourly temperature, relative humidity (%) and daily precipitation (mm) as input parameters. However, this model is location specific and needs to be calibrated for different agro-climatic conditions. Since then the model has been calibrated for predicting late blight appearance in Punjab (Arora *et al.*, 2012) for Tarai region of Uttarakhand (Pundhir *et al.*, 2014) and Chakraborty *et al.* (2015) for Plains of West Bengal. Chakraborty *et al.* (2015) have validated and concluded that modified version of this model could predict the late blight within 14 days with an accuracy of 100% if 7-days moving relative humidity (c 90%) period for ≥105 hours coupled with 7 days moving congenial temperate (7.2 to 26.6°C) for ≥150 hours prevail for 7 consecutive days or if 5-days moving relative humidity (≥90%) period for ≥65 hours, 5 days moving congenial temperature (7.2 to 26.6°C) for ≥105 for 5 consecutive days and sum of two consecutive days rainfall (≥2.5 mm) prevail.

Mehtha *et al.* (2001), Mishra *et al.* (2004) and Agarwal *et al.*, (2004) have utilized ordinal logistic model for qualitative data. They used this approach for adopting timely control measures to prevent pest or disease outbreak. They classified the data as 0/2 (two categories) or 0,1,2 (three categories) and probabilities of different categories were obtained thorough logistic regression. Mathematical representation of their approach is as follows

$$P(\mathrm{E} = 1) = \frac{1}{1 + \exp(-z)} + e$$

Where z is the function of weather variables and forecast rules are: If $P \geq 0.5$ more chance of epidemic occurrence and if $P < 0.5$ probability of occurrence of epidemic is minimum

Ramakrishhna *et al.* (2003) developed Southern Oscillation Index (SOI) to forecast crop yield.

Kapoor *et al.* (2004) had developed forecasting of rice blast in Kangra district of the Himachal Pradesh, India. They have concluded that moderate to high (10-30%) leaf blast severity in trap nurseries on any susceptible cultivar sown at 15 days intervals starting form 1st June or 5-10% leaf blast incidence on trap plants, continued high humidity (>80%), prevalence of low temperature (16-19°C) and maximum temperature (<28°C) for 6-8 days or cloudy weather and 5-6 rainy days in a week can be utilized as rice leaf blast rule for predicting this disease development.

Chattopadhay *et al.* (2005) developed forecasting for *Alternaria* blight of oilseed *Brassica* in India. According to them regional and cultivar specific model could predict the crop age at which *Alternaria* blight appeared on leaves and pods, the highest blight severity on leaves and pods and the crop age when blight severity was highest on leaves and pods at least one week ahead of first appearance of the disease on the corp.

On the same disease Kundal *et al.* (2006) have worked out on the machine learning techniques in disease forecasting. They demonstrated that supportive vector model (SVM) is better than existing machine learning techniques and convention linear regression approaches. They have developed a SVM based web server for rice blast prediction.

Desai *et al.* (2004) developed weather based forecasting model for *Brassica juncea* powdery mildew cause by *Erysiphe cruciferarum*. Through their weather based predicting model they have devised one week preciding date of observation for progression of this disease at S.K. Nagar and at Bhartpur.

Agarwal and Mehta (2007) have provided weather based forecasting of crop yields, pests and disease.

Dhar *et al.* (2007) predicted the infestation of *Helicoverpa armigera* in *Cajanus cajan* in Central Uttar Pradesh.

Maji *et al.* (2010) developed forecasting models of bacterial leaf spot disease of mulberry for Birbhum district of West Bengal. They provided that correlation between disease severity and meteorological parameters suggested that bacterial leaf spot disease severity significant positive correlated with maximum and minimum temperature, minimum relative humidity, rainfall and number of rainy days

Artificial Neural Network (ANN) is a machine learning technique (algorithms, neural networks, rough sets and fuzzy sets) for forecasting and classification purpose. ANNs are data driven self-adaptive methods in that there are few priori assumptions about models for problems under study. These learn from examples and capture subtle functional relationships among the data even if the underlying relationships are unknown or hard to describe. ANNs can often correctly infer the unseen part of a population event if data contains noisy information and as forecasting is performed *via* prediction of future behaviour from examples of past behaviour. Laxmi and Kumar (2011) used neural network approach for weather based forecasting for crop yield. They applied this approach using muti-layer preceptron for crop like rice, wheat and sugarcane yield with parameters like minimum and maximum temperatures, rainfall, morning relative humidity. They proposed that neural networks approach possess potential for prediction purposes. Venkatesh (2013) and Bhagawati *et al.* (2015) also utilized this approach for disease forecasting in grapes and for rice blast, respectively.

Chattopadhyay *et al.* (2011) developed forecasting models for white rust of *Brassica juncea*. They provided that regional and cultivar specific model devised could predict, at a few week after sowing the crop age at which white rust first appeared on the leaves, as staghead, the highest rust severity on leaves, staghead number and the crop age at peak rust severity on leaf, highest staghead numbers at least one week ahead of first appearance of the disease on the crop.

Dutta *et al.* (2012) developed disease forecasting model for leaf rust of mulberry (*Morus alba* L.) of Dimapur of North East India. They utilized meteorological parameters like maximum temperature, minimum temperature, maximum and minimum relative humidity and rainfall for prediction of disease severity thorough regression analysis.

Kumar *et al.* (2012) reported design and implementation of web-based aphid (*Lipaphis erysimi*) forecast system for oilseed Brassicas. Their statistical prediction was based on weather parameters (as independent variable) and crop age at time of first appearance of aphid on crop, peak number of aphid and crop age at peak population (as dependent variables).

Tilva *et al.* (2013) had calculated weather based plant disease forecasting using fuzzy logic.

5.10. COMPUTER SIMULATION MODEL FOR PREDICTING PLANT DISEASES

These are the computer programmes which utilize various primary information's like on weather, inoculums occurrence and process it and give future possible results. In a computer simulation of epidemic, the computer is given data describing the various subcomponents of the epidemic and control practices. The computer then provides continuous information regarding about spread and severity of disease over time as well also with regard to the final crop and economic losses likely to be caused by the disease under the conditions of the epidemic. First computer programme EPIDEM was developed in 1969 to simulate early blight of tomato and potato caused by *Alternaria solani*. Letter on various simulation models were developed like.

- LEAFBLAST (Choi *et al.*, 1988), EPIBLAST (Kim and Kim, 1993) and EPIBLA (Manibhusanrao and Krishnana, 1991) for Rice Blast.
- BLIGHTCAST for Late blight of potato in England
- JHULSCAST for Late blight of potato in India
- EPIDEMI for Insect and scab of Wheat
- CERCOS for Celert *Cercospora* leaf spot
- EPICORN for Sothern Corn leaf blight
- EPIVEN for Apple scab
- PLASMO for Grapes downy mildew
- MYCOS for Chrysanthemum leaf spot (*Mycosphaerella*)
- TOMCAST for Early blight of tomato
- EPICAST for Wheat leaf rust
- MARYBLIGHT for Apple fire blight
- EPIPREY for 6 foliar diseases and aphids of Wheat

5.10.1. NEGFRY

The NEGFRY is a computer based late bight forecasting model, developed in Denmark and is used for timing the initial fungicide application and scheduling of subsequent sprays throughout the growing seasons.

First part of this programme is based on the negative prognosis, that calculates the epidemic free period and then recommend the first spray at the end of this period.

The second part of the model is based on a method developed by Fry *et al* (1983) and calculates subsequent spraying intervals based on blight units.

Blight units are calculated using the length of any humid spell, the air temperature during the humid spell and the susceptibility of cultivar to late blight.

Model requires air temperature, relative humidity, rainfall, cultivar susceptibility anc crop emergence date as input for the calculations.

5.10.2. PROGEB

It is an integrated group of forecasting models developed for the main pests of potatoes and cereals n Germany.

One of their components, PHYTEB, forecast for *P. infestans* on potato. This component simulate symptotatic stages of the host (non-infected, latent, pre-infection, number of infection and amount of dead tissue) by means of state variables.

Variables are calculated every 3 hours

Temperature, relative humidity, rainfall, cultivar and plant protection actions taken are independent variables

PHYTRM consists of two sub models, SIMPHYT 1 and 2.

The first sub model forecasts the beginning of the epidemics 7-10 days ahead and it takes into account the cultivars, date of emergence and agro meteorological phenomena.

The second sub model simulates the course of epidemic for two cultivar classes and different fungicide application practice.

5.10.3. PROPHY

It is a computerized expert system developed in the Netherlands which recommend a first fungicide spray when the crop reaches a height of 15 centimetre in sensitive varieties and ten days later in resistant cultivar

A day is considered to be favourable when at least six hours with high relative humidity and at least two hours of leaf wetness or rain occur in a period from 8 PM on the previous day until 12 AM on the next day with temperature ranging from 8 to 25°C.

However, if viable lesions are observed calculation of favourable days is overruled and a spray is advised.

This formulates recommendations on the basis of the calculated protection periods and the weather forecast.

5.10.4. PHYTOPRE

It is a computer based information and decision support system (decision support systems are interactive computer based systems that help decision makers utilize and models to solve unstructured problems under complex and un-certain conditions).

PHYTOPRE contains a database of participants and their meteorological and disease records, a catalogue of epidemiological parameters, and the coordinates of all Swiss communities of about 3000 and the decision rules.

This model produces various kinds of output (a) field specific letters with application recommendations, lists with late blight records and graphs with IP, rainfall and fungicides applications, (b) weekly bulletins with the late blight records, rainfall etc.

5.11. AEROBIOLOGY

The air is swarming with microbial life. Some plant pathogen travel's thousands of kilometres from their source to successful final target. Atmospheric dispersal scale of these pathogen may varied from continental, regional, farm and at pathogen scale and knowledge of dispersal dynamic can enhance our disease management approaches.

Aerobiology is a scientific discipline that deals with the transport of organisms and biologically significant materials through the atmosphere (Isard and Gage, 2001 and 2005). Aerobiology also represents the generation, uptake, translocation, dispersion, viability, deposition and infection/infestation of seeds, viruses, fungi, bacteria and other agents including insects, such as aphids and mosquitoes, which act as virus vectors, as well as dealing with agriculturally significant insects like locusts, bush flies, and moths.

The notion of the aerobiological course has been developed into a predictive tool by integrating atmospheric transport models that function at nested spatial scales with phenological models that characterize the processes regulating the growth and development of plant and animal communities (Gage *et al.,* 1999). Atmospheric pathways are together to ecosystems that support mobile organisms before, during, and after the organisms' dispersal. Organisms are usually functionally adapted to the ecosystems in which they begin and end their airborne travel, as well as to the habitats where some aerobiota rest temporarily during their migration. Environmental conditions in source, resting, and receptor areas are typically very different. This general model of the aerobiological processes was used as the foundation of the integrated aerobiology modeling system (IAMS), built to forecast invasions of important species that use the atmospheric pathway.

Conceptual model of the soybean rust transport process as provided by Isard and Gage (2005) that involved the control of environmental and ecological controls on movement of fungal spores.

A frequent procedure assumed by agricultural meteorologists is to use seasonal meteorological indicators to signify the last stage, *i.e.*, the infection episode, rather than proceeding from the first phase, *i.e.*, the generation. When considering the end phase, *per se*, the inoculum is assumed to be present unless information to the opposing is provided by a plant pathologist. An index such as degree-days or heat units is sometimes used to indicate the phase where infection would probably occur if a suitable pathogen was present.

Criteria for the successful implementation of specific pest and disease management systems have also been given by Johnson (1987) as follows:

- A serious pest or disease problem must exist for which low-cost solutions, *e.g.* host resistance, are unavailable and unreliable;
- It should be likely to explain efficiently and forecast accurately the variations in the incidence of problems by means of a model;
- Facilities must exist for communication of model predictions so that timely control measures can be taken;
- Control measures must be available which are effective, economically justified and non-hazardous to the environment.

Angus (1988) mentions strategic methods, such as host resistance, crop rotation and fertilizer practices, along with tactical methods, such as the application of pesticides or fungicides, in response to model indications of infections or epidemics. The aim being to achieve prevention rather than containment of damage. The EPIPRE system (Djurle and Jonsson, 1985), among other models, considers the cost of application of a mixture of pesticides and fungicides in a single operation and the BLITECAST model (Krause *et al.*, 1975) provides a model for early disease control. Models used for aerobiological investigation could profit by adopting the above principles.

Climatic data is useful in the development of computer models to simulate outbreaks of pest and disease infection. The introduction of a new crop and its susceptibility to disease infection or pest infestation can be tested using a simulation model (Waggoner and Horsfall, 1969).

The trial with a simulator demonstrated that a lifetime of experiments in weather modification with regard to plant disease can be carried out in a matter of hours. Waggoner *et al.* (1972) also created a simulation of southern corn leaf blight. However, there is a compelling need for more computer simulator models for the important diseases and pests that cause epidemics or plagues of economic significance. Models like SIRATAC (Hearn and Brook, 1983; Ives *et al.*, 1984) and BLITECAST (Krause *et al.,* 1975) and the EPIPRE system (Djurle and Jonsson, 1985) are good prototypes.

Transport of plant pathogen through the atmosphere includes process like liberation (release and ascent), drift (passive horizontal transport), and deposition (landing). Liberation affects with by different ecological and environmental factors that directed the timing and mechanism of release; drift related with the passive, directed movement away from the ground surface in turbulent air and deposition involves settling (landing) at a new destination having a susceptible host.

Wind, temperature, radiation, precipitation, relative humidity, spore production, spore survival, incubation period and latent period are the major primary factors prerequisite for any aerobiological works.

Previously aerobiological works has concerted on liberation and deposition process with active (Burkard Volumetric Sampler) and passive samples like petri plates (Schmale *et al.*, 2008; Lin *et al.*, 2014).

Regional- and continental-scale atmospheric features known as Lagrangian coherent structures (LCSs) may shuffle plant pathogens along highways in the sky (Schmale and Ross, 2015). LCS is a key organizing structure of the motion of air parcels in the atmosphere, often a boundary, or front, between two air masses with different histories and biotic (pathogen) compositions. Nearby air parcels converge onto attracting LCSs and diverge away from repelling LCSs.

5.12. REFERENCES

Adiver, S.S., Anahosur, K.H., Giriraj, K. and Naik, S.I. 1998. Forecasting studies for foliar disease of Groundnut (*Arachis hypogaea* L.). *Karnataka Journal of Agricultural Sciences*, 11(1): 96–98.

Agrawal, R. and Mehta, S.C. 2007. Weather Based Forecasting of Crop Yields, Pests and Diseases – IASRI Models. The Indian Society of Agricultural Statistics, 61(2): 255–263.

Agrawal, Ranjana, Mehta, S.C., Bhar, L.M. and Kumar, A. 2004. Development of weather based forewarning system for crop pests and diseases - Mission mode project under NATP.

Andrade-Piedra, J.L., Forbes, G.A. Shtienberg, D., N.J. and Chacon, M.G 2005a. Qualification of a plant disease simulation model: Performance of the LATEBLIFHT model across a broad range of environments. *Phytopathology*, 95: 1412–1422.

Andrade-Piedra, J.L., Hijmans, R.J., Huarez, H.S., Forbes, G.A., Stienberg, D. and Fry, W.E 2005b. Simulation of potato late blight in the Andes. II: Validation of the LATEBLIGHT model. *Phytopathology*, 95: 1200–1208.

Angus, J.F., 1988. Modelling of disease epidemics. Proceedings of workshop on agrometeorological information for planning and operation in agriculture with particular reference to plant protection. Calcutta, India, pp. 307–312.

Arora, R.K., Sharma, S. and Singh, B.P. 2014. Late blight disease of potato and its management. *Potato Journal*, 41(1): 16–40.

Audsley, E., Milne, A. and Paveley, N. 2005. A folia disease model for used in wheat disease management decision support systems. *Annals of Applied Biology*, 147: 161–172.

Batista, D.C., Lima, M.A., Haddad, F., Maffia, L.A and Mizubuti, E.S.G. 2006. Validation of decision support systems for tomato early blight and potato late blight, under Brazilian conditions. *Crop Protection*, 25: 664–670.

Berrie, A.M. and Xu, X.M. 2003. Managing apple scab (*Venturia inaequalis*) and powdery mildew (*Podosphaera leucotricha*) using Adem. *International Journal of Pest Management*. 49 (3): 243-249.

Bhagawati, R., Bhagawati, K., Singh, A.K.K., Nongthombam, R., Sarmah, R. and Bhagawati, G. 2015. Artificial neural network assisted weather based plant disease forecasting system. *International Journal on Recent and Innovation Trends in Computing and Communication*, 3(6): 4168–4173.

Bhattacharya, S.K., Singh, D.S. and Khanna, R.N. 1982. Epidemiology and forecasting of *Phytophthora infestans* inciting potato late blight, *In*: *Phytophthora disease of tropical cultivated plants,* Nambiar, K.K.N. (ed.). Central Plantation of Crop Research Institute, Kasaragod, Kerala, pp. 94–101

Chakraborty, A., Singh, B.P., Ahmad, I. and Sharma, S. 2015. Forecasting late blight of potato in plains of west Bengal using JHULSACAST model. *Potato*, 42(1): 50–57.

Chattopadhay, C., Agarwal, R., Kumar, A., Bhar, L.M., Meena, P.D., Meena, R.L., Khan, S.A., Chattopadhay, A.K., Awasthi, R.P., Singh, S.N., Chakravarthy, N.V.K., Kumar, A., Singh, R.B. and Bhunia, C.K. 2005. Epidemiology and forecasting of Alternaria blight of oilseed in *Brassica* in India- a case study. *Journal of Plant Disease and Protection*, 112(4): 351–365.

Chattopadhyay, C., Agarwal, R., Kumar, A., Meena, R.L., Faujdar, K., Chakravarthy, N.V.K., Kumar, A., Goyal, P., Meena, P.D. and Shekhar, C. 2011. Epidemiology and development of forecasting models for white rust of *Brassica juncea* in India. *Archives of Phytopathology and Plant Protection*, 44(8): 751–763.

Choi, W., Park, E. and Lee, E. 1988. LEAFBLAST: A computer simulation model for leaf blast development on rice. *Korean Journal of Plant Pathology*, 4: 25–32.

De Wolf, E.D. and Isard, S.A. 2007. Disease cycle approach to plant disease prediction. *Annual Review of Phytopathology*, 45: 203–220.

Del Ponte, E.M., Fernandes, J.M.C. and Pavan, W. 2005. A risk infection simulation model for Fusarium head blight of wheat. *Fitopathology Brasilica*, 30: 634–642.

Desai, A.G., Chattopadhyay, C., Agarwal, R., Kumar, A., Meena, R.L. Meena, P.D., Sharma, K.C., Rao, M.S., Prasad, Y.G. and Ranakrishna, Y.S. 2004. *Brassica juncea* powdery mildew epidemiology and weather based forecasting models for India-a case study. *Journal of Plant Disease and Protection*, 111(5): 429–438.

Dhar, V., Singh, S.K., Kumar, M., Agarwal, R. and Kumar, R. 2007. Prediction of pod borer (*Helicoverpa armigera*) infestation in short duration pigeonpea (*Cajanus cajan*) in Central Uttar Pradesh. *Indian Journal of Agricultural Sciences*, 77(10): 701–704.

Djurle, A. and Jonsson, U. 1985. EPIPRE – a computerized pest and disease management system for winter wheat in Sweden. *Vaxtskyddsrapporter*, Jordbruk 32: 179–189.

Dutta, S.K., Ghosh, M.K., Borah, A. and Bindroo, B.B. 2012. Development of disease forecasting model for leaf rust of mulberry (*Morus alba* L.) of Dimapur of North East India. *Nature and Science*, 10(8): 66–72.

Gage, S.H., Isard, S.A. and Colunga-Garcia, M. 1999. Ecological scaling of aerobiological dispersal processes. *Agricultural and Forest Meteorology*, 97: 249–261.

Gent, D.H., Mahaffe, W.F., McRoberts, N. and Pfender, W.F. 2013. The use and role of predictive systems in disease management. *Annual Review of Phytopathology*, 51: 267–289.

Gilles, T., Phelps, K., Clarkson, J.P. and Kennedy, R. 2004. Development of MILLIONCAST, an improved model for predicting downy mildew sporulation on onions. *Plant Disease*, 88: 695–702.

Giosue, S., Spada, G., Rossi, V., Carli, G. and Ponti, I. 2000. Forecasting infections of the leaf curl disease on peaches caused by *Taphrina deformans*. *European Journal of Plant Pathology*, 106: 563–571.

Guillespie, T.J. and Sentelhas, P.C. 2008. Agrometeorology and plant disease management – A happy marriage. *Science Agriculture*, 65: 71–75.

Hearn, A.B. and Brook, K.D. 1983: SIRATAC – Case study in pest management in cotton. *In: New technology in field crop production* (Blythe, D.E., Foale, M.A., Mungomery, V.E. and Wallis, E.S. *eds.*). pp. 199-211.

Heger, M., Oerke, E.C., Dehne, H.W. and Hindorf, H. 2003. Evaluation of an action threshold based IPM wheat model in Rheinland (Germany) in 1999/2001. *OEPP/ EPPO Bulletin*, 33: 397–401.

Holtslag, Q.A., Remphrey, W.R. and Fernando, W.G.D. 2004. The development of a dynamic disease-forecasting model to control *Entomosporium mespili* on *Amelanchier alnifolia*. *Canadian Journal of Plant Pathology*, 26: 204–313.

Isard, S.A., Gage, S.H., Comtois, P. and Russo, J.M., 2005: Principles of the atmospheric pathway for invasive species applied to soybean rust. *BioScience,* 55: 851–861.

Ives, P.M., Wilson, L.T., Gull, P.O., Palmer, W.A. and Haywood, C. 1984: Field use of SIRATAC: An Australian computer-based pest management system for cotton. *Protected Ecology*, 6: 1–21.

Johnson, K.B. 1987: Role of predictive systems in disease management. *In*: Crop Loss Assessment and Pest Management Teng, P.S. *Ed.*, American Phytopathological Society, St. Paul, Minnesota, U.S.A. pp. 176–190.

Kang, W.S., Hong, S.S., Han, Y.K., Kim, K.R., Kim, S.G. and Park, E.W. 2010. A web based information system for plant disease forecast based on weather data at high spatial resolution. *The Plant Pathology Journal*, 26(1): 38–48.

Kapoor, A., Prasad, R. and Sood, G.K. 2004. Forecasting of rice blast in Kangra district of Himachal Pradesh. *Indian Phytopathology*, 57(4): 440–445.

Kaundal, R., Kapoor, A.S. and Raghava, P.S. 2006. Machine learning techniques in disease forecasting: A case study on rice blast prediction. *BMC Bioinformatics*, 7: 485–501.

Kim, C.K. and Kim, C.H. 1993. The rice blast simulation model EPIBLAST system approaches for agricultural development. Springer, pp. 309-321.

Krause, R. A. and Massie, L B. 1975. Predictive system: Modern approach to disease control. *Annual Review of Phytopathology*, 13: 31–47.

Kumar, V., Kumar, A. and Chattopadhyay, C. 2012. Design and implementation of web-based aphid (*Lipaphis erysimi*) forecast system for oilseed Brassicas. *Indian Journal of Agricultural Sciences*, 82(7): 618–622.

Kundal, R., Kapoor, A.S. and Raghava, P.S.G. 2006. Machine learning techniques in disease forecasting: A case study on rice blast prediction. *BMC Bioinformatics*, 7: 485–501.

Langston, D.B.J., Phipps, P.M. and Stripes, R.J. 2002. An algorithm for predicting outbreaks of Sclerotinia blight of peanut and improving the timing of fungicide sprays. *Plant Disease*, 86: 118–126.

Laxmi, R.R. and Kumare, A. 2011. Weather based forecasting model for crops yield using neural network approach. *Statistics and Application*, 9(1-2): 55–69.

Lin, B., Ross, S.D., Prussin, A.J. and Schmale, D.G.2014. Seasonal associations and atmospheric transport distances of fungi in the genus *Fusarium* collected with unmanned aerial vehicles and ground-based sampling devices. *Atmospheric Environment*. 94:385–91.

Llorente, I., Vilardell, P., Bugiani, R., Gherardi, I. and Montesinos, E. 2000. Evaluating of BSPcast disease warning system in reduced fungicide use programs for management of brown spot of pear. *Plant Disease*, 84: 631–637.

Magarey, R.D., Fowler, G.A., Borchert, D.M., Sutton, T.B., Garcia-Colunga, M. and Simpson, J.A. 2007. NAPPFAST: An internet system for weather based mapping of plant pathogen. *Plant Disease*, 91(4): 336–345.

Maji, M.D., Das, N.K., Chatterjee, S., Ghosh, A. and Bajpal, A.K. 2010. Forecasting models of bacterial leaf spot disease of mulberry for Birbhum district of West Bengal. *The Journal of Plant Protection Sciences*, 2(1): 92–95.

Manibhushanroa, K. and Krishnan, P. 1991. Epidemiology of blast EPIBLA: A simulation model and forecasting system for tropical rice in India. *Rice Blast Modelling and Forecasting*, IRRI, Manila, Philippines, pp. 31–38

Manter, D.K., Reeser, P.W. and Stone, J.K. 2005. A climate-based model for predicting geographic variation in Swiss needle cast severity in Oregon Cast Range. *Phytopathology*, 95: 1256–1265.

Mehta, S.C. Agrawal, R. and Singh V.P.N. 2000. Strategies for composite forecast. *JISAS*, 53(3).

Moschinie, R.C., Carranza, M.R. and Carmona, M.A. 2004. Meteorological based predictions of wheat head blight epidemic in the southern Argentinean pampas region. *Cereal Research Communication*, 32: 45–52.

Nagarajan, S., Joshi, L.M., Srivastava, K.D. and Singh, D.V. 1980. Epidemiology of brown and yellow rust of wheat over north India III. Impact of varietal change. *Plant Disease Reporter*, 62: 694–698.

Orlandin, S., Marta, A.D., D'Angelo, I. and Genesio, R. 2003. Meteorological based predictions of wheat head blight epidemic in the southern Argentinean pampas region. *Cereal Research Communication*, 32: 45–52.

Padmanabhan, S.Y. 1963. *Helminthosporium* disease of rice-VII: A study of the meteorological factor associated with the epiphytotic of 1942 in Bengal. *Oryza*, 1(2): 101–110.

Paul, P.A. and Munkvold, G.P. 2004. A model-based approach to pre-planting risk assessment for gray leaf spot of maize. *Phytopathology*, 94: 1350–1357.

Paul, P.A. and Munkvold, G.P. 2005. Regression and artificial neural network modelling for the prediction of gray leaf spot of maize. *Phytopathology*, 95: 388–396.

Peres N.A., Timmer L.W., and Chung K.R. 2009. Florida citrus pest management guide: Postbloom Fruit Drop. Publication No. PP-45, University of Florida, IFAS, EDIS, Gainesville, Florida.

Phipps, P.M., Deck, S.H. and Walker, D.R. 1997. Weather-based crop and disease advisories for peanuts in Virginia. *Plant Disease*, 81: 236–244.

Ponter, D.E.M., Canteri, G.C.V., Reis, M.G. and Yang, X.B. 2006. Models and applications for risk assessment and prediction of Asian soybean rust epidemics. *Fitopatologia Brasileria*, 31: 533–544.

Pundhir, V.S., Singh, B.P., Ahmad, I., Sharma, S., Kushwaha, J.S., Singh, V.K. and Joshi, V. 2014. Forecasting late blight of potato in Tarai region of Uttarakhand using modified JHULSACAST model. *Potato Journal*, 41(2): 95–104.

Ramakrishna, Y.S., Singh, H.P. and Rao, G.M. 2003. Weather based indices for forecasting foodgrain production in India. *Journal of Agrometeorlogy*, 5(1): 1–11.

Roget, D.K. 2001. Prediction modelling of soilborne plant disease. *Australian Plant Pathology*, 30: 85–89.

Rossi, V., Caffi, T., Giosue, S., Girometta, B. and Bugianie, R. 2005. Elaboration and validation of a dynamic model for primary infections of *Plasmopara viticola* in Northen Italy. *Review of Italian Agrometerology*, 3: 7–13.

Rouzet, J. and Jacquin, D. 2003. Development overwintering oospores of *Plasmopara viticola* and severity of primary foci in relation to climate. *OEPP/EPPPO Bulletin*, 33: 437–442.

Saxena, S.C. and Lal, S. 1984. Use of meteorological factors in prediction of *Erwinia* stalk rot of Maize. *Tropical Pest Management*, 30: 82–83.

Schmale, D.G., Dingus, B.R. and Reinholtz, C.F. 2008. Development and application of an autonomous unmanned aerial vehicle for precise aerobiological sampling above agricultural fields. *J. Field Robot.* 25: 133–47.

Schmale, D.G. and Ross, S.D. 2015. Highway in the sky: Scales of atmospheric transport of plant pathogen. *Annual Review of Phytopathology*, 53: 591–611.

Shtienberg, D., Shwartz, H. Oppenheium, D., Zilberstaine, M. and Herzong, A. 2003. Evaluation of local and imported fire blight warning system in Israel. *Phytopathology*, 93: 356–363.

Singh, B.P., Ahmad, I., Sharma, V.C. and Shekhawat, G.C. 2000. JHULSACAST: A computerized forecast of potato late blight in Western Utter Pradesh. *Journal of Indian Potato Association*, 27(1): 25–34.

Tilva, V., Patel, J. and Bhatt, C. 2013. Weather based plant diseases forecasting using fuzzy logic. *In*: International Conference on Engineering, Nirma University, November, pp.1-6. http://www.researchgate.net/publication/269272666

Turechek, W.W. and Wicox, W.F. 2005. Evaluating predictors of apple scab with receiver operating characteristic curve analysis. *Phytopathology*, 95: 679–691.

Twengstrom, E., Sigvald, R., Svensson, C. and Yuen, J. 1998 Forecasting Sclerotinia stem rot in spring sown oilseed rape. *Crop Protection*, 17: 405–411.

Venkatesh, H. 2013. A neural network approach for disease forecasting in grapes using weather parameters", *ICCCNT*, 2013, 2013 Fourth International Conference on Computing, Communications and Networking Technologies (ICCCNT), pp. 1-5,

Waggoner, P.E. and Horsfall, J.G., 1969 *EPIDEM: A simulator of plant disease written for a computer*. Bull. Conn. Agric. Exp. Sta., New Haven, No. 698.

Waggoner, P.E., Horsfall, J.G. and Lukens, R.J., 1972 *EPIMAY: A simulation of southern corn leaf blight*. Bull. Conn. Agric. Exp. Sta. New Haven, No. 729.

Wolf, P.F.J. and Verreet, J.A. 2005. Factors affecting the onset of *Cercospora* leaf epidemics in sugar beet and establishment of disease-monitoring threshold. *Phytopathology*, 95: 269–274.

Wolf, P.F.J. and Verreet, J.A. 2002. An integrated pest management system in Germany for the control of fungal leaf disease in sugar beet: The IPM sugar beet model. *Plant Diseases*, 86: 336–344.

Wu, B.M., Van Bruggen, A.H.C., Subbarao, K.V. And Scherm, H. 2002. Incorporation of temperature and solar radiation threshold to modify a lettuce downy mildew warning system. *Phytopathology*, 92: 631–636.

CHAPTER

6 Approaches for Epidemic Avoidance

6.0. INTRODUCTION

From epidemic point of view the basic objective of disease management is to keep disease intensity below an economic injury threshold (Nutter and Guan, 2001). Disease management involves the integration of tactics to achieve targeted goals like:

- Elimination of reduce initial inoculums (X_0)
- Reduce the rate of infection (r), and
- Reduce the time (t) that pathogen population and host population interact to reduce disease intensity (Nutter, 2007).

Based on pathogen and its disease cycle two strategies can be adopt

- If infection rate (r) is high then the main focus of disease management should be directed toward reduction of "r". For such condition disease risk (R) must be adequately lowered to a strategic threshold, below which the incorporation of other disease management principles like reduction of X_0 and t will provide an epidemiological benefit that reduces further disease risk.
- However, if infection rate (r) is low like in soil borne pathogen then the suitable strategy to reduce disease risk would be targeted toward reduction of initial inoculums (X_0).

Thus, appropriate methods are essential for correct and timely quantification of X_0, r and t. For such benchmark, selection of temporal model that best portray the temporal progress of plant epidemic involves contemplation of both subjective and objective standards.

Growth models commonly used are Monomolecular, Exponential, Logistic and Gompertz. In general, growth models that incorporate few variables to describe temporal disease dynamics have a good performance; however, this kind of models sometimes do not satisfy the acquiring process of key characteristics because they frequently ignore relevant variables that affect the epidemic development (Xu, 2006), *e.g.*, host growth, fluctuating environmental condition, length of latent and infectious period, etc.

Both logistic and Gompertz and models have a characteristic sigmoid form and an infection point that means secondary inoculation or plant to plant spread within the crop in comparison to monomolecular model which does not have infection point.

6.1. EXPONENTIAL

Also known as the logarithmic, geometric or Malthusian model.

Useful to describe early stages of most polycyclic epidemics.

This model suggested that the absolute rate of disease increase is directly proportional to disease intensity (X, *e.g.*, incidence or severity).

It can be write as $X=X_0e^{rt}$, where X_0 is the amount of initial inoculums/ initial disease intensity, e is the exponential and *r* is basic rate of infection and t is time.

This model is easy to extrapolate and interpolate (prediction), easy to see changes in initial X and the increase over time and easy to fit.

6.2. MONOMOLECULAR

Appropriate for modeling monocyclic epidemics

Also called negative exponential model.

In this model dx/dt is not proportional to x because inoculum produced by infected individual during the current epidemic does not lead to additional infection in the current epidemic. Thus, in other way this model is appropriate for modeling epidemics where there is no secondary spread within a growing season, meaning that the plant disease has a single cycle during growing season

However, the dx/dt is proportional to healthy tissue. The quantity of changes in disease (dy/dt) over time from a maximum at the beginning of epidemic and has a negative exponential probability density function.

The disease intensity is on a proportion scale. Thus, disease free (healthy) or level is 1-x

$$dx/dt = r_m (1-x).$$

In this model r_m is the product of amount of inoculums available for infection of disease free plants or units and the probability per unity time.

$$X=X_0+r_m t$$

6.3. LOGISTIC

Appropriate for most polycyclic diseases (having secondary spread within a growing season)

Most widely used for describing plant epidemic.

Particularly this model is most widely used for temporal analyses.

When the rate of infection is constant, the plot of dy/dt *versus t* increases initially, reaches a maximum (at 50% disease or infection point), and declines to zero

This curve that is symmetrical about the infection point can be linearized by using logarithmic transformation.

This model having features of both the exponential and monomolecular models.

$$X = 1/1+\exp \{-[1n * (X_0/1-X_0)+r *t]\}$$

6.4. GOMPERTZ

Appropriate for polycyclic diseases as an alternative to logistic models.

Gompertz model has an absolute rate curve that reaches a maximum more quickly and declines more gradually than the logistic models.

The curve for dy/dt *versus t* for Gompertz, unlike logistic model, is not symmetrical around the infection point.

The rate of infection approaches infection point more rapidly and declines more slowly than in the logistic model and the curve is skewed positively *i.e.*, a larger portion of the area under curve is to the right of the infection point.

$$X = e^{1n(X_0)e-r*t}$$

According to Berger (1981), the Gompertz transformation effectively linearized the DPC compared to logistic model and it can also avoid the curvilinearity which is commonly associated with logistically transformed values. He also found that this model more accurately quantified the epidemic rate, projection of future disease severity and determination of initial diseases.

The proportion of disease increase in a host plant, when plotted against time, gives a sigmoid curve though other types of curves are encountered. It is necessary to linearise such curves in order to evaluate rate reducing resistance for comparison of the epidemics in terms of the rate of disease increase. This could be achieved by applying either the logistic model or Gompertz model or Richards model or by Weibull model or Gaussian model. It is not necessary that all the models will not fit well to a specific plant-patho-system nor a specific model will fit in to all the plant-patho-systems or even diseases.

Mohapatra *et al.*, (2008) have evaluated the disease progress curve in the rice blast pathosystem and compared with the Logistic and Gompertz models. They had concluded that Gompertz model were more consistent among the studied slow-blasting genotypes over seasons, however, both Gompertz and logistic models fit well in to the rice blast patho-system.

The comparative utility of Gompertz, Exponential, Logistic and Monomolecular models in the analysis of disease progress and disease prediction was studied by Panicker and Ramraj (2010) and the transformations were fitted in the polynomial regression equation to get the absolute rate equation and linearized equation. Expression for the absolute rate of changes in y with (dy/dt) and the integrated linearized forms of these models are presented in Table 1.

For selection of most appropriate population growth model for disease progress datasets, several steps are involved like:

- First stem is to graph y (generally disease or pathogen incidence or severity) *versus* time t. The shape of disease/pathogen progress curve is very helpful in identifying the best model.
- After viewing these graphs, objective criteria for the acceptance or rejection of a model are provided by the estimated regression parameters and statistics, *i.e.,* r^2, coefficient of Variance (CV) and standard deviation (Campbell and Madden, 1990 and Nutter and Parker, 1997)
- Finally, the inspection of residual plots reveals if the model is satisfactory.

In above equations y = disease intensity; t= time; r= rate parameter (r_M = rate for monomolecular model; r_{E} = rate for exponential model; r_L = rate for the logistic model; r_G = rate for the Gompertz model) y_0 constant for integration, corresponding to y at t =0

Table 1: Comparative statements of growth models.

Model equation	***Integrated expression***	***Absolute rate of equation***	***Liberalized***
Monomolecular	$y=1-(1-y_0)$ *exp* $(-r_M$ *t)	dy/dy $-r_M$ * (1-y)	Ln[1/(1-y)] =1n[1(1-y_0)]+r_{m*}t
Exponential	$y=(y_0)$ *exp ($r*_{E*}$ t)	dy/dt =r_E*y	1n(y) = 1n *(y_0)+r_E*t
Logistic	y =1/{1+[(1-y_0)/y_0 * exp *(-r_L *t)	dy/dt =r_L y (1-y)	Ln[y/(1-y)] =1n[y/(1-y_0)]+r_L*t
Gompertz	Y=exp{1n(y_0) *exp (-$r_{G*t)}$}	Dy/dt =r_G[-1n(y)]	-1n[-in(y)] = 1n[-1n (y_0)] +r_G*t

6.5. CONTROL OF EPIDEMIC OTHER THEN CHEMICAL METHODS

By two mean (A) Cultural that includes sanitation, crop rotation, roughing, soil amendments, green manuring, removal of collateral and alternate host and vector control and (B) Host resistance like varietal diversification (at village level), multiline, gene deployment, gene pyramiding and gene cycling.

6.5.1. Sanitation

Sanitation is generally used for controlling soil borne and facultative parasite or saprophytes. Many obligate parasites also perennate through dormant structure in plant organs lying in or on the soil. Destruction of crop debris by burning in the field deceases these types of survival of pathogen. Burning the crop debris in the soil by soil turning plough also inactivates inoculums of many pathogens. Sanitation is very important when diseased crop residue is left on the field as a general practice by the farmer.

Sanitation is used for slow growing 'simple interest' type of diseases while it may not prove so, for other.

Efficiency of sanitation depends on the timely spotting of diseased plants and their removal.

Vanderplank (1963) defines sanitation as a process that reduces or completely eliminates or excludes the initial inoculums from which epidemics start.

From management point of view, cost effective disease reduction depends on knowledge about X_0 and *r*. A useful tool to help set strategic priorities related with the interactive effects of reducing initial inoculums (X_0) and the rate of disease development on plant disease epidemics is sanitation ratio

Sanitation ratio is the simple calculation provides an estimate of how infection r and X_0 interact to delay the time it takes to reach any specified level of disease intensity (*e.g.* disease onset, the time to reach 50% disease intensity, the time to reach an economic injury threshold). This ratio is calculated by quantifying the percent reduction in initial inoculums resulting from one or more sanitation practices that reduces X_0, along with an estimate of the rate of disease development (r)

$$SR\ (Delay\ in\ Time) = 1n \left[\frac{X_0}{X_{ws}}\right] x \frac{1}{r}$$

In this equation sanitation ratio is the estimated delay in time (t) required to reach a specified level of disease intensity, X_0 is the level of initial inoculums present without sanitation (*i.e.,* 100%), while, X_{ws} is the percentage of initial inoculums remaining after one or more sanitation practices performed (*e.g.,* X_{ws} = 100%-95% reduction in X_0 due to crop rotation = 5%) and r is the estimated rate of disease development.

For example if r is estimated to be 0.2 in units/day and a biological control agent is estimated to reduce initial inoculums by 90% then the estimated delay in time resulting from control measure would be

$$SR\ (Delay\ in\ Time) = 1n \left[\frac{100\%}{100\% - 95\%}\right] x \frac{1}{0.2\ 1n/day} = 12\ days$$

The another version of this equation is as follows

$$\Delta t = \frac{2.3}{r}\ x\ \frac{X_0}{X_{ws}}$$

According to this equation, both *Ät* and r are inversely proportion to the amount of initial inoculums and this relationships suggested that sanitation is helpful most when it needed less.

Since in systematic disease the r is low compared to non-systematic thus, such diseases are largely being control by sanitation. Again, for sanitation to be effective, the rate of an epidemic must be low, or slowed to a rate below which sanitation practices will effectively delay disease onset.

According to Nutter (2007) sanitation practices usually fall under one of three management principles *i.e.*, exclusion, avoidance or eradication (Fig. 1).

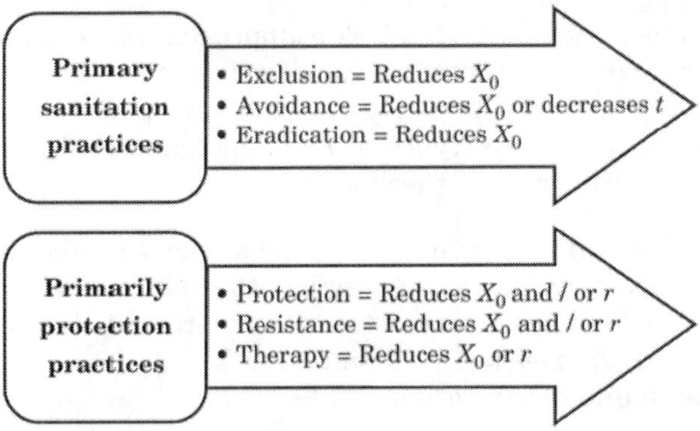

Fig. 1: Plant disease management approaches

Exclusion means keep the pathogen out of the crop production are so that initial inoculums (X_0) is kept at zero. This can be achieving through quarantine and seed certification.

In situation when plant disease threat cannot be kept out of crop growing area by exclusion, then avoid risk in the form of X_0 that is absent or very low, or the selection of planting sites in which environment limits the infection efficiency of the initial inoculums. Such management strategy known as avoidance that is "Use of management tactics that avoid high disease risk planting site, or exposure to environmental conditions that increase disease risk. Avoidance of disease can be achieved in space and time. *Fusarium solani* f. sp. *Phaseoli* is favored by dry soils and avoidance of such conditions will lower disease risk. Avoidance of pathogen population in time, like choice of planting date can be very effective strategy.

Sanitation approach is useful for controlling disease such as powdery mildew of wheat, barley, pea, downy mildew of peas and maize, red rot of sugarcane, Panama disease of banana, *Verticillium* wilt of cotton, Koleroga of *Areca* palms, foot rot and leaf rot *Piper betle*, brown rot of stone fruits, late blight of potato and angular leaf spot of cotton etc.

Eradication can be defined as any practice that reduces initial inoculums at the source. This practice involves burial, burning or removal of crop debris at the field. It can be achieve through various practices like crop rotation, removal of alternate and collateral host, rouging of diseased plants, pathogen eradication programs, flooding, soil solarization and soil fumigation and these approaches will discuss later.

Protection which can be defined as any tactic that provides a physical or chemical barrier to protect the crop. Through this tactics initial inoculums generally reduces however, protection is directed at the point of crop infection (*i.e.*, protection of potential infection courts), and not directly at the source of inoculums in the field. This approach can be achieve through use of physical barriers to protect crops, use of chemical barriers to protect crops, use of organic and reflective mulches and protection can also be achieve through resistance (types of resistance and their various forms will be discus in latter part of the chapter).

Working Example: Green ear disease of Bajara infection on March 12 was 3.4%. Following sanitation practices the disease severity reduces to 0.2% by March 20th. In a separate distance field without following sanitation, disease severity increases to 10.5% on March 20th. Calculate time delay in epidemic growth due to sanitation.

- As we know delay due sanitation can be quantified through

$$\Delta t = \frac{2.3}{r}\ x\ \frac{X_0}{X_{ws}}$$

- Thus according to this formula first we have to calculate the infection rate through using following formula

$$r = \frac{2.3}{t_2 - t_1}\ x \log \frac{X_2}{X_1} = \frac{2.3}{8}\ x \log \frac{0.105}{0.034} = \frac{2.3}{8}\ x \log 3.09\ = \frac{2.3}{8}\ x\ 0.49 = r$$
$$= 0.141$$

- Now put the r in sanitation formula i.e.,

$$\Delta t = \frac{2.3}{r}\ x\ \frac{X_0}{X_{ws}} = \Delta t = \frac{2.3}{0.141}\ x\ \frac{0.034}{0.002} = \frac{2.3}{0.141}\ x\ 1.23 = \frac{2.83}{0.141} = 20.1\ days$$

- Thus, by following sanitation practices a time delay of approximately 20 days accepting.

6.5.2. Crop Rotation

Crop rotation is the practice of growing a prescribed sequence of different crops in the same field, to discourage the build-up of pathogen population over time.

Crop rotation means planting a non host crop between the two successive season of main crops.

This tactics is more judgmental for soil invaders not for soil inhabitants

Thus, in other words soil-borne pathogen with shorter viability in the absence of host can be managed by rotation.

It provides a meaningful management benefit by reducing initial inoculums (X_0) in the field. This practice basically extends the time required for the build-up of inoculums.

Plant pathogens uses the plant residue both as a food base and as a refuge, and this tendency is known as "possession principal" (originally described by Bruhel 1975 and reported in Nutter, 2007). This suggested that pathogens that take possession of above ground plant parts have much less competition with soil microbes however, the crop is developing, but as crop mature and senesce, infected crop tissue above ground becomes infested crop residue that exist in on or below, the soil surface. The key for pathogen to maintain possession of a residue substrate until the next susceptible crop is present is a slow continued metabolism. However, a physical environment unfavorable for metabolism of plant pathogen in possession of a crop residue may not be unfavorable for potential colonist residing outside infested crop residue. And as crop residue degrades due to colonizing saprobes, pathogen survival rapidly declines.

Crop rotation is generally recommended with botanically unrelated crop as these will have different nutritional requirements and tend to be affected by different pathogen and pest.

Crop rotation helps to control many soil-borne diseases, such as wheat mosaic, wilt of arhar, wilt of linseed, finger and toe disease of crucifers, wilt of pea and gram, red rot and wilt of sugarcane, ergot and smut of bajra, wheat bunt, leaf smut and bunt of rice, flag smut of wheat and root rot of vegetables.

A three-year crop rotation is the standard recommendation for control of black rot (*Ceratocystis fimbriata*), stem rot (*Fusarium oxysporum*), and scurf (*Monilochaetes infuscans*) in sweet potatoes.

6.5.3. Removal of Alternate or Collateral Host

Many plant pathogens complete their life cycles on two hosts; one of them is generally a wild plant. The destruction of such a wild host plant brakes the chain in the completion of their life cycles of the pathogen.

Basically collateral host belongs to same family of the main host, while alternate host belongs to different family of the main host.

Nutter (2007) suggested alternate hosts as the hosts which the pathogen requires to complete its cycle while alternative hosts are defined as other host species on which a pathogen can complete its disease cycle.

The epidemiological importance of alternate and alternative hosts of pathogen is that they can serve as an over seasoning refuge (link between one crop growing season to the next time a susceptible crop is grown), as well as providing a local source of initial inoculums for the next susceptible crops.

Eradication of alternate host is effective controlling the local sources of initial inoculums (X_0).

Barberry eradication in temperate countries helps to control the black stem rust of wheat urdinia.

Pyricularia grisea can infect and survive on several wild species of riche and other grasses like *Panicum repens, Digitaria marginata* and *Setaria intermedia*.

Presents of alternative host serve as a source of initial inoculums by producing dispersal units (*e.g.* spores, sclerotia, viruliferous aphids) that can become infection units when they come in contact with other susceptible host and the environment is favorable for pathogen infection or its transmission.

Destruction and removal of collateral host is an important control measure. The yellow vein mosaic of lady finger persists on wild host (*Hibiscus tetraphyllus*) and systematic removal of this host effectively control the disease.

Removal of *Panicum antidotale* or *Cenchrus ciliaris* for pearl millet field contributes to lowering of ergot disease.

6.5.4. Roguing

Removal and destruction of plants as they become disease is called roguing, useful in many viral diseases.

If done properly it checks spread of the disease in healthy plants and also helps in production in disease free seed.

Rouging is practiced in many perennial plantation crops in the tropics. This is because roguing (in some pathosystems) can delay the epidemic and hopefully add years to the life of the plantation

Roguing is employed in such diseases as loose smut of wheat, loose and covered smut of barley, sorghum and maize, red rot of sugarcane, green ear disease of bajara, ergot of bajra, wilt of pigeon pea and Panama disease of banana.

Roguing delays plant disease epidemics by reducing the incidence of infection in vegetable and plantation crops. If r is high, then roguing in itself may not be beneficial.

6.5.5. Soil Solarization

Soil solarization is a method of controlling soil borne pests and pathogens by raising the temperature of the soil through application of transparent polyethylene sheet to a moist soil surface.

Basically it's a hypothermal method in which radiant energy capture by polythene sheets and transmit into soil.

Use of this method has been reported to reduce the population of many soil borne pathogens including fungi bacteria and nematodes as well as weeds (Katan *et al.*, 1983; Barbercheck *et al.*; 1986; Verma *et al.*; 2005).

This technology can easily be transmitted to the ordinary farmers and can be applied in large areas manually and mechanically. Thus, it is suitable for both developed and developing countries. It may have a long term effect, since effective disease control lasts for more than one season. This method has the characteristics of an integrated control, since physical, chemical and biological mechanisms are involved and because the control of a varieties of pests is achieved.

Solarization involves limitations, difficulties and potential negative side effects. It can only be used in regions where the climate is suitable (hot) and the soil is free of crops for about one month or more at a time of tarping with PE sheets.(a) It is too expensive for some crops and ineffective in the control of certain diseases, (b) Heat tolerant pathogens might develop after repeated application, though selection for tolerance to lethal agents is not likely to develop with disinfestation methods which are not target specific and (c) another possibility would be an increase in pathogen population due to a harmful effect on its antagonists.

Soil solarization has been demonstrated to control diseases caused by many fungal pathogens such as *Rhizoctonia solani*, *Fusarium* spp., *Pythium* spp., *Phytophthora* spp., *Verticillium* spp., *Sclerotium rolfsii* etc. in many crops (Katan *et al.*, 1983; Raoof and Rao,1997). Soil solarization has also been shown to significantly decrease the population of disease causing Agrobacteria and *Pseudomonas* (Raio *et al.*, 1997; Chellemi *et al.*, 1994). Many nematode diseases caused by *Meloidogyne* spp., *Heterodera* spp. etc. have been successfully controlled by this method (Rao and Krishnappa, 1995; Grinstein *et al.*, 1995).

6.5.6. Use of Host Resistance

Host resistance generally affects either initial inoculum by reducing X_0 to zero or the rate of epidemic (r).

Host resistance refers to "those characters or traits that enable host to avoid, tolerate or recover from attacks of insects under condition that would caused greater injury to other plants of the same species".

These are also reflected with heritable characteristics possessed by host that influence the ultimate degree of damage done by pathogen.

Resistance types in a host plants can be categorized on the basis of ecological point of view, genetic basis, based on biotype reactions, population or line based and based on evolutionarily concept.

Ecological resistance also known as pseudo resistance and it may be (a) *Host evasion*: It pertains with the capability of the host through which it may pass through the most susceptible stage rapidly or at a time which pathogen are less or evade injury by early maturating, (b) *Induced resistance*: It related with increased in resistance temporarily as a result of changed conditions of host or environment and (c) *Escape*: Absence of infestation or injury to the host plant due to transitory process like incomplete infestation. A special kind of resistance known as kludensity specifically uses for viral diseases in which host having the capability to avoid virus vectors (insects).

Resistance based on genes can be categorized into monogenic resistance: controlled by a single gene, easy to incorporate into host by breeding and they are also easy to break, oligogenic resistance: controlled by few genes, polygenic resistance: controlled by many genes, major gene resistance: controlled by one or few major genes also known as vertical resistance, minor gene resistance: controlled by many minor genes and their cumulative effect is also called adult resistance/mature resistance/field resistance. Also known as horizontal resistance.

Resistance based on biotype reactions is vertical resistance which is effective against specific biotype (specific resistance) and horizontal resistance which is equally effective against all known biotypes (*i.e.,* non-specific resistance).

Resistance based on population/line concept may be pure line: exhibited by lines which are phenotypically and genetically similar and multiline resistance: exhibited by lines which are phenotypically similar but genotypically dissimilar.

From evolutionary point of view resistance in host plant may be sympatric: acquired by coevolution of plant and insect (gene for gene) and it governed by major genes and allopatric resistance: not by co-evolution of plant and insect and governed by many genes.

Naturally, quantum of successful pathogen are rare, as the majority of host plant possess resistant to most pathogen species. This form of disease resistance is known as non host resistance (NHR). NHR can be defined as resistance exhibited by an entire plant species to all genetic variants of a non-adapted pathogen species or form a specialis (Stam *et al.*, 2014). NHR protect plants from pathogens adapted to other plant species. NHR mechanism overlap with that mediating host resistance to adapted pathogen and it can be based on both active and passive defense mechanism.

Resistance shown by an entire plant species to a specific parasite or pathogen is known as non host resistance and is expressed by every plants towards the majority of potentially pathogenic microbes.

Therefore, NHR is the most common form of disease resistance exhibited by plants and contrasts with host resistance, which expressed by plant genotypes within an otherwise susceptible host species

Host resistance is usually parasite-specific and commonly is expressed against specific pathogen genotypes.

Because of the durability of non-host resistance over time (pathogen have rarely altered their host species range over recorded history), it is commonly speculated that non-host resistance could be exploited by plants breeders seeking to improve disease resistance within host species.

NHR having two categories: Type 1: which does not result in visible cell death and Type II, which entails localized programmed cell death in response to non host pathogen (Mysore and Ryu, 2004).

Table 2 depicted the summary of types of resistance based on various ways of separation (Chen, 2013).

6.5.7. Vertical Resistance

Vertical resistance denotes the specific resistance, race specific resistance, qualitative resistance, true resistance, monogenic resistance and major gene resistance.

Vertical resistance is when a plant variety exhibits a high degree of resistance to a single race or strain of a pathogen and this quality is governed through one or few genes.

Host with this resistance quality hardly exhibits symptoms and resultant in the death of the infected cell and delimit the establishment of the pathogen.

Some of these genes may have multiple alleles like in leaf rust gene, Lr2 that provided resistance to *Puccinia recondite tritici*. In that locus, four genes namely Lr2a, Lr2b, Lr2c and Lr2d are linked together.

Table 2: Types of resistance based on ways for their aeparation

Ways of separation	*Resistance Type I*	*Definition*	*Durability*	*Resistance Type II*	*Definition*	*Durability*
Growth stage	All stages (seedling resistance)	Resistance can be quantified in the seedling stage however, remain effective throughout all growth stages	Usually not durable	Adult plant resistance	Plants with adult plant distance are susceptible in the seedling stage but become resistant in the latter stage	Usually durable
Testing condition	Greenhouse ble and field in the field	Resistance that is detected greenhouse is also effective	Usually not durable	Field resistance	Resistance can be detected in the field and during adult stage, however, cannot be detected in greenhouse	Usually dura- in
Specificity	Vertical resistance	Resistance is effective against some races, but not effective against other races	Usually not durable	Horizontal resistance	Resistance is similarly effective against all races	Usually durable
	Race specific resistance	Resistance is effective against some specific races, but not effective against other races	Usually not durable	Non-race specific resistance	Resistance is effective against all races	Usually durable
Degree of resistance	Complete resistance (immune)	Resistant plant do not show any visible symptoms	Usually not durable	Incomplete (partial resistance)	Resistance is not complete, showing various degree of reduction according to infection type and or severity	Usually durable
	Absolute resistance	Resistance is complete	Usually durable	Relative resistance	Resistance in incomplete	Depends
Sensitivity to	Hypersensi-	Resistant plants show necro-	Often not	Non-hypersen-	Either completely resistant	Depends

Table: 2 (*Contd...*)

Table: 2 (*Contd...*)

Ways of separation	*Resistance Type I*	*Definition*	*Durability*	*Resistance Type II*	*Definition*	*Durability*
pathogen infection	tive resistance	tic tissue result from cell death	durable	sitive response	or reduced severity, but susceptible infection types (*e.g.* slow rusting)	
Speed of symptom and or sign	Fast rusting (higher susceptible)	Rust develops fast and quickly reaches the highest level	Not applicable	Slow rusting	Plants have a susceptible infection type but rust develops slowly on them	Durable
Response to temperature	Temperature non-sensitive resistance	Resistance expenses at all range of temperature of which susceptible plants are susceptible	Usually not durable	Temperature sensitive resistance	Resistance expresses at some (for strip rust, high-temperature) range of temperature, but not the other range (low-temperature) of the temperatures	Usually durable
Inheritance Usually durable	Qualitative resistance	Resistance is inherited qualitatively, showing distinct two classes in a segregating population from a cross with a susceptible genotype.	Usually not durable	Quantitative resistance	Resistance is inherited quantitatively, showing continuous variation in a segregating population	
Effect of genes	Major gene resistance	Resistance is controlled by one or a few genes with a large effect	Usually not durable	Minor gene resistance	Resistance is controlled by one or more genes with minor effects	Usually durable
Number of genes	Monogenic resistance	Resistance is controlled by a single gene	Usually not durable	Polygenic resistance	Controlled by many genes	Usually durable
Durability	Non-durable resistance	Resistance is broken-down by some races	Not durable	Durable resistance	Resistance that remain effective during its prolonged and wide spread use	Durable
Molecular basis	NBS-LRR type resistance	Resistance is controlled by a gene with NBS-LRR domains in gene products	Usually not durable	Non NBS-LRR type resistance	Resistance is controlled by a gene without NBS-LRR domains in the products	Durable

6.5.8. Horizontal Resistance

This type of resistance protect the host against several strains of a pathogen, however it's in not complete in nature

Compare to vertical resistance, several genes are involved.

Horizontal resistances reduce the rate of disease spread and are evenly spread against all races of the pathogen.

The low incurable disease severity in horizontal resistance is assumed to result from polygenic resistance.

Partial resistances, dilatory resistance, lasting resistance are some other terms denoting horizontal resistance.

Morphological features like stomatal density and number, size, cuticle thickness, waxiness are influenced by the degree of resistance expressed.

Horizontal resistance does not protect plants from becoming infected, however, it slow down the development of individual infection loci on a host, slow down the spread of the disease and also slow down the development of epidemics in the field.

Characteristics differences between vertical and horizontal resistance are as follows:

Character	*Vertical resistance*	*Horizontal resistance*
Definition	Resistance that is effective against some pathotypes but not others	Resistance that operates to a similar extent against all pathogen types
Terminology	Pathotype specific resistance, major gene resistance, seedling resistance, monogenic and qualitative resistance	Non-pathotype specific resistance, generalized resistance, field resistance, adult plant resistance, quantitative resistance
Expression	Provide a high level of protection but is prone to breakdown as new virulent pathotypes evolve after which plants are fully susceptible	Provides lower levels of resistance and does not break down. Resistance often increases as plants mature.
Stage of Expression	Seedling to maturity	Expression increases as plant matures (adult plant)
Inheritance	Inherited as a single gene with identifiable effect. Liken with gene for gene concept	Reduced rate of degree of infection, colonization and sporulation results in a reduced rate of spore production.
Epidemiology	Reduces the amount of initial inoculums, delaying the start of an epidemic. Once epidemic starts, the disease progress curve is simi-	Slow down the rate of epidemic development due to its quantitative nature.

Table: (*Contd...*)

Table: (*Contd...*)

Character	*Vertical resistance*	*Horizontal resistance*
	lar to that of a fully susceptible cultivar	
Effectiveness	Most effective when a diversity of specific resistance genes occurs within and between crops. Usually used by breeders because it is easy to detect and manipulate	Most effective when there is genetic uniformity within and between crops. Often avoided by breeders because it is difficult to detect and manipulate.
Selection and Evaluation	Relatively easy	Relatively difficult

6.5.9. Plant Disease Resistance Gene (R)

Plant disease resistance (R) gene triggers defense mechanisms by detecting non-self or modified self-molecules (elicitors) which results in hypersensitive response (Sharma *et al.*, 2013).

Host possesses varying number of R genes with diverse recognition specificities which are activated in response to a variety of microbial pathogen.

In plants several R genes identified, cloned and characterized (Gururani *et al.*, 2012) and on the basis of their amino acid motif organization they can be categorized into eight classes (Table 3).

Majority of R protein consist the Leucine Rich Repeats (LRRs) and second major class encoding for cytoplasmic proteins with a nucleotide binding site (NBS), third class of resistance genes consists of cytoplasmic protein possessing NBS and LRR pattern.

The fourth major class of resistance gene family related to NBS motif, attached to a transmembrane (TM) domain. The fifth major class consist of eLRR, a transmembrane (TM), and a serine-threonine kinase domain (KIN). The sixth class contains genes which having putative extracellular LRR, with a PEST (Pro-Glu-Ser-Thr) domain for protein degradation.

Table 3: Different classes of R Genes

Classes	*Structure*	*R Gene*	*Plant*
1	Toxin reductase	*Hm-1*	Maize
2	NBS-LRR-CC	*RPM1, RPS 2, I2*	*Arabidopsis*, rice, tomato
3	NBS-LRR-TIR	*N, L6 and RPP 5*	*Tobacco,* Flax
4	eLRR-TM	*Cf-2,Cf-4, Cf-9 and FLS-2*	*Arabidopsis,* tomato
5	eLRR-TM-KIN	*Xa21*	Rice
6	eLRR-PEST-ECS	*Ve1, Ve2*	Tomato
7	TM-CC	*RPW8*	*Arabidopsis*
8	TIR-NBS-LRR-NLS-WRKY	*RPS1-R*	*Arabidopsis*

The first disease resistance gene cloned and characterized was *Hm1* in maize (Johal and Briggs, 1992) this provides resistance to maize fungal pathogen *Cochliobolus carbonum* by inactivating the HC toxin production of this fungus.

6.6. QUANTITATIVE RESISTANCE (QR)

Host resistance may be occur either in qualitative or in quantitative manner. With quantitative resistance, different genotypes in a population present as apparent phenotypes and generally controlled by a major genes.

QR defined by Vale *et al.* (2001) as a resistance that varies in continuous way between the various phenotypes of the host population, from almost imperceptible (only a slight reduction in the growth of the pathogen) to quite strong (little growth of the pathogen).

It presents at various levels to nearly all important pathogens in most cultivars and some of the examples are (a) Barley leaf rust: *Puccinia hordei*, (b) Rice-Bacterial blight: *Xanthomonas campestris* pv. *Oryzae*, (c) Wheat yellow rust, (d) wheat leaf rust and (e) barley powdery mildew.

It reduces the rate of epidemic development and reduces the severity of disease, but does not imply anything about the number of genes involved.

Unlike qualitative plant resistance, that confers immunity to disease, qualitative resistance confers only a reduction in disease severity and this can be nonspecific.

From plan epidemiology point of view, quantitative or partial resistance serves as an alternative to qualitative or total resistance (Singh *et al.*, 2004).

This type of resistance can be used to reduce the size of the pathogen population at the same time as also avoiding exerting high selective pressure (Sapoukhina *et al.*, 2013).

This resistance type can be quantified either by a relative measure determining a rank in comparison with a known standard cultivar, or by an absolute measure determining a point on a defined scale of continuous variations.

In polycyclic pathogens the epidemic development during each season depends on level of QR; higher the levels indicate the lower epidemic build up.

QR may reduce the chance of infection, resulting in fewer lesions and consequently it reduces the growth of pathogen.

Against pathogen the three components can be distinguish which are related with infection frequency, lesion size and sporulation rate per lesion. Associated with lesion size and sporulatoin rate is the latency period *i.e.,* the period between infection and first spore production. A short latent period is good for QR. Latent period can be recorded as the period between inoculation and appearance of symptoms on leaves. It can be decided when 50% of visible lesions developed on leaves (Patel and Kulkarin, 2006)

Infection frequency can be measure by counting the number of pustules per unit area (cm^2) and it can calculate on the same leaves with latent period measurement

Per cent Disease Index (PDI) can be quantified through following equation:

$$PDI = \frac{Sum\ of\ individual\ disease\ ratings}{Number\ of\ observations\ assessed}\ x\ \frac{100}{Maximum\ disease\ rating}$$

Categorize includes as resistant (R; PDI=0-5%), moderately resistant (MR; PDI = 6-10%), moderately susceptible (MS; PDI =11-20%), susceptible (S; PDI=21-50%) and highly susceptible (HS; PDI>50%).

For lesion size observation can be carried out at various time intervals like 3, 5 and 5 days based on length and width of lesion (cm^2).

In the prototype epidemiological model developed so far, we can consider four components of resistance: for infection efficiency (IE): RRIE; for sporulation (SP): RRSP; for latency period duration (LP): RRLP; and for infectious period duration (IP): RRIP.

Resistance increases with smaller IE, SP, and shorter IP, resistance decreases with shorter LP. Thus, the equations for relative resistances will vary depending on the direction with which decreasing observed values of IE, SP, LP, and IP will be. Operational definitions are meant to use observations, and what is observed by breeders is always relative. Large breeding programs always have a reference. When it comes to partial resistance, one can never be sure to have, within a given field experiment, the highest possible level of resistance. But what is available is the currently lowest level of resistance, a reference for susceptibility, which can be used as a control. In the case of partial resistance, the best practical reference therefore is a susceptible cultivar, c.

Assuming that the cultivar to be tested is denoted x and that the control is denoted c, the operational definitions for components of partial resistances therefore can be written as relative resistance terms (RR), which are functions of x and c:

RRIE(x) = 1 – [IE(x)/IE(c)];

RRSP(x) = 1 – [SP(x)/SP(c)]; and

RRIP(x) = 1 – [IP(x)/IP(c)].

These equations are based on the assumption that a cultivar *x* will have values of IE, SP, and IP smaller than (or equal to) the susceptible control. Note that, because IE(x) < IE(c), SP(x) < SP(c), and IP(x) < IP(c), all these terms follow the above condition: 0 < RR < 1. This is because, at the highest possible level of observed susceptibility: IE(x) = IE(c), SP(x) = SP(c), and IP(x) = IP(c).

In the case of latency period duration, however, resistance will correspond to higher LP values. Thus a slightly different equation:

RRLP(x) = 1 - [LP(c)/LP(x)].

As indicated earlier, RRIE, RRSP, RRIP, and RRLP are relative resistance terms, each of which corresponding to a unique, non-overlapping, step of the infection chain: therefore, they can be called components of resistance.

The question then arises on how to combine these components, that is, how to express the relative resistance of a host genotype that carries different levels of each of the separate, independent, components of resistance. The *combined* relative resistance, RR_c, of a variety carrying several components of resistance must meet the following conditions

- $0 \leq RR_c \leq 1$;
- If all the relative resistances (RR_i) corresponding to all components (i) are null, then the combined relative resistance (RR_c) is null;
- If any one of the values of the p components is equal to 1, then $RR_c = 1$;
- If any one of the values of the components is ≠ 0, then $RR_c \neq 0$.

Such a set of conditions are met by the following equation

$$RR_C = 1 - \left\{ \prod_{1}^{p} (1 - RR_i) \right\}$$

where p is the number of components of resistance involved. For instance: if:

- RRIE = RRSP = RRIP = RRLP = 1, then RR_c = 1 - {(1 - RRIE) * (1 - RRSP) * (1 - RRIP) * (1 - RRLP)}= 1 - {(1-1) * (1-1) * (1-1) * (1-1)} = 1 - 0 and: $RR_c = 1$

- if: RRIE = RRSP = RRIP = RRLP = 0, then RR_c = 1 - {(1 - RRIE) * (1 - RRSP) * (1 - RRIP) * (1 - RRLP)} = 1 - 1 and: RRc = 0 if:
- RRIE = RRSP = RRIP = RRLP = 0.5, then RR_c = 1 - {(1 - 0.5) * (1 - 0.5) * (1 - 0.5) * (1 - 0.5)} = 1 - {1 - 0.5^4} = 1 - 0.0625 and: RR_c = 0.9375
- if one of the components of resistance takes a value of 1 (the disease cycle cannot proceed), for example the first component: RRIE = 1, while RRSP = RRIP = RRLP = 0, then RR_c = 1 - {(1 - 1) * (1 - 0) * (1 - 0) * (1 - 0)} = 1 - {1 - (0 * 1 * 1 * 1)} = 1 - 1 and: RR_c = 0

QR can assess numerically in the laboratory as well as in field by measuring components of resistance like rate of infection and invasion of the host tissue, reproduction unit of host tissue per unit time and comparing these with the performance of known standards. It can also be evaluated by measuring the amount of tissue affected by the pathogen, and it's this measure that can be influenced by differences in the level of disease pressure.

QR trait can be link with plant and leaf age, inoculums density and light and temperature.

The first step towards efficient assortment for quantitative resistance is to determine which of the components of quantitative resistance has to be principally considered. There is no single model available because each host-pathogen interaction is different. A comprehensive description of the three major components includes reduced infection frequency or density, reduced lesion size or concentration, and reduced propagule production per unit of host tissue over a period of time.

In most host-pathogen interactions involving biotrophic pathogens, the major epidemiological components of quantitative resistance are infection efficiency (infection frequency), length of the latent period, and rate and duration of sporulation.

6.7. GENE FOR GENE HYPOTHESIS

Many major resistance genes operate in a gene-for-gene way.

For each resistance gene in the host there is a corresponding avirulence gene in the pathogen (Flor, 1971; *Linum usitatissimum-Melampsora lini*), and only the corresponding avirulence gene can initiate the hypersensitive reaction (HR) leading to incompatibility.

Resistance and avirulence inherit in most cases in a dominant manner, susceptibility and virulence in a recessive way.

The HR is now known to result from the specific interaction at the cellular level of the product of the resistance gene and the product of the avirulence gene. If one of the two products is absent, there is no incompatibility; the normal pathogenicity of the pathogen results in a compatible reaction (the host appears susceptible).

It occurs most clearly and frequently in pathosystems where a biotrophic, highly specialized (in terms of host range) pathogen is involved such as the cereals with the various rusts, smuts and bunts, and with the powdery mildew.

The resistance in these systems is typically race-specific and also very easily neutralized by new races of the pathogen; the resistance is highly elusive, with the exception of pathosystems involving viruses

According to this interaction of a resistance gene in the host and its corresponding avirulence gene in the pathogen are as follows

		Pathogen genes	
		A	*aa*
Host genes	R	Incompatible	Compatible
	rr	*Compatible*	*Compatible*

For two genes of host and pathogen this hypothesis can be explained as follows

		Pathogen genes			
		A1A2	*A1a2*	*a1A2*	*a1a2*
Host	**R1R2**	Incompatible	Incompatible	Incompatible	Compatible
genes	**R1r2**	Incompatible	Incompatible	Compatible	Compatible
	r1R2	Incompatible	Compatible	Incompatible	Compatible
	r1r2	*Compatible*	*Compatible*	*Compatible*	*Compatible*

This hypothesis is use for study of following relationships:

- The source of pathogenic variability in pathogens
- The mutability of resistance and virulence genes
- Why host resistance is expressed under one set of conditions and not others
- Prediction of putative genotypes
- Race nomenclature
- Genetic dissection of complex loci

- Cataloguing and storing of R genes in the form of plant seeds or cuttings and V genes in the form of pathogen strains
- Management and deployment of resistance genes in space and time
- Detection of linkage and allelic relationship
- Geographic distribution of R and V genes
- Synthesis of multilines and multigene cultivars.

This relationships have also been studied in Wheat-*Puccinia recondita, P. striiformis, P, graminis tritici*; Maize-*P. sorghi;* Potato-*Phytophthora infestans;* Rice-*Pyricularia oryzae;* apple-*Venturia inaequalis* and tomato-*Cladosporium fulvum*

Vander Plank (1978) elaborated protein for proteins hypothesis as a biochemical explanation of gene for gene interaction. The protein for protein hypothesis states that in gene -for -gene diseases the mutual recognition of host and pathogen is not by the genes themselves but by their coded proteins.

Vander Plank (1978) hypothesized that in susceptibility the pathogen excretes a protein (virulence for product) into the host cell which copolymenizes with a complementary host protein (resistance gene product). This co-polymerization interferes with one auto regulation of the host gene that codes for the protein and by so doing turns the gene on to produce more protein.

In resistance, the protein specified by the gene for avirulence in the pathogen and excreted into the host does not polymerize with the protein coded for by the gene for resistance. It is not recognized by the host at all.

6.8. GENETIC VULNERABILITY

Vulnerability, a qualitative terminology, warns against the possible ills of extensive and intensive cultivation of genetically similar crop over a large area.

This reduced diversity and increased uniformity of space enable rapid buildup of pandemics.

Widespread cultivation of varieties of similar resistance (pure line cultivars) largely eliminated the competition amongst the clones of the pathogen, thus, interfering with stabilizing tendencies and giving rise to new biotypes with selective advantages

Historically, there are many horrifying examples of the devastating effects of relying on such genetic bases. Most of the losses are historically more associated with pathogens than they are with insect pests as the latter have less chance of mutation as compared to the former (Singh, 2002).

- The Irish Potato Famine resulted from the widespread of genetically uniform clone (of a single variety called Lumpers) and the outbreak of potato late blight (*Phytophthora infestans*) caused 80% yield reduction. As a result, millions of Irish people starved to death and other two million emigrated (Keneni, 2012).
- The wheat stem rust (*Puccinia graminis*) which devastated wheat fields in 1917, the elimination of all oats derived from the variety Victoria by Victoria blight (*Cochliobolus victoriae*) in the mid 1940s and the southern corn blight (*Helminthosporium maydis*) resulted in crop loss estimated to 15% yield reduction in corn (Rubenstein *et al.*, 2005) or in a loss of one billion USD in the early 1970s could be mentioned as a few examples.
- A number of other examples are also available from Asian countries like the great Bengal famine in India in 1943 due a devastative disease (*Cochliobolus miyabeanus*) to rice (Asfaw, 2004).
- An excellent example of devastation of that scale by insect pests was encountered over a century ago in France when grapevine was totally wiped out by attacks on root stocks of *Phylloxera vitifoliae* until a resistant cultivar was introduced from the USA (Singh, 2002).

Some causes of genetic vulnerability are as follows:

6.8.1. Narrow Genetic Basis

Narrow genetic base of the cultivated varieties is considered as the direct cause of genetic vulnerability (de Boef *et al.*, 1996; Rubenstein *et al.*, 2005; Smolders, 2006).

It could be assumed that, in natural population, the resistant individual to the existing race of the pathogen, whatever it is, produces more progeny than the susceptible individual because of the relative reproductive advantages. Each cycle of generation advancement to the next progeny may, therefore, result in further genetic improvement and adaptation. In a genetically uniform population, however, no individual can have a relative reproductive advantage over the other as they all are identical and no genetic change is expected as far as there is no mutational change in race of the pathogen, the environment and/or the host itself

Crop production characterized by genetic diversity is normally stable as compared to those characterized by genetic uniformity for a number of reasons.

- It would be difficult for the pathogen to evolve matching genes for large number of resistance genes existing in genetically heterogeneous genotypes as it does for homogeneous genotypes.
- It may be assumed that mixture cultivars may be characterized by more efficient use of growth resources due to either spatial or temporal complementarities. Spatial and temporal complementarities may occur when the associated crops make efficient use of resources over space and time.
- The buffering effects of the components of a mixed cultivar reduce risks from stress as all components of a mixed cultivar may not become susceptible to a given stress at a time
- When one component of a mixed cultivar becomes susceptible to a given stress, there may be a partial or full compensation by the other components because more resources would be available for compensatory growth by the latter.
- The reproductive advantage of the components of a mixed cultivar that resisted a given stress may result in more number of them in consecutive generations and each cycle of generation advancement to the next progeny may, therefore, result in further adaptation.

6.8.2. Widespread of Dominant Varieties

The problem is aggravated by the wide-spread use of one or a few genetically uniform varieties over a large hectarage (spatial dominance), which is considered the other prerequisite for the widespread of disease and insect pests as it provides ideal condition and creates vulnerability.

It is misleading that uniform varieties with narrow genetic base may succeed for the first year or two following release, become established on large hectare and then may experience some serious loss such as unexpected disease epidemics.

The continuous production of a single variety of the same crop year after year (use of monocultivar) will also facilitate diseases epidemics particularly when host plant resistance is defeated by the counteractive mutation of disease pests and pathogens.

6.8.2.1. *Boom and bust cycle*

The breakdown of disease resistance in plants is based on two fundamental properties of microbial pathogens: their great genetic variability and their high rates of reproduction. Together these properties allow pathogens to rapidly change the frequencies of virulence genes in their populations and thus adapt to the presence of resistance genes in the host.

It may be expected that when a new variety carrying VR is developed, it is resistant to the prevalent pathotypes of the pathogen in question. Therefore, it becomes popular rapidly and occupies a large proportion of the growing area under the concerned crop. This period of boom (high popularity and cultivation in a large proportion of area) continues till the pathotypes specific to the VR gene (s) employed in the variety evolves and /or increases in frequency to cause an epidemic in the field occupied by the variety. The now susceptible variety loses popularity and area under it declines sharply thus the variety is effectively busted by the virulent Pathotypes. This has been called 'boom and bust' cycle.

When a variety goes bust, breeders develop new variety with a new VR gene, which makes it resistant to the predominant pathotypes. This variety also undergoes the 'boom and bust' cycle, and the breeders put forth a new variety and so on. The period for which such a variety may last is , on an average, -5 years in the case of air-borne fungal pathogens. But in many cases, this period is much shorter, in extreme cases, the variety go bust even before it reaches farmers fields.

The reason for boom and bust cycle line in the genetic system of the pathogen and the selection pressure exerted by the host on the pathogen pathotypes. When a resistant variety becomes popular it is cultivated on large areas.

The plants of this variety act as strong selection agents for the pathotypes specific to the new variety since only this pathotypes can cause disease in the plants of this variety. Therefore, the specific pathotypes increases rapidly in frequency in the pathogen population. This specific pathotypes may be newly evolved or may have already been present in the pathogen population, albeit, in a very low frequency.

Soon the frequency of the virulent pathotypes reaches a stage at which it causes an epidemic in the thus far resistance variety (Fig. 2).

Usually, the epidemic would occur after a couple of or more years following the first detection of the virulent pathotypes. The popularity of and the area under the variety will define rapidly following the epidemic. This will reduce or eliminate the selection pressure in favour of the pathotypes virulent on the variety. As a result, the frequency of this pathotypes in pathogen population will also decline.

Example of this phenomenon occurred in Festival variety of wheat in Australia and IR-26 in Kerala, India.

6.8.3. Failure of Vertical Resistance

Varietal resistance may fail in a shorter period than it takes for the development of a new variety.

Failure of vertical resistance is considered as another causes of genetic vulnerability connected with a narrow genetic base.

Vertical resistance is believed to be effected by a single or a few genes (monogenic or oligogenic) as opposed to horizontal resistance which is controlled by many genes (polygenic).

It is, therefore, highly specific and liable to changes in races of pathogens.

It is easier for a pathogen or an insect pest to overcome vertical resistance as it is a matter of defeating a single or a few genes of the host through the counteractive generation of the corresponding virulent genes through mutation.

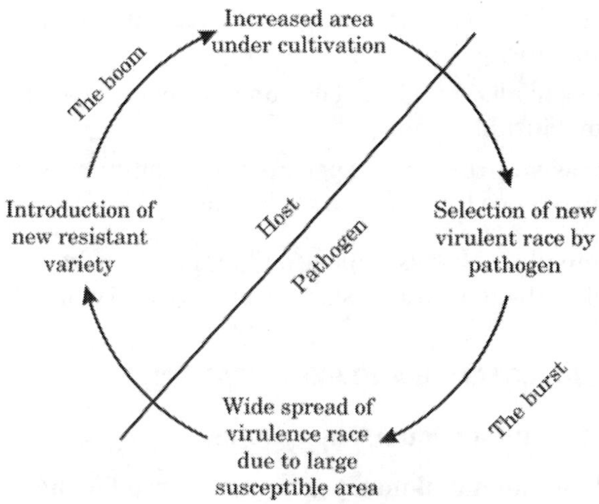

Fig. 2: Boom and Burst Phenomenon

6.8.3.1. Vertifolia effect

Vander Plank introduced the term Vertifolia effect derived from the German potato variety 'Vertifolia' having the late blight resistance genes R3 and R4. The variety became susceptible when the appropriate pathotypes P (3, 4) evolved in the pathogen leading to a nearly complete failure of the crop; such a total failure of vertical resistance leading to a disease epidemic is known as Vertifolia effect. Thus, Vertifolia effect refers to an epidemic development in a variety carrying vertical resistance genes and a low level of HR, leading to heavy economic losses.

The failure occurs because of the following two reasons: 1) The level of horizontal resistance in the varieties carrying oligogenes for resistance is usually low, and 2) The pathogen is able to evolve the virulent Pathotypes specific to the oligogenes for vertical resistance.

According to this a host must have horizontal and vertical resistance for avoiding the epidemic.

6.9. GENETIC DURABILITY

Resistance is considered durable when it remains effective for a considerable time, notwithstanding wide revelation and for example Rpg1gene is durable but not last forever.

According to Vale *et al.*, (2001) genetic durability having three major components.

- Resistance to pathogen with a wide host range are usually quantitative in nature and always durable.
- QR against specialist and based on some to several genes with additive effects seems durable, and
- Monogenic resistance against specialist of a non-hypersensitive nature and such resistance is often quite durable.

Details about durable resistance particular as a key to sustainable management of pathogens and pests can find with Mundt (2014).

6.10. VERTICAL RESISTANCE AND EXPLOITATION

6.10.1. Varieties with Individual Major Gene

In this method an individual major gene is referred for the prevalent race of pest info well adapted cultivar through backcrossing.

Resistant plants are selected from segregating population and backcrossed to the recurrent parent.

Generally, 5-6 backcrosses are sufficient to retain the genotype of recurrent parent with added resistance to specific race of pathogen.

This approach is widely used to control the prevalent race of pathogen. Transfer of major genes from one host to another is very simple.

The main disadvantages of this approach are that major gene may sometimes become susceptible to the minor prevalent races.

6.10.2. Gene Deployment

Another way of exploiting vertical resistance is gene deployment. Gene deployment refers to planned or strategic use of major genes in development of resistant cultivars for various geographical areas.

It is a guided distribution of genes in space and time.

This approach helps in preventing disease and maintaining the diversity for major genes in different geographical regions.

This approach requires a number of genes with similar effectiveness for control of prevalent races.

However, it is impossible to find out different genes with similar effectiveness. This is the main drawback of gene deployment approach.

Gould (1983) categorize the gene deployment strategies in to two broad categories as (a) spatial *i.e.* gene deployment at the farm level, at the field level and at the region level and (b) temporal.

6.10.2.1. *Gene deployment at the field level*

Spatial resistance gene deployment can be done within a small geographic area like a single field.

Gene deployment at the field level includes variety mixtures, species mixtures and multi blends.

6.10.2.1.1. *Variety mixtures*

It refers to a homogenous, spatial mixture of different genotypes of one plant species in a field.

Concept of mixtures is to protect individuals possess one particular resistance with a range of other possess other resistance genes.

Disease resistance in variety mixture relay on slowing down the development of the best- adapted race on each component.

The mechanism through which variety mixture reduces diseases involves (a) decrease in susceptible tissue and therefore a decrease in inoculums potential within the mixture, (b) an increase in distance for spores to move from one susceptible host to other, (c) the physical barrier of resistant plants and probably by cross-protection whereby the defensive mechanism of one component of the mixture may be activated by an a virulent isolate from another component.

6.10.2.1.2. *Development of multiline*

The term multiline variety was first defined by Jensen (1952) as a mix of multiple pure lines, each of which is of a different genotype.

Multiline variety is a mixture of several pure lines of similar phenotype(height, seed color flowering time, maturity time and various other agronomic characteristics) but have different genes for the character under consideration the disease resistance means these are isogenic lines. At the same time they do not reduce the yielding ability of each other when grown in mixture (*i.e.* compatible).

Multiline refers to seed mixture of isoline, related lines or unrelated lines. The seeds of genotypes with individual major gene are mixed together. This approach is used in self pollinated species. Multiline are mixtures of lines bred for phenotypic uniformity of agronomic traits while, cultivar mixtures are mixtures of agronomically compatible cultivars with no additional breeding for phenotypic uniformity

The resistant genes to different races of a pathogen are transferred into different genotypes by separate backcross programme.

Suppose there are five major genes each for a different race of pathogen. These five major genes will be transferred in five different backgrounds. At the end, seed of five resistant genotypes thus developed will be mixed together to constitute multiline. Thus multiline provides protection against several races of pathogen.

The new race of a pathogen whenever occurs will attack few genotypes of a multiline. All the component genotypes of a multiline will never become susceptible to new race of pathogen. The resistant components in a multiline will act as a barrier and delay the spread of the new race to susceptible plants.

Multiline varieties have many advantages for example, by using multiline we can exploit horizontal resistance, which is otherwise difficult to achieve because of the complexities associated with its genetic mechanism.

Multiline varieties proved to live longer than their corresponding pure stands because of stabilized selection.

Other advantageous features of this approach compare to variety mixtures is that all the component lines are phenotypically similar and the only difference being the gene for resistance.

Wheat variety Kalyan sona is the suitable example to explain the concept. This variety was originally resistant to brown rust. Later on became susceptible to new races of pathogen. Several pure lines with different resistance genes are produced through backcross breeding using one recipient or recurrent parent. The donor parents are the one with different genes for the disease resistance, every donor parent is used in separate back cross program. Because of this each line receives different gene for disease resistance according to the type of pathogen. Five to ten of such lines with different alleles for disease resistance are mixed to develop multiline variety. The lines to be mixed are determined by the races of the pathogen relevant to the area considered.

There are two main disadvantages of this method. First, several backcross programmes have to be launched to transfer each major resistant gene in a separate background. Second, the recurrent parent limits agronomic performance of the multiline. Moreover, it requires extensive testing against specific races.

6.10.2.2. *Gene deployment at the farm level*

A huge area with a single variety or many varieties with similar resistance parameters, would be vulnerable to virulent pathogen. The practical way to avoid such conditions is to implement diversification of varieties that possess different resistance for specific diseases.

For successful diversification breeder need complete knowledge about verity genetics and its durability.

This approach is primarily used in cereal varieties for reduction of mildew and yellow rust.

6.7.2.3. *Gene Deployment at the Regional Level*

Regional gene deployments are providing consistent resistance in different regional scale.

This approach have been proposed as a mean of controlling pathogens that disperse long distance over well-defined pathways and basically whose full geographic distribution has to be re-established annually.

For regional deployment strategies, the controls that would be needed around farmer-to-farmer seed trading and the use of farmer-saved seed, and agreements from breeders to forgo using some resistance resources are very substantial (Burdon *et al.*, 2014).

6.10.3. Gene Pyramiding

Gene pyramiding refers to incorporation of two or more major genes in the host for specific resistance to a pathogen.

This is differs from multiline, where each major gene is incorporated into separate genotype and then resistant genotypes are mixed together.

In pyramiding approach all major genes are incorporated in a single cultivar.

Combination of several major genes for specific resistance provides protection against several new races that may develop in the pathogen.

The new varieties can be developed with a contribution of 2, 3 or more major genes. Such varieties have durable resistance.

Objectives of gene pyramiding include:

- Enhancing trait performance by combining two or more complementary genes,
- Remedying deficits by introgressing genes from other sources,
- Increasing the durability of disease and/or disease resistance, and
- Broadening the genetic basis of released cultivars .

Traditionally, gene pyramiding is mainly used to improve qualitative traits such as disease and insect resistance. This is associated with the fact that the presence of target trait genes must be confirmed by phenotyping mostly at the individual level and that individual phenotypic performance is a good indicator of the genotype only if genes have a major effect on phenotypic performance and the error of phenotyping is minimal.

Main factors affecting gene pyramiding

Characteristics of the target traits/genes

- When the genes to be pyramided are functionally well characterized and markers used for selection equal to the gene itself (perfect markers), gene pyramiding will be more successful. For qualitative traits controlled by one or a few genes, the identification of the genes and tightly linked markers is easier provided phenotyping is carefully conducted. One or two markers per gene can be used for tracing the presence/absence of the target genes.
- When the target genes are QTL with moderate or small effects, pyramiding may be less successful due to the following reasons. Firstly,

the identified QTL may be more likely to be a false positive. Secondly, inaccurate QTL localizations result in the need to select for more marker loci covering large genomic segments to be certain that target QTL alleles are retained in selected. Thirdly, QTL effects may be specific to a particular genetic background.

Reproductive characteristics

- The propagation capability of a crop is determined by the number of seeds produced by a single plant.

Operating capital

All breeding programmes are operated within the limits of available operating capital. Therefore, reducing the overall cost is always an important consideration when choosing a strategy. In addition to the use of the most economic mating and testing approaches, other factors affecting the cost also need to be considered. In the context of gene pyramiding, cost affects both what can be achieved and how to achieve Increasing the number of generations (duration) wills it. Increasing the number of generations (duration) will reduce the pressure on population size required in each generation and may result in the reduction of the total cost. However, increasing the duration delays the release of the new cultivar and consequently reduced market share.

6.11. REFERENCES

Asfaw, A. 2004. Alternate approaches in deploying genes for diseases resistance in crop plants. *Asian Journal of Plant Sciences,* 3(5): 618–623.

Barbercheck, M.E. and Von Broembsen, S.L. 1986. Effect of soil solarization on plant parasitic nematodes and *Phytophthora cinnamomi* in South Africa. *Plant Disease*, 70: 945–950.

Berger, R.D. 1981. Comparison of the Gompertz and logistic equations to describe plant disease progress. *Phytopathology*, 71: 716–719.

Burdon, J.J., Barrett, L.G., Rebetzke, G. and Thrall, P.H 2014. Guiding deployment of resistance in cereals using evolutionary principles. *Evolutionary Application*, pp. 609–624.

Campbell, C.L. and Madden, L.V. 1990. Introduction to Plant Disease Epidemiology. John Wiley and Sons, New York, NY, USA.

Chellemi, D.O., Olson, S.M. and Mitchell, D.J. 1994. Effect of soil solarization and fumigation on survival of soil borne pathogens of tomato in Northern Florida. *Plant Diseases*. 78: 1167–1172.

Chen, X. 2013. High-temperature adult plant resistance, key for sustainable control of stripe rust. *American Journal of Plant Sciences*, 4: 608–627.

De Boef, W.S., T. Berg. and B. Haverkort. 1996. Crop genetic resources. *In*: Bunders, J., Haverkort, B. and Hiemstra, W. (*eds.*). Biotechnology; building on farmers' knowledge. Macmillan, London and Basingstoke, pp. 103–128

Gould, F. 1983. Genetics of plant herbivore system: Interaction between applied and based study. *In*: Derno, R.F. and McClure M.S. (*eds.*). Variable Plants and Herbivores in Natural and Managed Systems. Academic Press, New York, pp. 599–653.

Grinstein, A., Kritzman, G., Hetzroni, A., Gamliel, A. Mor, M. and Katan, J. 1995. The border effect of soil solarization. *Crop Protection*, 14: 315–320.

Gururani, M.A., Venkatesh, J, Upadhyaya, C.P., Nookaraju, A., Pandey, S.K. and Park, S.W. 2012. Plant disease resistance genes: Current status and future directions. *Physiological Molecular Plant Pathology*, 78: 51–65.

Jensen, N.F. 1952. Intravarietal diversification in oat breeding. *Agronomy Journal*, 44: 30–34.

Johal, G.S. and Briggs, S.P. 1992. Reductase activity encoded by the *Hm1* disease resistance gene in maize. *Science*, 258: 985–987.

Katan, J., Fishler, G. and Grinstein, A. 1983. Short and long term effects of soil solarization and crop sequence on *Fusarium* wilt and yield of cotton in Israel. *Phytopathology*, 73: 1215–1219.

Keneni, G., Bekele, E., Imtiaz, M. and Dagne, K. 2012. Genetic vulnerability of modern crop cultivars: Causes, mechanisms and remedies. *International Journal of Plant Research*, 2(3): 69–79.

Mohapatra, N.K., Mukherjee, A.K., Suriya, A.V. and Nayak, P. 2008. Disease progress curves in the rice blast pathosystem compared with the logistic and gompertz model. *ARPN Journal of Agricultural and Biological Science*, 3(1): 28–37.

Mundt, C. 2014. Durable resistance: A key to sustainable management of pathogens and pests. Infection, *Genetics and Evolution*, 27: 446–455.

Mysore, K.S. and Ryu, C.M. 2004. Non-host resistance: How much do we know? *Trends in Plant Sciences*, 9: 97–104.

Nutter, F.W., Jr. 2007. The role of plant disease epidemiology in developing successful integrated disease management programs. *In*: General concepts in integrated pest and disease management. Ciancio, A. and Mukerji, K.G. (*eds.*), pp. 45–79.

Nutter, F.W., Jr. and Guan, J. 2001. Disease losses. *In*: Encyclopedia of Plant Pathology. Maloy, O.C. and Murray, T.D. (*eds.*). John Wiley and Sons, Inc., New York pp. 340–351.

Nutter, F.W., Jr. and Parker, S.K. 1997. Fitting Disease Progress Curves Using EPIDEMIOL. Exercises in Plant Disease Epidemiology", Franc, L.J. and Neher, D.A. (*eds.*). APS Press, St. Paul, MN, pp. 24–88.

Panicker, S. and Ramraj, B. 2010. Studies on the epidemiology and control of *Ascochyta* blight of peas (*Pisum sativum* L) caused by *Ascochyta pinodes. Archives of Phytopathology and Plant Protection*, 43(1): 51–58.

Patel, V.S. and Kulkarni, S. 2006. Slow leaf blighting in wheat varieties caused by *Exserohilum hawaiiensi. Agricultural Science Digest*, 26(3): 157–161.

Raio, A., Zoina, A. and Moore, L.W. 1997. The effect of solar heating of soil on natural and inoculated agrobacteria. *Plant Pathology*, 46: 320–328.

Rao, V.K. and Krishnappa, K. 1995. Soil solarization for the control of soil borne pathogen complexes with special reference to *M. incognita* and *F. oxysporum* f.sp. *ciceri. Indian Phytopathology*, 48: 300–303.

Raoof, M.A. and Nageshwar Rao, T.G. 1997. Effect of soil solarization on castor wilt. *Indian Journal of Plant Protection*, 25: 154–159.

Rubenstein, D.K., Heisey, P., Shoemaker, R., Sullivan, J. And Frisvold, G. 2005. Crop genetic resources: An economic ap-praisal. United States Department of Agriculture (USDA). Economic Information Bulletin No. 2. (www.ers.usda.gov)

Sapoukhina, N., Paillard, S., Dedryver, F. and Vallavieille-Pope, C. 2013. Quantitative plant resistance in cultivar mixtures: Wheat yellow rust as a modeling case study. *New Phytologist*, 200(3): 888–897.

Sharma, T.R., Das, A., Thakur, S. and Jalali, B. 2013. Recent understanding on structure, function and evolution of plant disease resistance genes. *Proceeding Indian National Science Academy*, 80(1): 83–93.

Singh, B.D. 2002. Plant breeding: Principles and methods. Kalyani Publishers, New Delhi-Ludhiana

Singh, R.P., William, H.M., Huerta-Espino, J. and Rosewarne, G. 2004. Wheat rust in Asia: Meeting the challenges with old and new technologies. *In*: New directions for a diverse planet, Proceedings 4th Int Crop Science Cong, 26 September – 1 October 2004, Brisbane Australia, pp. 1–13.

Smolders, H. 2006. Enhancing farmers' role in crop devel-opment: Framework information for participatory plant breeding in farmer field schools. Centre for Genetic Re-sources, Wageningen University and Research Centre, The Netherlands.

Stam, R., Mantelin, S., McLellan, H. and Thilliez, G. 2014. The role of effectors in non-host resistance of filamentous plant pathogen. *Frontiers in Plant Science*, 5: 1–6.

Vale, F.X.R.D., Parleveliet, J.E. and Zambolim, L. 2001. Concepts in plant disease resistance. *Fitopatologica Brasileria*, 26: 577–589.

Verma, R.K. Singh, Y., Soni, K.K. and Jamalluddin, 2005. Solarization of forest nursery soil for elimination of root pathogens and weeds. *Indian Journal of Tropical Biodiversity*, 13: 81–86.

Xu, X. 2006. Modeling and interpreting disease progress in time. *In*: The epidemiology of plant disease. Cooke, B.M., Gareth Jones, D. and Kaye, B. (*eds.*). Springer, Dordrecht, The Netherlands, pp. 215–238.

CHAPTER 7 Academic Gains (Short, Descriptive and Objective Questions)

7.1. SHORT QUESTIONS

1. Provide comparative losses caused by historical plant diseases.
2. What do you understand from epidemic? What are the two essential criteria's for a disease termed as epidemic?
3. Epidemic may be partitioned into?
4. What is Van-derplank condition for epidemic (threshold theorem) and what is meaning of anti-epidemic?
5. What are the elements of epidemic? What do you understand from disease pyramid and tetrahedron? Describe how human activities involve in epidemic development.
6. Classify the disease on the basis of their geographic distribution.
7. What are the primary elements of an epidemic? Provide their comparative list.
8. List various collateral and alternate host for different rust disease of wheat.
9. What is inoculums and inoculums potential?
10. Provide your comments with example how introduction of new pathogen in an area aggravate epidemic development.
11. What do you mean by auto and allo-infection and how these two contribute in plant epidemic?
12. Describe in brief how environmental factor affecting epidemic development?
13. What is latent/incubation period and infection period.
14. Enlist some Indian works related with important epidemic occurred in country.

15. Enlist some of the computer programme related with epidemic simulation.
16. Describe contribution of J. J. Van-derplank in plant disease epidemiology
17. Describe contribution of Van Everdingen in plant disease epidemiology
18. Describe contribution of E. C. Large in plant disease epidemiology
19. Describe contribution of R. A. Krause in plant disease epidemiology.
20. Describe contribution of J. C. Zadok in plant disease epidemiology.
21. Describe contribution of J. Kranz in plant disease epidemiology.
22. Describe contribution of K. C. Mehta in plant disease epidemiology
23. Describe contribution of S. Nagrajan in plant disease epidemiology.
24. Describe contribution of Laurence Madden in plant disease epidemiology
25. Comment on disease development is a cyclic process.
26. How can we check high birth rate and low birth rate pathogen?
27. Comment on monocyclic pathogen.
28. Provide example of some soil borne and seed borne monocyclic pathogen.
29. What is polycyclic pathogen.
30. *Phytophthora infestans* is a polycyclic pathogen comment on it.
31. What are the essential criteria for a pathogen to call as polycylic?
32. Give an example for epidemic having both monocyclic and polycyclic in nature.
33. What is polyteic pathogen.
34. Explain the mathematical expression of monocyclic and polycyclic diseases with attributes like Q, R and T.
35. If $Q = 0.77$, $R = 0.0015$ and T is 80 days then what is epidemic progress if initiated by monocyclic pathogen.
36. If $X_0 = 35\%$, $T = 120$ days and $R = 0.12$ then find out the final incidence of decease.
37. How can we reduce the epidemic initiated by monocyclic ($X= QRT$) and Polycyclic ($X= X_0e^{rt}$) pathogens.
38. How the epidemic can be control through reducing amount of initial inoculums?
39. How the epidemic can be minimized through reducing the time available for its development?
40. How the epidemic can be delay by reducing the rate of disease incidence (r)?

41. What are the different models for quantification of temporal population growth?
42. Describe characteristic of monomolecular, exponential, logistic and Gompertz models.
43. What do you understand by simple interest and compound interest diseases?
44. What are the different phases during course of epidemic?
45. What are slow and rapid epiphytotic?
46. What is Disease Progress Curve (DPC) and what kinds of information's it provide?
47. Give examples of DPC of few diseases.
48. What is disease gradient curve?
49. What is difference between dispersal and disease gradient?
50. What is take off level and Zero date?
51. Explain epidemic rate (r) for various kind of diseases.
52. What are demerits of estimation of percentage infection?
53. Define Area Under Disease Progress Curve (AUDPC).
54. What are the two assumptions can be made through AUDPC model?
55. In plant pathological research, AUDPC can be widely use for two area, explain these two areas.
56. Explain AUDPC expression for two data point that is initial and final assessment.
57. What are the problems associated with AUDPC and provide their probable solutions?
58. What are three attributes of disease measurements?
59. Comments on disease incidence.
60. Comments on disease severity.
61. Define inoculum potential and disease potential.
62. Comment on disease severity.
63. What is epidemic threshold?
64. What do you mean by economic damage?
65. What is economic injury level?
66. What is warning threshold?
67. What is the literal meaning of yield loss?
68. What is damage curve and define its various attributes?
69. Comment about frequency of assessment for monocyclic and polycyclic diseases.

70. What types of relationships between two pathogen can be expected presented at same host?
71. Define pathogen spillover.
72. What is threshold theorem? Who coined this term and what its epidemiological significance?
73. What is apparent rate of infection and basic rate of infection. Describe them with their respective mathematical expressions.
74. Provide the explanation on why " toward the end of the epidemic there may be increase in r but decrease in R.
75. Describe exponential growth model.
76. What are the steps through which one can select appropriate population growth models?
77. What is Linked Differential Equation (LDE) and mentioned its advantages?
78. Provide a comparative account for compartmental models used in botanical epidemiology.
79. What are the demographic and environmental stochastic? and what are their significance in plant epidemic?
80. What do you understand from dispersal kernels?
81. From landscape point of view what are the factors affects the epidemic?
82. Draws diagram related with disease model dynamics.
83. What are the roles of quantification models in botanical epidemiology?
84. What are biological and dynamical issues need to consideration for modelling any botanical epidemiology?
85. What do you mean by compartmental models? Provide variables related with some compartmental models.
86. What is cryptic infection
87. What is the modified version of SIR model for soil borne plant pathogen? Explain its mathematical expression.
88. How can we incorporate cryptic infection in SIR model?
89. Explain SEIR model
90. What is basic reproductive ration (R_0)
91. What is SIDR model?
92. Describe compartmental model for vector borne plant epidemic. Give its equation and its interpretation criteria based on basic reproduction ration.
93. How do you quantify infection rate with particular example of *Pythium* and *Phytophthora*?

94. Colony diameter can be quantified through.
95. What is latent period and how it can be calculate?
96. How do you quantify spore released from a infected portion?
97. What is spore dispersal and what are factors affecting it?
98. How do you quantify lesion expansion?
99. What are simulations? Enlist some of the simulation model utilized in plant disease epidemiology?
100. What are hypothesis testing and its utility in plant disease epidemiology?
101. What are elements of the Null hypothesis and type I and II errors?
102. What are continuous and categorical data's?
103. What is alpha level?
104. On what basis hypothesis can be reject and what factors affecting it?
105. Describe sample size in plant pathology context
106. Describe and compare parametric and non-parametric tests.
107. What are advantages and disadvantages of non-parametric test?
108. Provide characteristic features of variable types.
109. What is Kruskal-Wallis test: Describe specific conditions for its applications
110. Describe Mann-Whitney U test
111. Describe Wilcoxon signed rank test
112. Describe Kendall's test of concordance
113. What are the drawbacks associated with the Null Hypothesis.
114. What is the Bayesian theorem? Describe essentials and how this approach is useful for botanical epidemiology.
115. What is ordination analysis and how it is differ from classification?
116. What are the Principal Component and Principal Coordinate Analysis?
117. What do you understand from Non-Metric Multi-dimensional Scaling (NMDS). How its better then linear ordination?
118. What is correspondence and canonical correspondence analysis?
119. Define spatial patterns and their advantages.
120. How spatial patterns help in plant disease studies?
121. Write specific role of spatial patterns in plant disease epidemiology
122. What are the different methods for characterization of spatial aggregation for disease incidence?
123. Describe different ways for collecting data's for spatial paternity in botanical epidemiology.

124. What are the issues related with spatial dynamics of epidemics?
125. Enlist spatial analysis tools generally utilize for epidemic studies.
126. Describes the uniform, the random and the cluster patterns.
127. What is dispersal gradient? List out few studies conducted for it.
128. Mentions the different approaches use for study of wind disperse spores.
129. Comments on spatio-temporal dynamics of airborne inoculums.
130. What is basic reproduction number?
131. Draw various steps for determining the appropriate spatial pattern analytical tool.
132. What do you understand from Complete Spatial Randomness (CSR)?
133. Define the conditions to carry out CSR test.
134. What is first and second order characteristic?
135. What is the heterogeneous possion process?
136. Enlist various tools for testing CSR.
137. What is the Diggle tests of spatial randomness?
138. What is the quadrat counting test and provide factors that affecting its power?
139. Comments on Kolmogorov- Smirnov test.
140. What are the frequency distribution for spatial pattern analysis?
141. What do you understand from Spatial Analysis by Distance Indices (SADIE)?
142. Write note on Blocked Quadrat Variance (BQV), Two Term Local Quadrat Variance (TTLQV), Paired Quadrat Variance (PQV) and Random Paired Quadrat Variance (RPQV).
143. Define spatial autocorrelation. What are the advantages of this approach over negative binomial and Poisson distribution?
144. Poisson probability distribution provides information about?
145. What are the different steps for Poisson probability distribution?
146. Write a note on uses of binomial and negative binomial distribution in plant pathology.
147. Define conditions on which negative binomial models applies.
148. What are the two important parameters associated with negative binomial?
149. What are the parameters associated with binomial and beta binomial distribution?
150. Draw a flow diagram for selecting appropriate equation for estimation of negative binomial exponent (K).

151. Provide a comparative statement of various binomial distributions.
152. Enlist spatial indices for natural sampling
153. What is the Index of Dispersal (I_d) and its interpretation criteria?
154. What is the index of clumping?
155. What is the Green index?
156. What is the Lloyd's mean crowding?
157. What is the Morisita index of dispersion?
158. What do you understand from Standardized Morisita Index (MIS)?
159. Describe the Taylor power low on the background of disease incidence data.
160. What do you mean by Doublet and Run analysis?
161. What is the spatial point pattern analysis?
162. What is the first order effect for point pattern?
163. What is the Kernel estimation?
164. Describe various methods for dealing with edge effect.
165. What is the meaning of quadrant count method?
166. Describe types of quadrat variance method.
167. Provide comparison between Blocked Quadrat Variance and Paired Quadrat Variance methods.
168. What is Block Quadrat Variance, its types and limitations?
169. Provide a comparative account of 3TLQV, 4TLQV, 9TLQV and TTLQV
170. What is T-square method?
171. Explain limitations of indices (like T-square index, Hopkins index) based upon nearest neighbor and random point to host/pathogen distance.
172. What is individual-centre analysis? Does it related with species association analysis?
173. What is O-ring statistics and its advantages?
174. What is the Kernel estimation of Intensity?
175. What is nearest neighbor analysis?
176. Define spatial autocorrelation
177. What is negative and positive spatial autocorrelation?
178. What do you understand from redundant information?
179. What is the Moran coefficient and Geary ratio?
180. What are factors responsible for spatial autocorrelation?
181. What is global and local autocorrelation?

182. Define contiguity. Draw the Rook, the Bishop's and t he Queen moves.
183. Enlist different tests for global autocorrelation.
184. Define Lag.
185. Describe Gamma index of spatial autocorrelation.
186. What is Moran's I index?
187. Compare Moran's I and Geary's C indices
188. What is correlogram
189. Define the variogram
190. What is semivariogram
191. What are basic variogram models?
192. What is nugget and sill?
193. Define Kriging
194. Defines forecasting and what are its advantages
195. Enlist the information's requires for developing forecasting systems
196. Plant disease forecasting having applied values for what types of diseases?
197. Appropriateness of any forecasting system depends on.
198. What are the different methods for forecasting disease?
199. What is predictive system and how its differ from forecasting?
200. What are mechanistic and empirical approaches of predictive system?
201. What is Disease Pressure Index?
202. What are attributes of forecasting models?
203. Write a note on weather based plant disease prediction system.
204. What are the "Four Dutch Rules" for late blight of potato?
205. Give a comparative account about forecasting system developed for late blight of potato.
206. What is BLITECAST and how it work?
207. Classification of crop-weather models.
208. Differentiate between primary and secondary inoculums.
209. What is indexing? Provide index host for different pathogen.
210. Draw a flow chart related with forecasting method based on amount of inoculums
211. Provide details about forecasting based on primary inoculum.
212. Forecasting based on secondary incoculum.
213. Give details for forecasting for pea root rot.

214. Forecasting for the fire blight of apple and pear.
215. Forecasting for Apple scab.
216. Forecasting for Leaf spot of peanut.
217. Forecasting for early blight of celery.
218. Forecasting for blister blight of tea.
219. Forecasting for sugar beet yellow.
220. What is the Indian Stem Rust Rules?
221. Write strategies for disease management particularly when infection rate (r) is high and during low situation.
222. What is sanitation. How sanitation is related with epidemic management?
223. What is sanitation ration?
224. What is exclusion, avoidance, eradication and protection
225. Charcoal rot disease (caused by *Macrophomina phaseolina*) of pearl millet infection on January 12th was 3.4%. Fowling sanitation practices the disease severity reduces to 0.2% by January 20th. In a separate distance field without fallowing sanitation, disease severity increase to 10.5% on January 20th. Calculate the time delay in epidemic growth due to sanitation.
226. What is crop rotation?
227. How removal of alternate and collateral hosts helps in reducing epidemics.
228. What is rouging and how its differ from sanitation
229. What is soil solarisation and for what kind of pathogen its is most meaningful?
230. How host resistance is helpful in reducing epidemic
231. Define non-host resistance
232. What are the different classes of R gene?
233. What is quantitative resistance?
234. What is percent disease index?
235. What do you mean by Gene-for Gene hypothesis
236. What is protein for protein hypothesis?
237. What is boom and Bust cycle?
238. What is the Vertifolia effect?
239. What is gene deployment?
240. What are multiline?
241. What is gene pyramiding?

7.2. DESCRIPTIVE QUESTIONS

1. How different host factors contribute in epidemic development
2. What are predisposing factors?. Describe some of them for few fungal and bacterial diseases.
3. List various pathogenic factors essential for epidemic development.
4. Describe various milestone related with past epidemiology. Provide list of some important epidemic, their casual agent and geographical extent.
5. Explain with example that environmental or physical factors are crucial for determining the pathogen as a monocyclic.
6. What do you understand by Area Under Disease Progress Curve? and explain its mathematical expression
7. Describe epidemiological forms in view of its quantitative perspective and enlist the various quantitative tools available for its measurement.
8. What are the challenges associated with plant disease modeling? Provide their present approaches and possible future thrust.
9. Provide basic of attribute of various growth models utilize for quantification of disease progress curve.
10. What do you understand by SIR model? Explain its mathematical express. Based on its result, how can we interpret about epidemic probability? What is the role of SIR model in management practices?
11. How test association helps in plant pathological studies. What variables are needed for its quantification?
12. Provide mathematical expression and interpretation's related with the Ochiai index, Dice index and with the Jaccard index. Explain basic differences between them.
13. What are the different methods of ordination analysis? Provide their basic approaches and merits and demerits.
14. What is the spatial heterogeneity? What are the factors affecting it?
15. Provide details account on first and second order test.
16. What are distance models? Briefly describe the various methods linked with this approach.
17. What is Ripley's K function? Provide details about its quantification part. How it's better then nearest neighbor distance method. What are its limitation?
18. Provide details about information's needed for disease forecasting.
19. Describe pathogen disease cycle and its role in forecasting.
20. Give a comparative account of forecasting works carried out in the India

21. What is computer simulation model for predicting plant diseases. Give suitable examples.
22. What is aerobiology. Provide some milestones.
23. Provide a compressive account on control of epidemic other then chemical methods.
24. Provide a comparative account of types of resistance on the basis of their separation
25. What is genetic vulnerability and factors causes for genetic vulnerability?
26. What are different approaches for genetic durability?

7.3. OBJECTIVE QUESTIONS

1. Who coined term simple interest disease

   A. Van-derplank B. L. R. Jones

   C. E. C. Large D. K. C. Mehta

2. Low birth pathogen nature are

   A. Non- Systemic B. Systemic

   C. Both D. Obligate

3. Low birth rate pathogen effectively control through

   A. Chemical methods B. Sanitary practices

   C. Biological methods D. None

4. Nature of high birth rate pathogen is

   A. Nacrotrophs B. Systemic

   C. Non-systemic D. Both

5. High birth rate pathogen effectively control by

   A. Chemical method B. Biological method

   C. Physical methods D. By host resistance

6. Rapid epiphytotic develops through

   A. Simple interest disease B. polyetic

   C. Both A and B D. Compound interest disease

7. Example of high birth rate pathogen is

   1. *Phytophthora infestens,* 2. *Venturia inaequalis,*

   3. *Verticillium,* 4. *Pythium*

   A. 1. B. 2

   C. 1 and 2 D. 3 and 4

8. Slow epidemic occur in

   A. Annuals B. Perennials

C. Both A and B D. None of the above

9. Rapid epidemic occur in
   A. Annuals B. Perennials
   C. Both A and B D. None of the above
10. Example of polyetic pathogen are
    A. *Ceratocystis ulmi* B. *V. inaequalis*
    C. None of them D. Both A and B
11. Disease Progress Curve (DPC) is draw between
    A. Disease and time B. Time and environment
    C. Environment and pathogen D. Pathogen and host
12. For monocyclic pathogen DPC curve is
    A. Bell type B. Saturation type
    C. Sigmoid D. Logistic
13. For polycyclic pathogen DPC curve is
    A. Bell type B. Saturation type
    C. Sigmoid D. Logistic
14. Disease which infect multiple plant parts known as
    A. Polyteic B. Bi-model
    C. Monocyclic D. Polycylic
15. On zero date the epidemic rate is
    A. Low B. Zero
    C. High D. Negative
16. Fungicidal application is generally advisable
    A. At high disease intensity B. Before zero date
    C. At high disease severity D. After zero date
17. Epidemic rate of disease is denote by
    A. R B. r
    C. γ D. Δt
18. Epidemic rate curve for potato late blight
    A. Sigmoid B. Unsaturated
    C. Symmetrical or bell D. Saturated
19. Epidemic rate curve for apple scab
    A. Asymmetrical B. Unsaturated
    C. Sigmoid D. Saturated
20. Epidemic rate for polycyclic diseases are
    A. Equal to monocyclic B. Lesser than monocyclic

C. Greater than monocyclic D. Negative

21. Example of polyetic epidemics are
    A. Apple Scab and elm yellow
    B. Potato blight and apple scab
    C. Chestnut blight and elm yellow
    D. Peach leaf curl and chestnut blight
22. During early stage of epidemic disease incidence is ——while disease severity is ——.
    A. Low and negative B. High and Low
    C. Low and high D. High and positive
23. Which growth model is appropriate for most polycyclic diseases
    A. Logistic B. Sigmoid
    C. Bell D. Unsaturated
24. Which growth model is appropriate for modeling monocyclic disease
    A. Monomolecular B. Gompertz
    C. Logistic D. Exponential
25. In SIR Model $R_0 > 1$ indicates
    A. Pathogen arrive and infection occurs
    B. Represent pathogen invade and epidemic cause
    C. Pathogen establish but latent stage
    D. Systematic in nature
26. A non-parametric test useful for
    A. For normally distributed data
    B. For Non normally distributed data
    C. Parametric data
    D. Both for normally and non-normally distributed data's
27. Which is dividing hypothesis between regular and aggregated patterns
    A. CSR B. Pattern test
    C. Monte Carlo D. K-S test
28. Kolmogorov-Smirnov test (K-S test) is a
    A. Parametric test B. Non parametric test
    C. Both A and B D. Conditional test
29. Monte-Carlo procedure can be use for
    A. Point Pattern B. Negative binomial
    C. Departure from CSR D. SIR test

30. What is the common coefficient in spatial autocorrelation.
    A. Moran's I B. Getis
    C. Geary C D. Both A and C
31. For Poisson distribution variance (α^2) is
    A. Equal to mean B. Lower than mean
    C. Greater than mean D. No relation
32. For negative binomial variance is
    A. Lesser than for clumped B. Greater than for clumped
    C. Negative than for clumped D. No relation
33. For positive binomial variance
    A. Lesser than for uniform B. Greater than for clumped
    C. Negative than for clumped D. Greater than for uniform
34. Negative binomial distribution mathematically related to
    A. Chi-Square B. Poisson and Normal
    C. Poisson and Logarithmic D. Normal and exponential
35. Normal binomial distribution is most commonly use probability for
    A. Random pattern B. Uniform Pattern
    C. Clumped Pattern D. Both A and B
36. Poisson distribution is the most commonly used probability for
    A. Random pattern B. Uniform Pattern
    C. Clumped Pattern D. Both B and C
37. Which test is useful for determining appropriateness of beta-binomial distribution compare to binomial distribution
    A. K exponent more than one B. K exponent less than 2
    C. K exponent equal to 0 D. K exponent is in negative form
38. Low and positive value of K exponent of negative binomial distribution represent
    A. Uniform B. Clumped
    C. Uniform and random D. Highly aggregated
39. K exponent value equal to 8 represent
    A. Moderate Aggregated B. Random Pattern
    C. Uniform Pattern D. Uniform and clumped patterns
40. K exponent values more than 8 represent.
    A. More Aggregated B. Random Pattern
    C. Uniform D. None of these

41. Lloyd's index more than one indicates.
    A. Random Pattern
    B. Aggregated Dispersion
    C. Uniform
    D. Both A and C
42. Lloyd's index less than one indicates.
    A. Regular dispersion
    B. Uniform
    C. Random
    D. B and C
43. Value of the Morisita index equal to one represents.
    A. Uniform
    B. Random Pattern
    C. Uniform and Random
    D. None of above
44. Value of the Morisita index more than one represents.
    A. Random
    B. Uniform
    C. Both A and B
    D. Aggregated
45. Value of the Morisita index less than one represents
    A. Uniform
    B. Clumped
    C. Random
    D. None of the above
46. A positive one value of standardized Morisita index (MIS) indicates
    A. Maximum randomness
    B. Maximum dispersion
    C. Maximum aggregation
    D. A and B
47. A zero value of standardized Morisita index indicates
    A. Random distribution
    B. Clumped Distribution
    C. Uniform distribution
    D. B and C
48. A negative one value of standardized Morisita index represents
    A Perfectively clumped pattern
    B Perfectly uniform distribution
    C More dispersion
    D None of these
49. Taylor power low is based on the.
    A. Empirical relationships between mean and variance that are related by exponential function
    B. Mechanistic relationship between mean and variance that are related by exponential function
    C. Mechanistic relationship between mean and variance that are related by logarithmic function
    D. Empirical relationship between mean and variance that are related by logistic function
50. Value of variance to mean ratio more than one indicates
    A. Uniform
    B. Random
    C. Clumped
    D. Negative binomial

51. Value of variance to mean ratio less than one indicates
    A. Uniform B. Clumped
    C. Random D. Negative binomial
52. Value of variance to mean ratio equal to one indicates
    A. Uniform B. Random
    C. Negative binomial D. Clumped
53. Value of Johnson and Zimmer index less than two represents
    A. Uniform B. Clumped
    C. Random D. Negative binomial
54. Value of Johnson and Zimmer index more than two represents
    A. Uniform B. Negative binomial
    C. Clumped D. Random
55. Value of Johnson and Zimmer index equal to two represents
    A. Negative binomial B. Random
    C. Clumped D. Uniform
56. Which index can be use for two species case
    A. Pielou index of segregation B. Simpson Index
    C. Green Index D. Lloyd's Index
57. Which spatial index provides information about species combination diversity
    A. Pielou index of segregation B. Simpson Index
    C. Individual-centre analysis D. Lloyd's Index
58. In point pattern analysis F denotes
    A. Empty space function
    B. Nearest-neighbor distance
    C. Reduced second moment function
    D. Pearson coefficient
59. In point pattern analysis G denotes
    A. Reduced second moment function
    B. Pearson coefficient
    C. A empty space function
    D. Nearest-neighbour distance
60. In point pattern analysis K denotes
    A. Empty space function
    B. Reduced second moment function
    C. Nearest-neighbour distance

D. Pearson coefficient

61. Moran's I formula related to
    A. Nearest-neighbor distance
    B A empty space function
    C. Reduced second moment function
    D. Pearson coefficient
62. Father of the Indian aerobiology
    A. K. C. Mehta   B.
    C. C. D. Mayee   D. S K Bhattacharya
63. Who carried out the role of Barbery and Mahonia species in annual recurrence of wheat rust in India
    A. S.K. Bhattacharya   B K. C. Mehta
    C. G. V. Rao   D. S. Nagrajan
64. "Further Studies in Cereal Rust in India" wrote by
    A. S K Bhattacharya   B. G. V. Rao
    C. K. C. Mehta   D. K. Muralidharan
65. *"Epidemiology and Plant Disease Management"* wrote by
    A. J.C. Zadoks   B. J. Kranz
    C. E.C. Large   D L.R. Jones
66. Who studied spread of fungal spore with in canopy
    A. R.A. Krause   B. J.C. Zadoks
    C. L.R. Jones   D. E.C. Large
67. "Plant Disease: Epidemic and Control" wrote by
    A. L. R. Jones   B. J. C. Zodoks
    C. J E Vanderplank   D. J Kranz
68. ***"Principal of Plant Infection" written by***
    A. S. Nagrajan   B. J. C. Zodoks C K.
    C. Mehta   D. J E Vanderplank
69. Vertifolia phenomena studied on
    A. Vertifolia variety of potato   B. Vertifolia variety of apple
    C. Vertifolia variety of wheat   D. Vertifolia variety of rice
70. Vertifolia effect related with
    A. Erosion of vertical resistance   B. Erosion of horizontal resistance
    C. Multiline varieties   D. Gene Pyramiding
71. Vertifolia effect studied by
    A. J Kranz   B. C. D. Mayee

C. J.E. Vanderplank D. V Rao

72. Threshold theorem provided by
    A. J.E. Vanderplank B. Van Everdingen
    C. R.A. Krause D. Madden
73. EPIDEM program related with
    A. Early blight of tomato and potato
    B. Apple scab
    C. Late blight of potato
    D. Downy mildew of grapes
74. MYCOS computer programme related with
    A. Late blight of potato
    B. Mycosphaerella blight of *Chrysanthemum*
    C. Root rot of papaya
    D. Rice blight
75. EPIVEN programme related with pathogen
    A. *Alternaria solani* B. *Venturia inaequalis*
    C. *Helminthosporium maydis* D. *Plasmopara viticola*
76. EPICORN computer programme related with
    A. *Plasmopara viticola* B. *Helminthosporium maydis*
    C. *Phytophthora infestans* D. *Alternaria solani*
77. BLIGHTCAST programme developed for
    A. Rice blast B. Wheat rust
    C. Potato late blight D. Green ear disease of Bajara
78. TOM-CAST related with
    A. *Plasmopara viticola* for downy mildew of grapes
    B. Early blight of tomato
    C. *Phytophthora infestans* on potato foliage
    D. *Taphrina deformans* for peach leaf curl
79. PLASMO programme related with
    A. *Taphrina deformans* for peach leaf curl of grapes
    B. Early blight of tomato
    C. *Phytophthora infestans* on potato foliage
    D. *Plasmopara viticola* for downy mildew
80. PLASMO developed by
    A. Rosa *et al.* (1993) B. Jorge *et al.* (2005)
    C. Phtbaldo (1992) D. Krause *et al.* (1985)

81. TOM-CAST developed by
    A. Pfnder and Upper (2015) B. Phtbaldo (1992)
    C. Jorge *et al.* (2005) D. Krause *et al.* (1985)
82. LATE BLIGHT developed by
    A. Jorge *et al.* (2005) B. Phtbaldo (1992)
    C. Krause *et al.* (1985) D. Pfnder and Upper (2015)
83. Rice blight warning system with trap plots developed by
    A. S. Nagrajan B. J.C. Zodoks
    C. K.C. Mehta D. Muralidharan and Rao, 1981
84. In India Late Blight forecasting rules were developed by
    A. S. Nagrajan B. V. Rao
    C. Bhattacharaya D. Muralidharan and Rao
85. In India forecasting system for Groundnut rust developed by
    A. C.D. Mayee B. S. Nagarajan
    C. Shanta and Shukla D. V. Verma
86. Casual organism for ergot of rye
    A. *Claviceps purpurea* B. *Hemileia vastatrix*
    C. *Plasmopara viticola* D. *Ceratocystis ulmi*
87. Pathogen responsible for fire blight of pear and pome fruits
    A. *Plasmopara viticola* B. *Cryphonectria parasitica*
    C. *Erwinia amylovora* D. *Claviceps purpurea*
88. Pathogen associated with the Irish Famine
    A. *Erwinia amylovora* B. *Hemileia vastatrix*
    C. *Tilletia indica* D. *Phytophthora infestans*
89. *Phytophthora* infestans is a
    A. Nacrotrophs B. Facultative saprophytes
    C. Obligate parasite D. Obligate and bisexual parasite
90. Coffee rust caused by
    A. *Ceratocystis ulmi* B. *Fusarium* spp.
    C. *Hemileia vastatrix* D. *Plasmopara viticola*
91. Pathogen responsible for downy mildew of grapes
    A. *Erwinia amylovora* B. *Plasmopara viticola*
    C. *Puccinia recondita* D. *Cochliobolus miyabeanus*
92. Panama disease of the banana initiated by
    A. *Fusarium oxysporum* f. sp. *Cubense*
    B. CMV virus

C. Clavibacter

D. *Alternaria solani*

93. Chestnut blight is a polyetic epidemic and it's cause by

A. TMV | B. Potato virus X and Y
C. *Cryphonectria parasitica* | D. *Ceratocystis ulmi*

94. Dutch elm pathogen *Ceratocystis ulmi* present in which host tissue

A. Epidermal | B. Ground tissues
C. Cuticle | D. Vascular tissues

95. Casual organism for Bengal famine of rice

A. *Tilletia indica* | B. *Cochliobolus miyabeanus*
C. *Claviceps purpurea* | D. *Claviceps microspherella*

96. Maize leaf blight casual organism is

A. *Tilletia indica* | B. *Alternaria* spp.
C. *Claviceps purpurea* | D. *Helminthosporium maydis*

97. Casual organism for Leaf rust of wheat

A. *Puccinia recondita* | B. *Puccinia triticii*
C. *Puccinia striiformis* | D. None of these

98. Karnal bunt casual organism is

A. *Pythium* spp. | B. *Fusarium* spp.
C. *Claviceps* spp. | D. *TIlletia indica*

99. Term "Crytic error" provided by

A. S. Nagrajan | B. J.C. Zodoks
C. K.C. Mehta | D. J.E. Vanderplank

100. Predisposing factors related with

A. Pathogen | B. Host
C. Environment | D. Host and pathogen

101. Diseases triangle components includes

A. Host, environment and time
B. Host, human and pathogen
C. Host, pathogen and environment
D. Environment, time and pathogen

102. Anti-epidemic denote with

A. $IR_c<1$ | B. $IR_c = 0$
C. IRc >1 | D. $IR_c = -2$

103. A disease is said to pandemic when

A. When a disease is more or less constantly occurring year after year in a moderate to severe form in a country or locality

B. Diseases which occur at irregular intervals over limited areas or locations

C. Disease is constantly present in a locality but it assumes severe form only on occasions.

D. When an epidemic disease spreads over continents or subcontinents and involves mass mortality

104. A disease is said to Endemic when

A. When an epidemic disease spreads over continents or subcontinents and involves mass mortality

B. Disease is constantly present in a locality but it assumes severe form only on occasions.

C. When a disease is more or less constantly occurring year after year in a moderate to severe form in a country or locality

D. Diseases which occur at irregular intervals over limited areas or locations

105. A disease is said to Sporadic when

A. Diseases which occur at irregular intervals over limited areas or locations

B. Diseases which occur at irregular intervals over limited areas or locations

C. When an epidemic disease spreads over continents or subcontinents and involves mass mortality

D. Disease is constantly present in a locality but it assumes severe form only on occasions.

106. Alternate host belongs to

A. Different family B. Same family

C. Both A and B D. None of these

107. Collateral host belongs to

A. Different family B. Same family

C. Both A and B D. None of these

108. Alternate host of Black rust of wheat

A. *Berberis* species B. *Mahonia species*

C. *Mahoberberis* D. All of these

109. Alternate host of Yellow rust of wheat

A. *Isopyrum fumarioides* B. *Thalictrum species*

C. *Clematis* species D. All of these

110. Alternate host of Orange rust of wheat
    A. *Isopyrum fumarioides*
    B. *Berberis* species
    C. *Thalictrum species*
    D. Not known
111. Collateral host of Black rust of wheat
    A. *Puccinia graminis tritici*
    B. *Puccinia recondita*
    C. *Puccinia striiformis*
    D. Not known
112. Collateral host of Yellow rust of wheat
    A. *Puccinia graminis tritici*
    B. *Puccinia recondite*
    C. *Puccinia striformis*
    D. Not known
113. Collateral host of Orange rust of wheat
    A.. *Puccinia striiformis* and *Bromus coloratus*
    B. *Puccinia recondita* and *Bromus catharticus*
    C. *Puccinia graminis tritici* and *Aegilops* species
    D. *Hilaria jamestii* and *Aegilops squarrosa*
114. For development of a predictive system two essential conditions are
    A. Knowledge about host
    B. Knowledge of inoculums release and time required for the pathogen to cause systems development after infection
    C. Knowledge about control methods
    D. Information's about current environmental conditions
115. Which model is use for Asian Soybean rust epidemics disease
    A. CLIMEX
    B. GENECROP
    C. EPIDERMES
    D. PROCULTURE
116. For Gray leaf spot of corn the prediction model has been developed by
    A. Roget
    B. Berrie and Xu
    C. Paul and Munkvold
    D. Wolf and Verreet
117. For leaf spot of sugarbeet the prediction model has been developed by
    A. Wolf and Verret
    B. Paul and Munkvold
    C. Roget
    D. Berrie and XU
118. For take all of apple (*Gaeumannomyces graminis*) the prediction model has been developed by
    A. Wolf and Verret
    B. Roget
    C. Berrie and XU
    D. Paul and Munkvold
119. A prediction model for *Podosphaera leucotricha* has been developed by
    A. Wolf and Verret
    B. Roget
    C. Paul and Munkvold
    D. Berrie and XU

120. Powdery mildew of grapes cause by
   A. *Uncinula necator* B. *Septoria tritici*
   C. *Erwinia amylovara* D. *Plasmopara viticola*

121. Severity of the Tikka disease of groundnut (*Cercospora arachidicola* and *C. persontata* depends on (High humidity and low temperature
   A. High temperature B. Low temperature
   C. High humidity and high temperature D. High humidity and low temperature

122. Which weather acts as predisposing factors for severe rice blast in India.
   A. Combination of 20-24^{0}C temperature and 90% RH
   B. Combination of 23-28^{0}C temperature and 80% RH
   C. Combination of 24-35^{0}C temperature and 99% RH
   D. Combination of 20-28^{0}C temperature and 99% RH

123. Rice blast cause by
   A. *Septoria tritici* B. *Pyricularia oryzae*
   C. *Bulmeria graminis* D. *Plasmopara viticola*

124. Prediction of Erwinia stalk rot of maize based on
   A. Sunshine hours and rainfall
   B. Relative humidity and temperature
   C. Leaf wetness period
   D. Rainfall and air temperature

125. 10:24 rules are associated with
   A. Apple scab B. Downy mildew of grapes
   C. Wheat rust D. Fire blight of apple

126. In 10:24 rules 10 represents
   A. 10^{0}C temperature with 24 mm of rainfall in 24 hours period
   B. 10^{0}C temperature with 24 mm of rainfall in 10 hours period
   C. 10^{0}C temperature with 10 mm of rainfall in 24 hours period
   D. 10^{0}C temperature with 100 mm of rainfall in 10 hours period

127. Epidemic of downy mildew onion caused by
   A. *Peronospora pannosa* B. *Peronospora destructor*
   C. *Uncinula necator* D. *Podosphaera leucotricha*

128. BLIGHTCAST relay on which two models
   A. Temperature
   B. Temperature and humidity
   C. Humidity and leaf wetness period

D. Favourable day, humidity and temperature

129. Stewart wilt of corn cause severe disease during
    A. December to February temperature more than or equal to +7^{0}C
    B. December to February temperature more than or equal to +10^{0}C
    C. December to February temperature more than or equal to +23^{0}C
    D. December to February temperature more than or equal to –4^{0}C
130. Minimum or mild disease of Stewart wilt of corn
    A. If temperature is <11^{0}C B. If temperature is <15^{0C}
    C. If temperature is >11^{0}C D. If temperature is <–11^{0}C
131. Pea root rot cause by
    A. *Aphanomyces euteiches* B. *Alternaria solani*
    C. *Bremia lactucae* D. *Magnaporthe grisea*
132. Firelight of apple and pear caused by
    A. *Bremia lactucae* *B. Erwinia amylovora*
    C. *Aphanomyces euteiches* D. *Alternaria solani*
133. Apple scab forecasting based on
    A. Humidity and ascospore release
    B. Temperature and perithecia present on lived leaves
    C. Humidity and temperature during ascospore maturation
    D. Leaf wetness period and severity is directly proportional to amount of perithecia present on dead leaves
134. Mills chart related with
    A. Wheat rust forecasting B. Rice blast forecasting
    C. Apple scab forecasting D. Grapes downy mildew
135. Early blight of celery caused by
    A. *Stemphylium vesicarium* *B. Cercospora apii*
    C. *Septoria tritici* D. *Cercospora* species
136. Blister blight of tea cause by
    A. *Exobasidium vexans* *B. Cercospora apii*
    C. *Septoria tritici* D. *Cercospora* species
137. Blister blight of tea severity
    A. High at 87% RH B. High at 90% RH
    C. High at 99% RH D. High at 83% RH
138. Indian stem rust rules provided by
    A. Nagarajan and Singh, 1975 B. Murlidharan and Rao (1981)
    C. Bhattacharya *et al.* (1982) D. Shukla and Anjaneyulu (1992)

139. In India Late blight forecasting system was developed by
    A. Shukla and Anjaneyulu (1992) B. Nagarajan and Singh (1975)
    C. Murlidharan and Rao (1981) D. Bhattacharya *et al.* (1982)
140. In India Rice blast forecasting was developed by
    A. Bhattacharya *et al.* (1982) B. Shukla and Anjaneyulu (1992)
    C. Murlidharan and Rao (1981) D. Nagarajan and Singh (1975)
141. EPIBLA is use for
    A. For forecasting of rice blast in the China
    B. For forecasting of rice blast in the America
    C. For forecasting of rice blast in the Philippines
    D. For forecasting rice blast in the India
142. Rice tungro forecasting system in India developed by
    A. Shukla and Anjaneyulu (1992) B. Nagarajan and Singh (1975)
    C. Murlidharan and Rao (1981) D. Padmanabhan (1963)
143. Meteorological factors associated with Bengal famine studied by
    A. Nagarajan and Singh (1975) B. Shukla and Anjaneyulu (1992)
    C. Padmanabhan (1963) D. Murlidharan and Rao (1981)
144. JHULSACAST is
    A. Computerized system developed by Singh *et al.* (2000)
    B. Computerized system developed by Padmanabhan (1963)
    C. Computerized system developed by Murlidharan and Rao (1981)
    D. Computerized system developed by C. D. Mayee (1982)
145. JHULSACAST system developed
    A. For late blight B. For Peach leaf curl
    C. For Apple scab D. For rice blast
146. Southern Oscillation Index (SOI) utilize for
    A. Forecasting disease severity
    B. Forecasting disease incidence
    C. Forecasting primary inoculums amount
    D. Forecasting crop yield
147. Forecasting for Altenraria blight of oilseed developed by
    A. Nagarajan and Singh (1975) B. Shukla and Anjaneyulu (1992)
    C. Chattopadhay *et al.* (2005) D. Murlidharan and Rao (1981)
148. LEAFBLAST, EBIBLAST and EPIBLA programmes were developed for

A. Rice blast B. Late blight of potato
C. Early blight of potato D. Fire blight of apple

149. BLIGHTCAST related with
A. Rice blast in India B. Late blight of potato in England
C. Early blight of potato in Brazil D. Late blight of potato in India

150. JHULSACAST related with
A. Early blight of potato in Brazil B. Rice blast in India
C. Late blight of potato in India D. Late blight of potato in England

151. EPIDEMI is use for
A. Insect and scab of wheat B. Scab of potato
C. Charcoal rot D. Downy mildew of grapes

152. CERCOS is related with
A. *Cercospora* leaf blight of cereals B. Powdery mildew of cereals
C. Foot rot of papaya D. High and low bunt of cereals

153. EPICORN is related with
A. Foot rot of papaya B. Southern corn leaf blight
C. Northern corn leaf blight D. Southern blight of onion

154. CERCOS is related with
A. *Cercospora* leaf blight of cereals B. High and low bunt of cereals
C. Powdery mildew of cereals D. Apple scab

155. PLASMO is related with
A. Downy mildew of pea B. Downy mildew of rose
C. Downy mildew of grapes D. Downy mildew of bajara

156. MYCOS is related with
A. *Chrysanthemum* leaf spot B. Wheat leaf spot
C. Apple fire blight D. Early blight of potato

157. TOMCAST is related with
A. Early blight of Groundnut B. Early blight of tomato
C. Early blight of potato D. *Chrysanthemum* leaf spot

158. EPICAST is related with
A. Early blight of potato B. Early blight of crucifers
C. Wheat leaf rust D. Wheat stem rust

159. MARYBLIGHT is related with
A. Early blight of Groundnut B. Fire blight of apple
C. Early blight of potato D. Rice blight

160. From epidemic point of view the basic objective of disease management is to keep density below

    A. An economic injury level
    B. An epidemic threshold
    C. An economic injury threshold
    D. Quantifiable with yield loss

161. Effective management practices includes

    A. Timely quantification of X_0, r and t
    B. Quantification of initial inoculums only
    C. Quantification of epidemic rate of infection only
    D. Quantification of time at which disease incidence is highest

162. Exponential growth model suitable for

    A. For monocyclic disease
    B. For polyetic disease
    C. For both mono and polycyclic diseases
    D. Most polycyclic diseases

163. Exponential model suggest that

    A. Absolute rate of disease increase is indirectly proportional to disease severity
    B. Absolute rate of disease increase is directly proportional to disease intensity
    C. Absolute rate of disease decrease is directly proportional to disease severity
    D. Rate of disease increase is not related with disease severity or intensity

164. Exponential growth model written as

    A. $X=X_0ert$
    B. X=XRT
    C. $X=X_0$
    D. $X=X_0/rt$

165. Monomolecular model is most appropriate for

    A. For monocyclic disease
    B. For polyetic disease
    C. For both mono and polycyclic diseases
    D. Most polycyclic diseases

166. Monomolecular model also called

    A. Positive exponential model
    B. Negative exponential model
    C. Logistic Model
    D. Gompertz model

167. Monomolecular model written as

    A. $X=X_0/rm$
    B. $X=X_{0*}t$
    C. $X=X_0+rmt$
    D. $X=X_{0RT}$

168. Logistic model is most appropriate for
    A. Polycyclic disease having secondary spread with in a field only
    B. Polycyclic disease having secondary spread with in canopy
    C. Polycyclic disease not having secondary spread with in growing season
    D. Polycyclic disease having secondary spread with in growing season
169. Gompertz
    A. Appropriate for monocyclic disease as an alternate to logistic models
    B. Appropriate for polycyclic disease as an alternate to logistic models
    C. Appropriate for polycyclic disease as an alternate to exponential models
    D. Appropriate for monocyclic disease as an alternate to exponential models
170. Sanitation is an useful for controlling epidemic initiated by
    A. Monocyclic pathogen B. Systematic pathogen
    C. Polycyclic pathogen D. Both A and B
171. Sanitation is generally use for controlling
    A. Soil borne and facultative parasite
    B. Obligate pathogen
    C. Soil transient pathogen
    D. Non-systematic pathgoen
172. Sanitation is generally useful for
    A. Compound interest disease B. Simple interest disease
    C. Both A and B D. None of these
173. Efficiency of sanitation depends on the
    A. Use healthy planting material
    B. Use of resistant varieties only
    C. Timely application of chemical spray
    D. Timely spotting of diseased plants and their removal
174. Crop rotation is most useful for
    A. Soil invaders only
    B. Soil inhabitants
    C. Soil invaders not for soil inhabitants
    D. Both for soil invader and for soil inhabitants

175. Soil solarisation is a
   A. Chemical process
   B. Physical method
   C. Mean of biological control
   D. Hypothermal process
176. Hm-1 gene isolated from
   A. Maize
   B. Rice
   C. Tomato
   D. Potato
177. PTO gene isolated from
   A. Tomato
   B. Rice
   C. Potato
   D. Wheat
178. AUDPC denotes
   A. Area under disease progress curve
   B. Area use for disease progress curve
   C. Area under disease progress curve
   D. None of these

ANSWER KEY

Question Number	Answer	Question Number	Answer	Question Number	Answer	Question Number	Answer
1	A	49	A	97	A	145	A
2	B	50	C	98	D	146	D
3	B	51	A	99	D	147	C
4	C	52	B	100	B	148	A
5	A	53	A	101	C	149	B
6	D	54	C	102	A	150	C
7	C	55	B	103	D	151	A
8	B	56	A	104	C	152	A
9	A	57	C	105	A	153	B
10	D	58	A	106	A	154	D
11	A	59	D	107	B	155	C
12	B	60	B	108	D	156	A
13	C	61	D	109	D	157	B
14	B	62	A	110	D	158	C
15	C	63	B	111	A	159	B
16	D	64	C	112	C	160	B
17	B	65	A	113	B	161	AA
18	C	66	B	114	B	162	D
19	A	67	C	115	A	163	B
20	C	68	D	116	C	164	A
21	C	69	A	117	A	165	A
22	B	70	B	118	B	166	B
23	A	71	C	119	D	167	C
24	A	72	A	120	A	168	D
25	B	73	A	121	D	169	B
26	D	74	B	122	A	170	D
27	A	75	A	123	B	171	A
28	B	76	B	124	A	172	B
29	C	77	A	125	B	173	D
30	D	78	B	126	C	174	C
31	A	79	D	127	B	175	D
32	B	80	A	128	D	176	A
33	A	81	B	129	A	177	B
34	C	82	C	130	D	178	A
35	C	83	D	131	A		
36	A	84	C	132	B		
37	B	85	A	133	D		
38	D	86	A	134	C		
39	A	87	A	135	B		
40	B	88	D	136	A		
41	B	89	D	137	D		
42	A	90	C	138	A		
43	B	91	B	139	D		
44	D	92	A	140	C		
45	A	93	C	141	D		
46	C	94	D	142	A		
47	A	95	B	143	C		
48	B	96	D	144	A		

7.4. SOME EXPLANATION

Q.1. What is the epidemiological consequence of reducing the initial inoculum when the apparent infection rate is high?

Ans. A reduction in initial inoculum (or initial disease incidence) results in an apparent delay in the epidemic. Note that in the early part of the epidemic, before the logistic factor (reduced amounts of susceptible tissue) has much of an influence, every point along the disease curve is reduced by the proportion of the reduction in initial disease incidence.

Q. 2. To provide effective control of an epidemic with a high apparent infection rate, how much must the initial inoculum be reduced?

Ans. To have any significant impact on disease development in a polycyclic epidemic with a high apparent infection rate, the initial inoculum must be reduced to exceedingly low levels.

Q. 3. What is the epidemiological consequence of reducing the initial inoculum when the apparent infection rate is low?

Ans. The reduction in disease incidence throughout the epidemic is proportional to the reduction in initial inoculum, just as it is for the early part of the epidemic when **r** is high.

Q.4. What is the epidemiological consequence of reducing the apparent infection rate?

Ans. Unlike reducing initial inoculum, which merely delays the onset of the epidemic, a reduction in the apparent infection rate reduces the rate of disease development throughout the epidemic.

Q.5. Which has a greater impact on a polycyclic epidemic, reducing the initial inoculum or reducing the apparent infection rate?

Ans. A reduction in the apparent infection rate has a far greater impact on the epidemic than a comparable reduction in the initial inoculum.

Q.6. What is the epidemiological consequence of an integrated approach that reduces both the initial inoculum and the apparent infection rate?

Ans. Small reductions in the initial inoculum or the apparent infection rate may not seem adequate to control an epidemic when used individually, but when combined into an integrated program, satisfactory disease control can be achieved.

Q.7. Calculate the apparent infection rate (r) if disease increases from 0.1 to 5.0% in 7 days

Ans. Here, x_1 =0.1% (0.001) and x_2 = 5.0% or (0.05) and time in advancement of disease (t_2-t_1 =7 days), therefore applying the following formula for r

$$r = \frac{2.3}{t_2 - t_1}\left[\log 10\left(\frac{x_2\,(1-x_1)}{x_1\,(1-x_{12})}\right)\right]$$

Putting values in this formula = (2.3/7=0.3285); (1-0.001=0.999; 1-005=0.95)

$$= 3.285\ x\left[log10\left(\frac{0.05\ x0.99}{0.001\ x0.95}\right)\right] = r = 3.285\ x\left[log10\left(\frac{0.0495}{0.00095}\right)\right] = r$$

$$= 3.285\ x\,[log10\,(52.105)]$$

$$r = 3.285\ x\ [1.716] = 0.564\ or\ 57\%$$

Q. 8. In a field experiments it was found that rust infection of sorghum was 0.45 per cent on April 10th on susceptible cultivar and which was increase up to 0.78 per cent at 18th of April, find out the apparent infection rate

$$r = \frac{2.3}{t_2 - t_1}\ x\ log\frac{x_2}{x_1}$$

$$r = \frac{2.3}{8}\ x\ log\frac{0.78}{0.45} = \frac{2.3}{8} x \log 1.733$$

$$\frac{2.3}{8} x \log 1.733 = \frac{2.3}{8} x0.23 = 0.2875\,x\,0.238 = 0.0686$$

Q. 9. In a research work related with strip rust of wheat, initial level of disease was 1.1 per cent on May 15th and remained the same till 27 May as the incubation (p) under that weather was for 12 days. Since then the severity increased from 13.4 per cent on January 8th when $2p$ was completed. Calculate the R value

Ans. R can be quantify through following formula

$$R = \frac{1}{x_0(t_2 - t_1)}\ x\left(1n\frac{1}{1-x_2} - 1n\frac{1}{1-x_1}\right)$$

In present, X_0 and X_1 is 1.1 per cent (*i.e.*, 0.011) X_2 is 13.4 per cent or 0.134 and t_2-t_1 is 12 days, thus,

$$R = \frac{1}{0.011x(12)} x \left(1n \frac{1}{1-0.134} - 1n \frac{1}{1-0.011}\right)$$

$$R = \frac{1}{1.32} x \left(1n \frac{1}{0.866} - 1n \frac{1}{0.989}\right)$$

$$R = \frac{1}{1.32} x \,(1n\, 1.154 - 1n 1.011)$$

$$R = \frac{1}{1.32} x \,(0.14387 - 0.0110)$$

$$R = \frac{0.1328}{1.32} = 1.01\, units/day$$

Q. 10. *Phytophthora* infestation infection of potato was 7.9 percent on 8th May and it increased 45.2 per cent on 20th May. During this time total green leaf areas also increased from 425 cm^2 to 900 cm^2 Calculate the incubation or *p* value

Ans. Incubation period can be quantified by following equation

$$p = \frac{1}{t_2 - t_1} log_{10} \frac{ax_2(1 - x_1)}{x_1(1 - x_2)}$$

In this problem time period is 12 days; a is 900, x_2 is 0.452, x_1 is 0.079

$$p = \frac{1}{12} log10 \frac{900\, x0.452\, x\, (1 - 0.079)}{425\, x\, 0.079\, x\, (1 - 0.452)}$$

$$p = \frac{1}{12} log10 \frac{900\, x0.452\, x\, (0.921)}{425\, x\, 0.079\, x\, (0.548)}$$

$$p = \frac{1}{12} log10 \frac{374.66}{18.399}$$

$$p = \frac{1}{12} log10\, 20.36$$

$$p = \frac{1}{12} x1.3088$$

$P = 1.05$ per unit/day

QUESTIONS

University Of Kwazulu-Natal School of Agricultural, Earth & Environmental Sciences Discipline Of Plant Pathology Supplementary Examination: July 2013

1. Define the term "Epidemiology". Discuss and show graphically the difference between an epidemic and an endemic disease. Epidemic diseases fall into two (2) main classes. Define these classes, given an example of each.
2. Explain Vanderplank's correction factor and when would you use it?
3. A third year Plant Pathology student from University of KwaZulu-Natal monitoring powdery mildew on baby marrows in 2004 collected the following data

Date	*% leaf area infected*
July 1	0.005
July 8	0.07
July 15	4.5
July 22	32.0
July 29	75.0

The latent period in 7 days and there is an infection period of 7 days.

- Calculate the logistic, basic infection and corrected basic infection equations.
- List and show in a graphical form the three stages of an epidemic

4. Discuss the differences between the logarithmic, logistic and basic infection and the corrected basic infection equations.
5. Define the following term. Include diagrams where appropriate
   - Vertical resistance
   - Horizontal resistance
   - Boom and bust cycle of cultivar production
   - Stabilizing selection pressure
   - Directional selection pressure
   - Allo and auto infections

6. Discuss and show graphically the differences between an epidemic and an endemic disease. Epidemic disease fall into two main classes. Define these classes given examples of each.
7. If rust on germinations is at 0.5% leaf area infected on 1st April and 35% on the 30th April, what is the rate of disease progress?
8. What is a compound interest disease and what is the equation used for calculating compound interest disease progress? Give an example of a compound interest disease.
9. What are the useful aspects of the logistic transformation, and what are its limitations?
10. Describe how one would breed for VR and HR
11. Discuss the advantages and disadvantages of VR and HR
12. What is vertifolia effect? Discuss the underlying genetics in terms of selection pressure and genetic load. Why it's a problem with VR but not with HR breeding programmes?
13. Comment on
    - Auto-infection and allo-infection
    - Crop uniformity and crop vulnerability
    - Disease incidence, disease severity and disease count
    - Constant ranking and apparent horizontal resistance
    - Differential interaction and epidemiological competence
    - Interplot interference
    - Gene for Gene relationship
    - Epidemic and endemic disease
    - Discontinuous and continuous pathogen-systems
    - Qualitative and Quantitative resistance
    - Interplot interference
    - Apparent vertical resistance
14. Describe in detail, the One Pathotype and Two Pathotype techniques that may be used to inactivate/eliminate functional Vertical Resistance genes in a host Horizontal Resistance. Contrast the two techniques.
15. "Horizontal resistance is durable and durable resistance is common" Discuss this statement, providing theoretical as well as practical examples to substantiate your answer.
16. Vanderplank introduced the terms "cryptic error", or "interplot interference". What do these terms mean and what is the evidence that this phenomenon is significant in Plant Pathology trials?
    - What value are guard rows in avoiding these problems?

- What problems do unsprayed plots introduce into fungicide trials?
- How can one avoid cryptic errors in cultivar trials for disease resistance?
- What effect does cryptic errors having in choosing for VR and HR in cultivar trials?
- What is the effect of the size and shape of plots on cryptic errors?
- When do untreated plots not present a problem

Subject Index

P

Q

R